# COMPLEMENT REGULATORY PROTEINS

# COMPLEMENT REGULATORY PROTEINS

## B. Paul Morgan and Claire L. Harris

*Complement Biology Group, Tenovus Building,*
*Department of Medical Biochemistry,*
*University of Wales College of Medicine,*
*Heath Park, Cardiff CF4 4XN, UK.*

ACADEMIC PRESS

Harcourt Brace & Company, Publishers

San Diego  London  Boston  New York  Sydney  Tokyo  Toronto

ACADEMIC PRESS
525 B Street, Suite 1900, San Diego,
California 92101–4495, USA
http://www.apnet.com

ACADEMIC PRESS
24–28 Oval Road
LONDON NW1 7DX
http://www.hbuk.co.uk/ap/

Library of Congress Catalog Card Number: 98–89879

A catalogue record for this book is available from the British Library

ISBN 0–12–506965–0

Typeset by Phoenix Photosetting, Chatham
Printed in Great Britain by the University Printing House, Cambridge
99 00 01 02 03 04 CU 9 8 7 6 5 4 3 2 1

# CONTENTS

# ABOUT THE AUTHORS

Paul Morgan and Claire Harris are in the Complement Biology Group, Department of Medical Biochemistry, University of Wales College of Medicine. Professor Morgan has been involved in complement research since the early 1980s. After qualifying in Medicine, he undertook a PhD degree in Cardiff, studying the structure and function of the membrane attack complex of complement. After post-doctoral stints in the laboratories of Professor Manfred Mayer and Professor Alfred Esser in the US, he returned to Cardiff, first as a lecturer in Medical Biochemistry and then as a Wellcome Senior Clinical Fellow and Senior Lecturer. He was promoted to a Personal Chair by the University of Wales in 1995. He continues to be supported by The Wellcome Trust. Dr. Harris has come into the field more recently. Her PhD degree was in Professor Peter Lachmann's unit in Cambridge where, under the supervision of Dr. Richard Harrison, she undertook studies of the biochemistry of C3. She joined Professor Morgan's laboratory immediately upon completion of her PhD in 1993. She retains a strong interest in the biochemical aspects of the complement system and in the regulators which control complement at the level of C3.

# PREFACE

The complement system provides a powerful defence against invading organisms, acting either directly to destroy the target or indirectly by activating and guiding phagocytes to destroy. Complement is a cascade system and generation of active products from the cascade during activation on pathogens presents a considerable risk to the host. Our own cells are protected from this potent cytotoxic and cytolytic system by a battery of fluid-phase and cell surface complement regulatory proteins, the majority of which have been identified during the last 20 years, which act to restrict complement activation on the host cell surface. This once tiny area of research has undergone an information explosion over the last decade to the point where it has become difficult to assimilate from the primary papers and reviews of individual regulators the complex interplay between these proteins. No comprehensive account of current knowledge of the complement regulatory proteins has been produced and the interested reader must search the literature to find the required information. The time is thus ripe for a book which provides a thorough review of all the complement regulatory proteins and places them in context with one another in maintaining control of complement. We have written this book in order to fill this need.

The book provides in the first two chapters a succinct introduction to complement and its control, aimed at those new to the field. Chapters 3 and 4 contain the meat of the subject – comprehensive reviews of each of the individual regulators. The remaining chapters contain what we believe to be most interesting, discussions of the roles of the complement regulators *in vivo* in health and disease. Deficiencies of the various regulators, interactions of pathogens with the regulators, roles of complement regulators in reproduction and utilization of the regulators in therapy are all discussed in depth. A section on complement regulators in other species provides information essential for interspecies comparison and highlights important differences from control in humans. We hope that the text will highlight the relevance of the complement regulators to homeostasis and their contributions to disease processes. This is a young and growing area which will undoubtedly contribute in the future to the understanding and therapy of disease.

This two-person enterprise benefited enormously from discussion and interactions with other members of the Complement Biology Group in Cardiff, particularly Philippe Gasque, Carmen van den Berg, Sim Singhrao, Rhian Morgan and Stewart Hinchliffe. Financial support for this enterprise and most other output from the Complement Biology Group has been provided over the last 11 years by The Wellcome Trust; we are deeply indebted to Dr. David Gordon, Dame Bridget Ogilvie and others at the Trust for their unstinting support. We thank Tessa Picknett at Academic Press for stepping so enthusiastically into the breach and for her subsequent help.

B. Paul Morgan and Claire L. Harris

# ABBREVIATIONS

AI, anaphylatoxin inactivator
AP, alternative pathway
C, complement
C1inh, C1 inhibitor
C3i; C3($H_2O$), $C_3$ hydrolysed at the thioester
C4bp, C4b binding protein
CCPH, complement control protein homologue
CD, cluster of differentiation
Cho, carbohydrate
CHO, Chinese hamster ovary
CMV, cytomegalovirus
CP, classical pathway
CPN, carboxypeptidase N
CR, complement receptor
CRP, complement regulatory protein
DAF, decay-accelerating factor
E, erythrocyte
EBV, Epstein–Barr virus
EGF, epidermal growth factor
fD, factor D
fH, factor H
fI, factor I
fIM, factor I module
fJ, factor J
GPI, glycosyl phosphatidylinositol
HAE, hereditary angiedema
HfHL, human factor H-like
HfHR, human factor H-related
HIV, human immunodeficiency virus
HRF, homologous restriction factor

Ig, immunoglobulin
LHR, long homologous repeat
MAC, membrane attack complex
MCP, membrane cofactor protein
$M_r$, relative molecular mass
NMR, nuclear magnetic resonance
P, properdin
PNH, paroxysmal nocturnal haemoglobinuria
R, receptor
RCA, regulators of complement activation
RGD, Arg-Gly-Asp
SCR, short consensus repeat
SDS-PAGE, sodium dodecyl sulphate-polyacrylamide gel electrophoresis
SP, signal peptide
STP, Ser/Thr/Pro
TCC, terminal complement complex
tm, transmembrane
TSR, thrombospondin repeat

# Chapter 1

# THE COMPLEMENT SYSTEM:
## a brief overview

## INTRODUCTION

Complement (C) was discovered a little over a century ago as a heat-labile component of blood plasma which conferred bactericidal properties and the capacity to lyse erythrocytes from other species [1]. The multi-component nature of the system was first recognized at the turn of the century [2] but it was not until the 1960s that the individual components were isolated and characterized, beginning an exciting period during which the intricacies of this fascinating system have emerged [3]. Despite this long history, new discoveries continue to be made and this super-ficially simple system continues to surprise.

The C system consists of a group of 12 soluble plasma proteins which interact with one another in two distinct enzymatic activation cascades (the classical and alternative pathways) and in the non-enzymatic assembly of a cytolytic complex (the membrane attack pathway) (Figure 1.1; Table 1.1). A third activation pathway, termed the lectin pathway, has recently been described [4, 5]. Control of these enzymatic cascades is essential to prevent the rapid consumption of C in response to trivial stimuli and is provided by ten or more plasma- and membrane-bound inhibitory proteins which act at multiple stages of the system to regulate activation.

C plays a central role in innate immune defence. The physiological roles of C are to provide an innate system for the rapid destruction of a wide range of invading micro-organisms and to mediate the solubilization and clearance of immune complexes. C, via interactions with other components of the immune system, also plays a role in the development of immunity. The mechanisms by which C augments the immune response to antigen are only now becoming clear and provide a fascinating link between the systems of innate and acquired immunity. C also contributes to pathology in a broad range of diseases, predominantly through the inflammatory effects of many of the products of C activation.

In order to appreciate the necessity for control of C, it is essential first to understand the basics of the system. The first two chapters of this text set

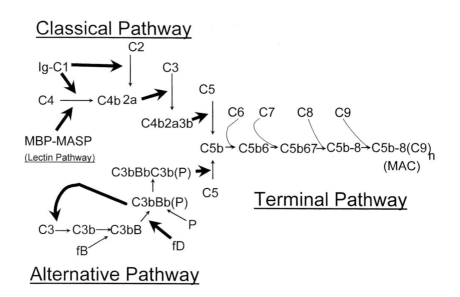

**Figure 1.1. The complement system.**

The constituent pathways of the C system and the component proteins are shown. Enzymatic cleavages are represented by thick arrows. The lectin pathway differs from the classical pathway (CP) only in that the MBP–MASP complex replaces the C1 complex. (MBP, mannan-binding protein; MASP, MBL-associated serine protease.)

the scene for what is to come and are intended as an introduction. The aficionados of C may wish to skip this chapter (and perhaps the next).

## ACTIVATION OF C

### The classical activation pathway

The classical activation pathway (CP) is so called because it was the first pathway to be described, dating back over a century to the original descriptions by Paul Ehrlich and others of the interactions between heat-stable anti-toxin (antibody) and the heat-labile serum component which 'complemented' the effects of immune serum[6,7]. The prototypic activator of the CP is, indeed, antibody bound to particulate antigen. However, it is now clear that many other substances, including components of damaged cells, bacterial lipopolysaccharide and nucleic acids, can initiate the CP in an antibody-independent manner and that these non-antibody activators may be of considerable physiological and pathological significance[8].

**Table 1.1. *The component proteins of the complement system.***

The proteins which constitute the classical, alternative and membrane attack pathways are listed together with their approximate concentration in plasma and the location of the gene (where known). Abbreviations: MHC, major histocompatiblity complex; Chr, chromosome.

| Component | Structure | Plasma concentration (mg/l) | Gene location |
|---|---|---|---|
| **Classical pathway** | | | |
| C1 | Complicated molecule, composed of 3 subunits, C1q (460 kDa), C1r (80 kDa), C1s (80 kDa) in a complex (C1qr$_2$s$_2$) | 180 | C1q A, B and C chains, 1p34-36; C1r, C1s; 12p13 |
| C4 | 3 chains (α, 97 kDa; β, 75 kDa, γ, 33 kDa); from a single precursor | 600 | 2 genes (C4A & B) in MHC Chr 6 |
| C2 | Single chain, 102 kDa | 20 | MHC Chr 6 |
| **Alternative pathway** | | | |
| fB | Single chain, 93 kDa | 210 | MHC Chr 6 |
| fD | Single chain, 24 kDa | 2 | ? |
| Properdin | Oligomers of identical 53 kDa chains | 5 | X |
| **Common:** | | | |
| C3 | 2 chains: α, 110 kDa, β, 75 kDa | 1300 | 19 |
| **Terminal pathway** | | | |
| C5 | 2 chains: 115 kDa, 75 kDa | 70 | 9 |
| C6 | Single chain, 120 kDa | 65 | 5 |
| C7 | Single chain, 110 kDa | 55 | 5 |
| C8 | 3 chains: α, 65 kDa, β, 65 kDa, γ, 22 kDa | 55 | α, β, Chr 1 γ, Chr 9 |
| C9 | Single chain, 69 kDa | 60 | 5 |

**Binding and activation of C1**

The CP is initiated by the binding of the C1q component of the hetero-oligomeric C1 complex to an activator [9–11]. Activation *in vivo* involves binding of C1q to aggregated or immune complex bound IgG or IgM antibody. C1 is a large hetero-oligomeric complex (molecular weight approx. 800 kDa) consisting of a single molecule of C1q and two molecules each of C1r and C1s associated non-covalently in a Ca$^{2+}$-dependent complex

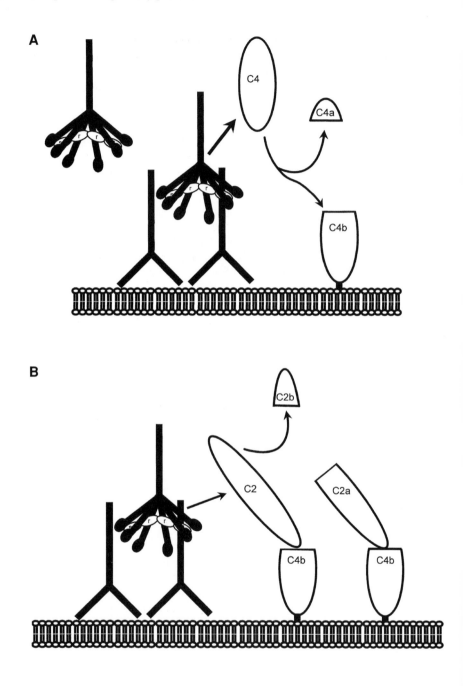

(Figure 1.2) [12, 13]. C1q is itself a complex molecule, composed of six copies of each of three related but distinct polypeptide chains (A, B and C). Each of the six subunits consists of single copies of the A, B and C chains, wound around each other in a triple helix containing an extended, collagen-like tail and a large globular 'head' region [14–16]. The three chains constituting each subunit are highly homologous and are encoded by three closely linked genes located on chromosome 1p in the order A-C-B [14, 17]. In the intact C1q molecule the collagen tails of the six triple helices are tightly associated along the amino terminal half but then diverge to form six connecting strands, each bearing a globular head. C1q binds via its globular heads to the Fc portion of IgG or IgM. C1q contains no enzymatic activity, but conformational changes occur in C1q upon binding of multiple heads of the C1q molecule by aggregates of IgG which trigger activation of the other components of the C1 complex. Among the human IgG subclasses the ranking of efficiency for binding and activation of C1q is IgG3 > IgG1 > IgG2; IgG4 does not activate C1q. IgM is a multivalent molecule and can thus activate C1q efficiently without the need for aggregate formation. However, fluid-phase IgM does not bind C1q or activate C; conformational changes occurring within the molecule on binding particulate antigen appear to be essential for exposure of the C1q-binding site [18]. Activation of C by IgA has long been an area of dispute. It is now widely accepted that IgA does not bind C1q and trigger CP activation, but polymeric IgA, once associated with particulate antigen, appears to be an efficient activator of the alternative pathway (AP) of C [19–21].

The enzymatic activity of the C1 complex is provided by C1r and C1s. These are both single-chain molecules of molecular weight 80 kDa, encoded by closely linked genes on chromosome 12 and sharing a high degree of homology. In the presence of $Ca^{2+}$, C1r and C1s associate with each other to form an elongated $C1r_2–C1s_2$ complex which binds between the globular heads of C1q (Figure 1.2) [22, 23]. Binding of C1q through the globular heads to Fc regions of aggregated IgG triggers activation through

---

**Figure 1.2. Activation and activities of C1.**

The C1 complex consists of one molecule of C1q and two molecules each of C1r and C1s. These latter components associate in a heteromeric complex and bind between the globular 'heads' of C1q. Activation of C1 occurs when at least two of its six 'heads' associate with immunoglobulin Fc regions and involves the sequential cleavage/activation of C1r and C1s to form the active C1 esterase.
**A.** The active esterase cleaves C4 from the fluid-phase, the small fragment, C4a, being released and the larger fragment, C4b, binding via a thioester group to the membrane in the vicinity of the initiating esterase.
**B.** Membrane-associated C4b binds C2 and presents it for cleavage by the C1 esterase to form the next enzyme in the cascade, C4b2a.

conformational changes which require that multiple heads of C1q are simultaneously engaged. This requirement prevents the activation of C1 by monomeric IgG and imposes constraints on antibody density for efficient activation. The conformational changes occurring within C1q upon binding antibody trigger the auto-activation of the proenzyme C1r, a process which involves cleavage at a single site within the molecule, thereby exposing the buried active site. C1r then activates C1s in the complex, again by cleaving at a single site in the molecule[22]. C1s in the activated C1 complex will enzymatically cleave and activate the next component of the CP, C4.

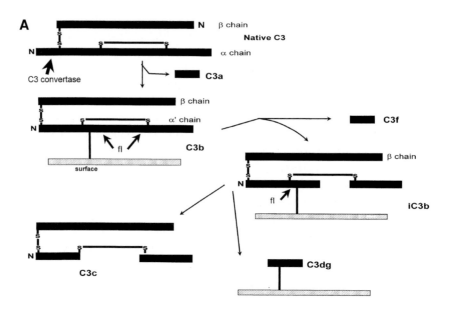

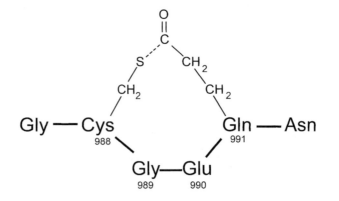

## Binding and cleavage of C4

C4 is a large, plasma protein (200 kDa) containing three disulphide-bonded chains ($\alpha$, $\beta$ and $\gamma$) [24, 25]. C4 is synthesized as a single-chain precursor which is processed intracellularly to give the three-chain molecule. C4 is encoded by two closely linked genes in the class III region of the major histocompatibility complex (MHC) on the short arm of chromosome 6 which give rise to the two isotypic variants, C4A and C4B [26, 27]. These variants differ by only six amino acids but these small changes cause significant differences in function, C4A binding preferentially to amino groups after cleavage and C4B to hydroxyl groups (see below). C4B is also more efficient in propagating continued activation of C.

Activated C1, bound through IgG to a surface, cleaves plasma C4 at a single site near the amino terminus of the $\alpha$ chain, releasing a small fragment, C4a (Mr approx. 9 kDa) and, in the process, exposing a reactive thioester group in the $\alpha$ chain of the large fragment, C4b. The thioester group, discovered by Law and Levine 20 years ago [28], is crucial to the function of both C4 and C3. It consists of an intramolecular bridge between a Cys and a Glu residue, separated by two 'spacer' amino acids. In native C4, the thioester is buried deep within a hydrophobic pocket in the molecule, rendering it non-reactive. Once exposed in C4b the thioester can form covalent amide or ester bonds with exposed amino or hydroxyl groups respectively on the activating surface (Figure 1.3). Exposure of the thioester group thus bestows upon C4b the capacity to bind covalently and irreversibly to membranes or other surfaces [29]. The

---

**Figure 1.3. Activation and decay of C3 or C4.**
**A.** Activation of C3 or C4 involves a single cleavage in the $\alpha$-chain. For C4, the cleaving enzyme is activated C1 and for C3 it is the C3-convertase of the CP or AP (alternative pathway). A small fragment, C4a or C3a respectively, is released and a thioester group is exposed in the large fragment, C4b or C3b respectively. Decay of C3b is mediated by fI which, in the presence of an appropriate cofactor (fH, CR1 or membrane cofactor protein (MCP)), cleaves the molecule into inactive fragments. FI cleaves C3b at two sites in the $\alpha'$ chain, releasing a 3-kDa fragment, C3f, and yielding iC3b (i = inactive). A further cleavage by fI, catalysed by CR1 but not the other cofactors, releases a large fragment, C3c, leaving the fragment C3dg (41 kDa) attached to the activator surface. C3dg is promptly further proteolysed by serum proteases, releasing a small fragment (C3g, 5 kDa) and leaving C3d bound to the membrane. Decay of C4b follows a similar pattern except that no equivalent of the intermediate iC3b exists. In the presence of an appropriate cofactor (C4bp, MCP or CR1), fI cleaves at two sites in C4b on either side of the thioester, releasing the large fragment, C4c, and leaving C4d on the membrane. **B.** Structure of the internal thioester group formed by residues 988–992 in the C3d region of the $\alpha$ chain of C3. A similar thioester moiety is also found in C4 and in the structurally related protein, $\alpha$2-macroglobulin.

precise chemistry of thioester bond reactivity and the molecular basis for the different substrate specificities of the thioesters in the isotypes C4A and C4B have only recently been elucidated [30, 31]. The thioester group, once exposed, is extremely labile due to its propensity to inactivation by hydrolysis. As a consequence, C4b binding to surfaces is a very inefficient process, restricting C4b binding to the immediate vicinity of the activating C1 complex. The bulk of the C4b formed is hydrolysed at the thioester and becomes rapidly degraded in the fluid phase. The role of membrane-bound C4b is to provide a receptor for the next component of the CP, C2.

### Binding and activation of C2

C2 is a single-chain plasma protein of molecular weight 102 kDa which, like C4, is encoded by a gene in the class III region of the MHC on chromosome 6. The gene for C2 is closely linked to that for factor B (fB), its functional homologue in the AP (see below). C2 and fB are the most temperature-sensitive of the C components, the only ones to be significantly denatured by brief incubation at 56°C. This thermal instability can be used to selectively deplete serum of C2. Membrane-bound C4b expresses a binding site which, in the presence of $Mg^{2+}$ ions, binds the proenzyme C2 near its amino terminus and presents it for cleavage by C1s in an adjacent C1 complex to yield a 30-kDa amino-terminal fragment, C2b, and a 70-kDa carboxy-terminal fragment, C2a [32]. The C2b fragment may be released or remain loosely attached to C4b but is not required for enzymatic activity. The C2a fragment remains attached to C4b to form the C4b2a complex, the next enzyme in the CP. The enzymatic activity in this complex resides entirely in C2a, C4b acting to tether C2a to the activating surface. The C1 complex plays no further part in C activation.

### Binding and cleavage of C3

The penultimate component of the CP, C3, is the most abundant (1–2 mg/ml in serum) and most important of the C components. It is essential for activity of both the CP and AP. C3 is a large (185 kDa) molecule composed of two chains ($\alpha$, 110 kDa and $\beta$, 75 kDa) held together by disulphide bonds [33]. Like C4, with which it shares sequence homology and many structural features, C3 is synthesized as a single-chain precursor molecule and cleaved intracellularly prior to secretion. The gene for C3 is on chromosome 19.

C3 binds non-covalently to C2a in the C4b2a complex and is then cleaved by the C2a enzyme at a single site in the $\alpha$ chain (between Arg77 and Ser78), releasing the small fragment, C3a (9kDa), from the amino terminus and exposing in the large fragment, C3b, a labile thioester group and binding sites for several C receptors and regulatory proteins (see below). The thioester group, which is essentially identical to that in C4b, confers upon C3b the capacity to bind covalently to the activating C4b2a complex [34–37]. C3b may also be released from C4b2a and bind to the adjacent membrane through its thioester group, as described above for C4b (Figure 1.3). Binding is inefficient and the bulk of the C3b formed decays through hydrolysis of the thioester in the fluid phase and is subsequently degraded. Only C3b bound to the activating C4b2a complex takes any further part in activation, although C3b bound further afield has other important roles in mediating interactions with phagocytic cells. The enzyme so formed, C4b2a3b is the C5 cleaving enzyme (convertase) of the CP.

### Binding and cleavage of C5

The final component of the CP, and also the first component of the membrane attack pathway, is C5, a two-chain plasma protein of 190 kDa molecular weight, encoded on chromosome 9, which is structurally related to C3 and C4. All three of these molecules have probably evolved from a common precursor. Like C3 and C4, C5 is synthesized as a single-chain precursor and cleaved intracellularly prior to secretion. However, C5 does not contain a thioester group and thus cannot bind directly to surfaces [29]. Instead, C5 binds non-covalently to a site on C3b in the C4b2a3b convertase and is presented for cleavage by C2a in the complex. The C4b2a complex can only cleave C5 bound to C3b; C3b is thus an essential component of the C5 convertase of the CP. Cleavage occurs at a single site (after residue 74) in the $\alpha$ chain of C5, releasing a small amino-terminal fragment, C5a (approx. 10 kDa), and exposing in the larger fragment, C5b, a labile hydrophobic surface binding site and a site for binding C6 [10]. C5b remains attached to C3b on the surface during the early stages of membrane attack complex (MAC) assembly.

It has been suggested that C4b2a can cleave C5 even in the absence of C3 – the so-called C3 bypass pathway [38, 39]. While it is of interest for understanding the cleavage events which occur during C activation, the pathway is unlikely to be of physiological relevance. Also of no physiological relevance but useful as a laboratory tool, is the fact that C5 can be 'activated' *in vitro* by freeze-thawing or by incubation at acidic pH, manipulations which presumably cause a limited degree of denaturation and expose binding sites hidden in the native molecule [40, 41]. C5 activated in this manner will bind C6 and the complex thus formed can initiate MAC formation.

## The alternative activation pathway

The alternative pathway (AP) provides a rapid, antibody-independent route for C activation and amplification on invading micro-organisms and other foreign surfaces. The history of the discovery of the AP reads like a Greek tragedy[42]. Pillemer in 1954 described the binding and activation of C3 on yeast cell walls and from this initial observation went on to propose the existence of a C activation pathway separate from the CP and activated in an antibody-independent manner[43]. Pillemer termed this novel pathway the 'properdin pathway' and the discovery generated enormous scientific and media interest. Others challenged this proposal, attributing the observed activation to the presence of natural antibodies reactive with yeast components[44]. The resulting battle ended with Pillemer's untimely death, a sad episode in the history of research. Pillemer's belief in his system was ultimately vindicated by his colleagues who identified all the components of the pathway and convincingly showed independence from the CP[45, 46].

### Initiation of the AP

C3 is the key component of the AP but three other proteins, unique to the AP, are also required, factor B (fB), factor D (fD) and properdin. Factor B is a single-chain 93-kDa plasma protein which shares some 40% sequence identity with C2, the two proteins being encoded by closely linked genes in the MHC on chromosome 6. C3b, in the fluid phase or bound through its thioester to an activating surface, binds fB in an $Mg^{2+}$-dependent manner. Binding C3b renders fB susceptible to cleavage by fD, a 26-kDa serine protease enzyme present in plasma in its active form, which cleaves fB at a single site, releasing a 30-kDa fragment, Ba, and exposing a serine protease domain on the large (60-kDa) fragment, Bb[47]. Factor D can only cleave fB which is bound to C3b. The C3bBb complex thus formed is the C3-cleaving enzyme (C3 convertase) of the AP and cleaves C3 at a site identical to that utilized by the C4b2a enzyme in the CP. Factor D is an essential component of the AP and is unusual among C components in that it is not made in the liver. Instead, fD, also termed *adipsin*, is synthesized almost exclusively by adipose tissue and, besides its role in the C system, has important roles in fatty acid handling by adipose tissue[48]. The third component unique to the alternative pathway is *properdin*, the protein discovered by Pillemer. Properdin binds and stabilizes the C3bBb complex, extending the lifetime of the active convertase three- or four-fold[49, 50]. Properdin is a basic glycoprotein made up of oligomers – mainly dimers, trimers and tetramers – of a 53-kDa monomer[51]. The oligomers are formed by non-covalent interactions between the monomers, which are each composed of six tandemly arranged 60-amino-acid repeats known as thrombospondin repeat (TSR) modules [52].

The AP requires $Mg^{2+}$ ions for assembly of the C3bBb complex while the CP requires both $Mg^{2+}$ ions (for assembly of the C4b2a complex) and $Ca^{2+}$ ions (for assembly of the C1 complex). This provides a most useful means of distinguishing the two pathways in serum samples. EDTA, by chelating both ions, will block both pathways, while EGTA (with supplemental $Mg^{2+}$) chelates only $Ca^{2+}$ and specifically blocks the CP.

Initiation of the AP on a surface requires that C3b be already present and in a conformation which allows fB to bind and be cleaved by fD. The origin of the initiating C3b presents an interesting conundrum. Limited activation of the CP may, by depositing C3b, trigger activation of the AP. This is likely to be the most important trigger for activation of the AP *in vivo* where the two pathways act in concert, the AP acting as an amplifier. Nevertheless, it appears that the AP can be initiated independently of the CP. In biological fluids, C3 is continuously hydrolysed at a slow rate to form a metastable $C3(H_2O)$ molecule which has many of the characteristics of C3b. $C3(H_2O)$ binds fB in solution and renders it susceptible to cleavage by fD. Bb in the fluid-phase C3 convertase thus formed $(C3(H_2O)Bb)$ cleaves C3 to form C3b which can then deposit on adjacent surfaces [29, 53]. As a result of this 'tickover' phenomenon, C3b is continuously deposited in small amounts on all cells, foreign and self, in the body.

### Amplification on activator surfaces

'Tickover' deposition of C3b occurs on all host cells exposed to C but does not result in continued activation because the surface features do not favour efficient binding of fB (non-activator surfaces) and bound C3b is rapidly degraded, a process catalysed by fluid-phase and membrane inhibitors (Chapter 2). In contrast, the surface features of many microorganisms and foreign cells favour amplification (activator surfaces). On activator surfaces, C3b binds fB in an $Mg^{2+}$-dependent manner and presents it for cleavage by fluid-phase fD. The AP C3 convertase thus formed, C3bBb, cleaves more C3 (Figure 1.4). Activating surfaces thus rapidly become coated with C3b molecules, each one of which can itself recruit fB to form more convertases and further amplify activation. The precise nature of the surface features which determine whether activation and amplification will occur are still not clear. However, differences in surface carbohydrates, particularly a relative deficiency of sialic acid on activating surfaces, appear to be important [54-56].

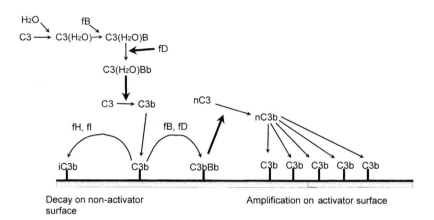

**Figure 1.4. Initiation and amplification in the AP.**

Deposition of C3b is a prerequisite for activation of the AP on a surface. The initiating C3b may be generated by limited CP activation or by a number of other mechanisms which generate C3b-like C3, such as limited cleavage by serum proteases, hydrolysis etc. Conformational changes in hydrolysed C3 [C3(H$_2$O)] expose the binding site for fB and permit the assembly of fluid-phase C3-convertase C3(H$_2$O)Bb which cleaves C3 and causes deposition of C3b. This 'tick over' deposition of C3b occurs continuously on all surfaces exposed to plasma.

What happens next depends on the nature of the surface. On non-activator surfaces the deposited C3b is rapidly inactivated by fI acting in concert with membrane and plasma CRP. On activator surfaces which lack regulators, stable convertases form which cleave more C3, triggering a positive feedback loop which rapidly coats the surface with C3b.

### Cleavage of C5

As in the classical activation pathway, bound C3b acts as an essential receptor for C5, permitting the cleavage of C5 by Bb in an adjacent C3bBb complex. The AP C5 convertase is thus composed of two molecules of C3b, one binding Bb in the C3bBb complex and an adjacent (or attached) C3b acting as a receptor for C5. The convertase hence has the shorthand notation C3bBbC3b (Figure 1.4), nomenclature which has contributed to the widespread misapprehension that C is 'difficult'! The site of cleavage in C5 is identical to that utilized by C2a in the CP convertase (C4b2a3b). C5a is released and labile membrane- and C6-binding sites are exposed on C5b as described above for the CP.

## The lectin pathway

The lectin pathway represents a recently described activation pathway which provides a second antibody-independent route for activation of C on bacteria and other micro-organisms. Recognition of this novel pathway came from a series of insightful studies on an opsonic defect commonly found in infants with repeated infections [4, 5, 57]. Turner and others demonstrated that the yeast opsonization activity defective in these infants was, in normal sera, C-dependent. However, the infants had apparently normal C activity in the known pathways, suggesting that yeast opsonization might involve atypical activation of C.

The key component of this novel pathway is mannan binding lectin (MBL; also termed mannan binding protein, MBP), a high molecular weight serum lectin made up of multiple copies of a single 32-kDa chain [5, 58]. Infants with the yeast opsonic defect have low or absent expression of MBL. Deficiencies of MBL are frequently found in the normal population and in adults MBL deficiency appears to be without consequence. The role of MBL *in vivo* is to bind the mannose and N-acetyl glucosamine residues present in abundance in bacterial cell walls. MBL has a structure similar to that of C1q – a multimeric molecule with globular binding regions and a collagenous stalk. Indeed, MBL *in vitro* associates with C1r and C1s to form a functional C1-like molecule which can initiate CP activation. However, it appears that the association of MBL with C1r and C1s is an artefact of the *in vitro* situation and MBL *in vivo* associates with a novel serine protease termed MBL-associated serine protease (MASP), a 100-kDa protein which is homologous with C1r and C1s [59]. A second MBL-associated protease, termed MASP-2 has recently been described which may also be an essential component of the C1-like complex [60]. The surface-bound MBL–MASP complex activates C4 in much the same manner as the activated C1 complex. MBL thus provides a rapid, antibody-independent means of activating the CP on bacteria. A capacity to activate C is not unique to MBL; various other lectins have also been shown to activate C, leading to the proposal that lectin-triggered activation represents a distinct C pathway specifically tailored to target carbohydrate residues expressed predominantly on bacteria [58]. C-reactive protein is a lectin of the pentraxin family present in plasma at vanishingly low concentrations, but increasing a thousand-fold or more in response to injury or inflammation. Indeed, measurement of C-reactive protein as an index of inflammation (acute phase) is common in clinical practice. C-reactive protein binds the C-polysaccharide present on the pneumococcus and a variety of other bacteria and, once bound, activates C in a manner similar to MBP, providing a lectin pathway which is switched on in response to inflammation [61].

## The membrane attack pathway

Cleavage of C5 by the C5 convertase of either the CP or AP is the final enzymatic step in the C cascade. The membrane attack pathway involves the non-covalent association of C5b with the four terminal C components, all of which are hydrophilic plasma proteins, to form an amphipathic membrane-inserted complex (Figure 1.5). While still attached to C3b in the convertase C5b binds C6, a large (120-kDa) single-chain plasma protein. Binding of C6 stabilizes the membrane binding site in C5b and exposes a binding site for C7, a 110-kDa single-chain plasma protein which is homologous to C6 (see below). Attachment of C7 causes conformational changes in the complex which result in its release from the convertase to the fluid phase. The hydrophobic membrane binding site is unstable and susceptible to inactivation by hydrolysis or by interaction with numerous plasma proteins. If the C5b67 complex encounters a membrane during the brief lifetime of the membrane binding site it associates tightly with the membrane. However, the complex does not at this stage penetrate deeply into the membrane and does not disturb the integrity of the lipid bilayer. In addition to the natural decay of the membrane binding site, attachment of C5b67 is further restricted by the tendency of the

**(a)**

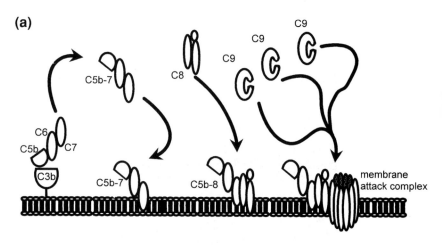

**Figure 1.5. The membrane attack pathway.**
**(a).** Cleavage of C5 is the final enzymatic step of the C system. C5b bound to the convertase sequentially binds C6 and C7. The C5b67 complex is released from the convertase and associates with adjacent membrane via a labile hydrophobic binding site in the C6 component. The C5b67 complex stably associated with membrane then binds C8 and finally, multiple copies (up to 12) of C9.
**(b).** Monomers of C9 unfold and associate with each other to form a barrel-like structure with a central pore which traverses the membrane.
**(c).** Electron micrograph of ring lesions on C-lysed sheep erythrocyte membranes. The internal diameter of the pores is about 0.1 μm (scale bar 1μm).
**(d).** Electron micrograph of polymerised C9 to illustrate the MAC-like complexes which spontaneously form *in vitro* (scale bar 1 μm).

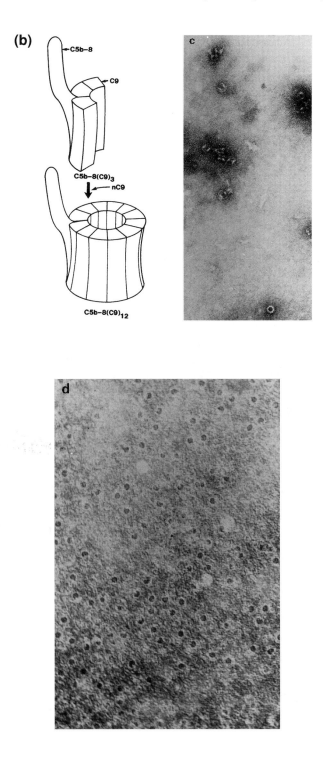

**(b)**

C5b–8

C9

C5b–8(C9)$_3$

nC9

C5b–8(C9)$_{12}$

c

d

complex to aggregate and by the presence of multiple fluid-phase inhibitors (see below). Deposition of C5b67 is thus limited to the target cell. A small proportion of complexes may attach to the membranes of closely apposed host cells and cause damage or lysis, so-called 'innocent bystander' lysis[62–64].

The penultimate component of the membrane attack pathway, C8, is a complex molecule made up of three chains, α, β and γ (molecular weights 65 kDa, 65 kDa and 22 kDa respectively), encoded by separate genes, α and β, which are closely linked on chromosome 1, and γ on chromosome 9. The α and β chains are homologous with each other and with C6 and C7 whereas C8γ displays no homology with any of the C components. The α and γ chains are covalently linked whereas the β chain is non-covalently associated in the complex. The β chain in C8 binds C7 in the C5b67 complex and the resulting complex, C5b-8, becomes more deeply buried in the membrane and forms small pores, causing the cell to become slightly leaky[65].

The final component in the pathway, C9, is a single-chain plasma protein (molecular weight 69-kDa) which is homologous with C6, C7, C8α and C8β. The first C9 molecule to enter the C5b-8 complex binds to C8α and undergoes a major conformational change from a globular, hydrophilic form to an elongated, amphipathic form which traverses the membrane and exacerbates membrane leakiness. Unfolding of C9 also exposes binding sites which enable additional C9 molecules to bind, unfold and insert in the membrane. The pore thus grows with the recruitment of additional C9 molecules, individual complexes (membrane attack complexes, MACs) containing as many as 18 C9 molecules. MACs containing multiple C9 molecules can be visualized in the electron microscope as ring-like structures enclosing a 10-nm pore (Figure 1.5). Evidence that the pore is formed mainly from C9 molecules is provided by the observation that C9 incubated *in vitro* in the presence of $Zn^{2+}$ ions forms identical ring structures which are composed of 12 or more C9 monomers (Figure 1.5)[64, 66–68]. The precise mechanism by which the MAC causes lysis is still the subject of debate between those who contend that the MAC ring surrounds a rigid, transmembrane pore and those who consider that the MAC induces areas of lipid perturbation (leaky patches) in the membrane[69, 70]. This latter concept is supported by evidence that incorporation of only one or two C9 molecules into the C5b-8 complex is sufficient to produce a lytic complex. Whatever the mechanism, the MAC forms functional pores in cell membranes through which ions and small molecules pass, bringing about osmotic lysis of the cell. This strategy of pore formation as a means of killing targets is widely used in nature, notably by bacterial toxins and by mammalian cytotoxic T lymphocytes[71].

The MAC component proteins C6, C7, C8α, C8β and C9 are all highly homologous and have clearly arisen from a common ancestor by gene reduplication (Figure 1.6). C6 and C7 are closely linked on chromosome 5 with C9 more distant on the same chromosome [72]. C8α and C8β are closely linked on chromosome 1. These duplication events have probably arisen to increase the efficiency of targeting and of cytolysis by the MAC.

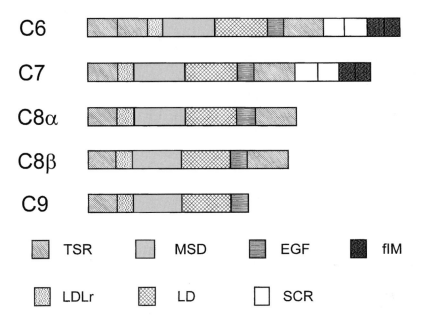

Figure 1.6. **Structural similarities between terminal C components.**

The component proteins of the membrane attack complex (MAC), C6, C7, C8α, C8β and C9, share many structural features and have clearly arisen by gene reduplication. The lytic domain (LD) is present in each of the proteins and is also present in the cytolytic protein perforin, indicating that it has a role in the interactions of these proteins to form the pore. Other shared domains include the thrombospondin repeat module (TSR), the LDL receptor module (LDLr), the MAC-specific domain (MSD) and the epidermal growth factor module (EGF). C6 and C7 each contain two short consensus repeat (SCR) domains, found in the receptors and regulators which bind C3, C4 or C5.

## PHYSIOLOGICAL ROLES OF C

The physiological importance of C is dramatically demonstrated in individuals deficient in components of the system. The main problems encountered by such individuals are recurrent and severe bacterial infections and immune complex disease, vividly demonstrating that the principal roles of C are to mediate killing of invading bacteria and to solubilize immune complexes [73, 74]. C also plays an important physiological role as an initiator or enhancer of inflammation at sites of infection or injury, and in recent years it has become evident that C also makes an important contribution to the induction of antibody responses [75–77]. These roles are all mediated by the products of C activation and the C receptors with which they interact.

## Cell membrane receptors for C

Cells express surface receptors specific for C1q, for the large fragments of C3 and C4 generated during C activation and for the small fragments of C3, C4 and C5 (Table 1.2). Almost all of the effects of C are mediated through interactions with cell membrane receptors and it is impossible to understand the various effects of C without some knowledge of the receptors. Here we will provide a brief overview of the various receptors with which C proteins, complexes and fragments interact to bring about the physiological and pathological effects.

### Receptors for C1q

Evidence that C1q bound to lymphocytes was first described in the early 1970s [78]. Binding was later shown to be specific and saturable, indicating the presence of a cell surface receptor [79]. The lymphocyte receptor was isolated and partially characterized [80] and receptors for C1q (C1qR) were subsequently described on cells of the monocyte/ macrophage lineage and endothelial cells in the early 1980s [81–83]. The receptor purified on C1q-sepharose had a molecular mass of around 60–70 kDa, was widely distributed and had roles in the modulation of activation of B cells and phagocytes [84]. The receptor bound C1q through the collagen tail and also bound other collagenous molecules, including lectins [85]. The ability to bind these other molecules has led to the adoption of the term 'collectin receptor' (i.e. collagen/lectin receptor) for this molecule. Detailed structural analysis has revealed that C1qR is either very closely related to or identical with the intracellular $Ca^{2+}$-binding protein, calreticulin [86, 87]. The process by which a calreticulin-like molecule appears at the cell surface to act as a receptor for C1q remains unknown.

A C1q-binding protein of molecular mass 100 kDa has been isolated from monocytes and suggested to be a second receptor for C1q [88]. This highly acidic molecule is expressed on neutrophils and monocytes and, like the collectin receptor, binds C1q through its collagenous tail.

**Table 1.2.** *Receptors for C components, fragments and complexes.*

The known cellular receptors for C are listed together with their CD assignment (where known), natural ligands, molecular characteristics and cell distribution. Abbreviations: sc, single chain; tm, transmembrane; SCR, short consensus repeat; E, erythrocyte; FDC, follicular dendritic cell; NK, natural killer; nd, not determined.

| Receptor | Ligand | Characteristics | Distribution |
|---|---|---|---|
| cC1qR | C1q, collagenous region | 70 kDa; calreticulin-like | Broad |
| cC1qR (100) | C1q, collagenous region | 100 kDa; acidic glycoprotein | Monocytes, neutrophils |
| gC1qR | C1q, globular heads | 33 kDa; 80 kDa | Leukocytes; platelets |
| fHR | factor H | nd | Broad (?) |
| CR1 (CD35) | C3b/C4b | 180–200 kDa, sc, tm, approx. 30 SCRs | E, B cells, neutrophils, monocytes etc. (broad?). |
| CR2 (CD21) | C3d (and EBV) | 145 kDa, sc, tm, 15 or 16 SCRs | B cells, FDCs, epithelia, glia, ? others? |
| CR3 (CD11b/18) | iC3b (and matrix) | Heterodimer | Myeloid and NK cells |
| CR4 (CD11c/18) | iC3b, C3dg | Heterodimer | Myeloid, macrophages |
| C5aR (CD88) | C5a | 40 kDa; 7-tm-spanning | Neutrophils, macrophages, mast cells, muscle, ?others? |
| C3aR | C3a | Approx. 60 kDa; 7-tm-sp. | As above |
| C4aR (?) | C4a (?) | nd | nd |
| C5b-7R | C5b67 | nd | nd |

A third receptor for C1q, initially distinguished by its molecular mass (33 kDa) and the observation that it bound C1q through the globular heads (gC1qR), was isolated from Raji cells and shown to be expressed on leucocytes and platelets[89]. The gC1qR is a highly charged, acidic glycoprotein. Neutrophils expressed this receptor and a second gC1qR with a molecular mass of 80 kDa[90]. The precise roles of these multiple receptors for C1q in the regulation of C activation on cells and in other C1-mediated cell activation events remain poorly understood.

**Receptors for the large fragments of C3 and C4**
Multiple receptors exist for the large products of cleavage of C4 (C4b) and C3 (C3b) generated during C activation and for the fragments generated during subsequent cleavage of C4b and C3b.

*C receptor 1* (CR1; CD35) has dual roles as C regulator and C receptor and is described in detail in Chapter 3. The ligands for CR1 are C3b and C4b which bind to separate sites on the large CR1 molecule[91]. CR1 on erythrocytes plays a key role in the transport of immune complexes (coated with C3b and C4b) through the circulation to the sites of clearance in spleen and liver.

*C receptor 2* (CR2; CD21) is the receptor for C3d, the surface-bound fragment which is the end-product of the cleavage of C3b by fI and serum proteases (Figure 1.7)[92]. It is also the receptor for the Epstein–Barr virus, the consequences of which are discussed elsewhere in

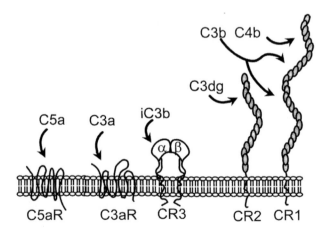

**Figure 1.7. Receptors for fragments of C3.**
Fragments of C3 mediate their effects by binding specific receptors expressed on cells (Table 1.2). Illustrated are the various receptors so far identified for C3 fragments and the C fragments with which they interact.

this volume. CR2 is highly homologous to CR1 and the genes are adjacent in the regulators of C activation (RCA) cluster on chromosome 1. Nevertheless, CR2 does not function as a C regulator. The structure and function of CR2 will be discussed in Chapter 3. It is now becoming apparent that CR2 on the B-cell surface plays a major role in the induction of antibody responses [reviewed in 92, 93]. The primary role of CR2 expressed on other cell types is undefined.

C *receptor 3* (CR3; CD11b/CD18) is the receptor for iC3b (Figure 1.7) and also binds numerous extracellular matrix proteins. It is a member of the β-2 integrin family of cell adhesion molecules and, like other members of this family, is a heterodimer composed of one molecule of the common integrin β chain, CD18, and one molecule of a specific α chain which in CR3 is CD11b [94–97]. CR3 is expressed on myeloid and natural killer cells and functions primarily as an adhesion molecule, mediating binding to matrix and to C-opsonized surfaces.

C *receptor 4* (CR4; CD11c/CD18) binds iC3b and C3dg. Like CR3, it is a β-2 integrin with wider roles in cell adhesion.

**Receptors for the small fragments of C4, C3 and C5**
Small (approx. 10-kDa) peptides are generated from cleavage of C3, C4 and C5 during C activation, and termed C3a, C4a and C5a respectively. C3a and C5a express important biological activities through interaction with specific receptors; no receptor for C4a has been identified and current evidence suggests that C4a is without function in man. C3a and C5a are short (respectively 77 and 74 amino acids) highly cationic peptides. Removal of the carboxyl-terminal Arg residue by serum carboxypeptidase N (anaphylatoxin inactivator) reduces or abrogates receptor binding for each anaphylatoxin. The presence of specific and distinct membrane receptors for C5a and C3a on neutrophils and macrophages was first demonstrated by classical methods using radiolabelled or fluorescent ligands [98, 99]. The C5a receptor (C5aR; CD88) was characterized as a 40-kDa G-protein-associated molecule and its cloning from U937 cells [100] placed C5aR in the family of 7-transmembrane-spanning receptors which includes many cytokine receptors and the fMLP receptor. C5aR is abundantly expressed on neutrophils, macrophages and monocytes. There is a growing body of evidence that C5aR is expressed, albeit at lower copy number, on many other cell types [101, 102]. The importance of C5aR in immune defence has been illustrated by the recent engineering of a C5aR knockout mouse; the animals display a severe impairment of mucosal immunity [103].

The receptor for C3a (C3aR) has proved more elusive and was finally cloned only in 1996 [104]. C3aR is also a member of the 7-transmembrane-spanning receptor family and is highly homologous with C5aR, the major difference being the large size (172 amino acids) of the second extracellular loop, which is short in C5aR and in all other members of this receptor family. The cDNA sequence predicts a mature protein of 482 amino acids and a predicted molecular mass of 54 kDa. However, the sequence includes two putative N-glycosylation sites in the large loop and the finding that C3aR has an apparent molecular mass of 65 kDa on SDS-PAGE suggests that it is heavily glycosylated in this loop [105].

It had been suggested that C3a and C4a shared a common receptor. However, recent studies using the expressed protein show that this is not the case. Given the failure to demonstrate any biological activity for C4a in humans it seems unlikely that a specific, functional C4aR remains to be discovered.

## Bacterial killing by C

Most bacteria, with the important exception of Gram-negative organisms belonging to the genus *Neisseria*, are resistant to the lytic effects of the MAC, a consequence of their thick cell walls and capsules. Nevertheless, C contributes to bacterial killing by attracting phagocytic cells to the site of infection and coating the bacteria in readiness for phagocytosis. The surfaces of most bacteria activate C via the AP; bacteria will also activate C via the CP in the presence of specific antibodies and via the lectin pathway if the relevant sugar groups are expressed on the organism. As a consequence, the bacterial surface rapidly becomes coated with C fragments, particularly C3b and its breakdown products iC3b and C3d [106]. This process, known as opsonization, enables phagocytic cells to recognise bacteria by binding to fragments of C3 through CR1 and CR3 on the phagocyte membrane. The bacterium can then be efficiently internalized, killed and digested within the cell [107]. In the presence of specific antibody, opsonization can additionally be achieved by bacterium-bound antibody interacting with Fc receptors on phagocytes.

C activation at the site of infection also releases the small fragments, C3a, C4a and C5a which act to attract phagocytic cells to the site and activate them for efficient killing. The fragments C4a, C3a and C5a are all very similar in structure, all have an Mr of 9–10 kDa (74–77 amino acids) and are derived by cleavage at an Arg-X peptide bond from the N-terminus of the α-chain of the parent molecule [108]. C3a and C5a have similar biological effects, although their potencies differ, C5a being the most active. No convincing evidence for function of C4a in humans has been produced. C3a and C5a bind to specific receptors (described above) on mast cells and basophils, triggering degranulation with the consequent release of vasoactive amines which mediate vasodilatation and increased vascular permeability (anaphylaxis). Both contain an Arg residue at the carboxy terminus which is essential for optimal activity. Removal of this residue by a constitutively active plasma enzyme carboxypeptidase N (anaphylatoxin inactivator) yields the desArg metabolite and completely abrogates anaphylactic activity of these fragments, thus restricting their activity to the site of activation[108]. C5a and to a lesser extent, C3a, also attract phagocytic cells to the site of activation (chemotaxis). This activity is mediated through C5aR and C3aR expressed on phagocytes. Binding of C5a stimulates cell adhesion to and migration through vessel walls and the concentration gradient of C5a guides the cells towards the inflammatory site. In addition, C5a also causes activation of phagocytic cells, priming them for bacterial killing. Importantly, C5adesArg retains some affinity for the phagocyte C5aR even after loss of the carboxy-terminal Arg and is thus able to signal chemotaxis at some distance from the activation site[109]. As noted above, many other cell types express C5aR and C3aR. Of particular interest is the recent observation that C5aR is highly expressed on hepatocytes and that binding of C5a triggers activation of the cells for synthesis of acute phase proteins[110]. This finding provides a fascinating bridge between activation of C and initiation of the acute phase response which, among other things, increases the synthesis of C components and activators.

### Immune-complex solubilization by C

Immune complexes consist of aggregates of antigen and antibody which form in the plasma under physiological and pathological conditions. Growth of immune complexes occurs as more molecules of antibody and antigen are recruited into the expanding matrix, a consequence of the bivalent nature of antibody and the broad, polyclonal response with antibodies directed against multiple epitopes on the antigen (Figure 1.8). As the complex grows, its solubility decreases; large, insoluble complexes will precipitate in capillary beds and cause inflammation. C plays several crucial roles to prevent this occurring:

i.   The complexes efficiently activate the CP and hence become coated with C fragments (opsonized); coating with C masks antigenic sites, limiting further growth of the immune complex.

ii.  Large immune complexes activate the alternative pathway and become coated with C3b which effects disaggregation, converting large, insoluble complexes to small soluble ones.

iii. C3b and C4b on the immune complex binds the major erythrocyte C receptor, CR1, effectively removing the immune complex from the circulation by tethering to the erythrocyte.

Erythrocytes are a central component of the immune complex clearance system (Figure 1.8). Erythrocyte-bound complexes are carried to the liver and spleen where transfer of the complex to fixed tissue macrophages occurs with subsequent internalization and digestion [111-113]. Binding of immune complexes to erythrocytes is a dynamic event; C3b once bound to the cofactor CR1 is rapidly cleaved by serum fI to yield iC3b which is bound only weakly by CR1, causing release of the complex. Another C3b on the complex will then bind the CR1 and the cycle of binding, cleavage and release is repeated (Figure 1.8). As a consequence of this 'on-off' binding, delivery of the immune complex to fixed macrophages in the tissues does not retard or damage the erythrocyte. Individuals with deficiencies of early components of the CP, particularly those with deficiencies of components of the C1 complex, do not activate C on immune complexes which therefore grow large and do not bind erythrocytes. Large, insoluble complexes will form and deposit in the tissues, causing immune complex disease.

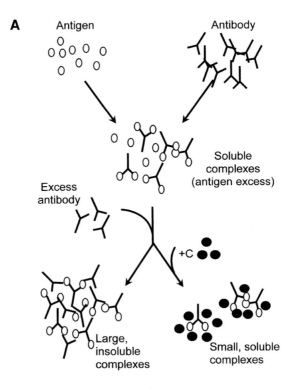

**A**   Antigen                                    Antibody

Soluble complexes (antigen excess)

Excess antibody

+C

Large, insoluble complexes

Small, soluble complexes

**B**     <u>Circulation</u>

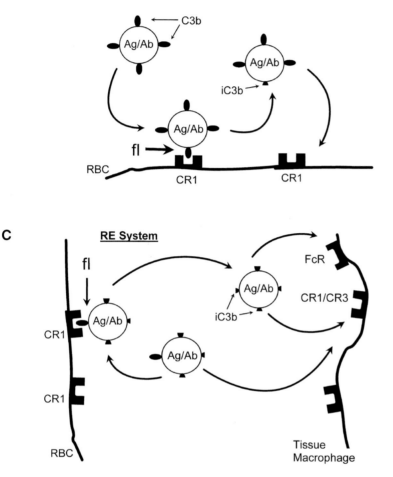

**Figure 1.8. Immune complexes and C.**
**A.** C interferes with the growth of immune complexes. As immune complexes form they activate C through the CP. Coating of the immune complex with C fragments masks antigen, preventing the recruitment of more antibody and growth of the complex.
**B.** C involvement in immune complex transport by erythrocytes (red blood cell; RBC). C3b on the immune complex binds to CR1 on the RBC surface, tethering the complex and effectively removing it from the circulation. However, CR1 is a cofactor for fl cleavage of C3b; the cleavage product, iC3b, binds CR1 very weakly and the complex is thus released. The complex will rapidly reassociate through another C3b molecule. This on-off cycle provides an efficient and dynamic system of transport.
**C.** In the organs of the reticuloendothelial (RE) system (spleen and liver), immune complexes released from RBC are taken up and destroyed by tissue macrophages.

## Cell activation by C

Several of the products of C cause priming of phagocytic cells for more efficient bactericidal activity at the site of inflammation. Binding of C1q to
the C1qR on phagocytes enhances the ability of the cells to mount an oxidase response. Binding of C5a or C3a to their respective receptors on
phagocytes and other cell types can trigger oxidase responses and other
cell activation events.

The MAC forms pores in membranes and can thus bring about lysis of
some target cells. However, MAC is also important as a cell activator *in
vivo*. Metabolically active nucleated cells at the inflammatory site are resistant to lysis by the MAC and are activated to release proinflammatory
molecules, thus enhancing the inflammatory response. These important
non-lytic effects of the MAC are discussed in several recent reviews[114, 115].

## C and the immune response

Several of the complexes and fragments generated during C activation
influence the cellular immune response. Most work has focused on C3 and
its fragments[75, 76]. C3 and the fragments C3b and iC3b have been reported
to enhance the mitogen-induced proliferative response of B and T lymphocytes, probably acting through C receptors (CR1, CR2 and CR3) on
phagocytic cells and lymphocytes[116]. *In vivo*, fragments of C3 play an
important role in the trapping of antigen-containing immune complexes
in lymph nodes. Follicular dendritic cells (FDC) express C receptors
which bind C3 fragments on the complexes. The immune complexes are
retained on the surface of the FDC where they can interact with B and T
cells in the follicle to trigger an antibody response. Individuals deficient in
C3 trap antigen on FDC very poorly and as a consequence do not mount a
good immune response[77, 117]. As well as this indirect effect on antibody
generation by modulating the development of T-cell help, C3 fragments
also have a direct role at the surface of the B cell. Antigen binding to the
B cell receptor (membrane IgM) may be sufficient to trigger a B-cell
response with proliferation and antibody production, but, as with activation events in other lymphocytes, it appears that a second signal is necessary for an optimal response. Work pioneered by Fearon, Tedder and others
over the last 8 years has shown that C3 fragments bound to the antigen
may provide this second signal[92, 93]. Antigen which has activated C will
bear C3 fragments at various stages of breakdown. The final covalently
associated fragment, C3d, is the ligand for CR2, a C receptor expressed on
B cells in a complex with, among other components, CD19. The antigen
may then bind the B cell receptor and, through covalently bound C3d,

may co-ligate CR2 (Figure 1.9). The CR2/CD19 complex, once ligated with the B cell receptor, will deliver a signal which will markedly enhance the response of the B cell to antigen. The potential applications of this novel pathway to adjuvant design have not escaped the attention of the discoverers[118, 119].

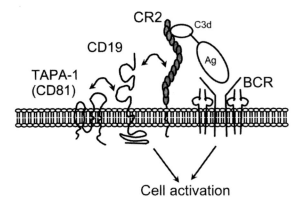

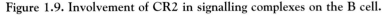

Cell activation

**Figure 1.9. Involvement of CR2 in signalling complexes on the B cell.**
CR2 is constitutively associated with CD19 and TAPA-1 on the B cell surface. Antigen coated with C3dg will cross-link the B cell receptor (BCR; mIgM) and CR2 in the CR2/CD19/TAPA-1 complex. Ligation of the BCR triggers a first activation signal and the co-ligated CR2/CD19/TAPA-1 complex delivers a second signal, dramatically increasing the B cell response to antigen.

The anaphylatoxins may also play a role in the regulation of immunity. C3a has been shown to inhibit lymphocyte activation induced by mitogen or antigen. These effects are probably mediated through C3aR on T lymphocytes. C3adesArg (C3a which has lost its carboxy-terminal Arg) has no inhibitory effect. Other fluid-phase fragments of C3 have also been demonstrated to influence the cellular immune system *in vitro* although the physiological relevance of these findings is unclear. C5a indirectly enhances antibody responses by stimulating cytokine release from macrophages. Importantly, C5adesArg, which is much more abundant and stable than C5a *in vivo*, retains this activity. As a consequence, the stimulatory effect of C5a and C5adesArg is likely to outweigh the inhibitory effect of C3a and the net result is that the small, soluble products of C activation cause enhanced humoral immune responses[120, 121].

## INVOLVEMENT OF C IN PATHOLOGY

The above account provides a cogent case for the physiological import-ance of C, but there is within the system an enormous potential for harm – a double-edged sword. In general, harm will result when C activation occurs in an uncontrolled manner and/or at an inappropriate site; the end result of this will be inflammation and tissue destruction. In C deficiencies, the converse situation applies. Here we see the consequences of the loss of the physiological functions of C. Here we will first review briefly the pathologies associated with the various C deficiencies and then provide an even shorter account of some situations *in vivo* where inappropriate C activation causes disease.

### Deficiencies of C

Deficiencies of almost every C protein and regulator have been described and more detailed accounts of the various C deficiencies can be found in several recent reviews [73, 74, 122]. Deficiencies of the regulators of C will be discussed in Chapter 5. The clinical consequences of deficiency of a single component are dependent on the particular pathway of C activation com-promised by the loss (Figure 1.10).

Deficiencies of components of the CP (C1, C4 or C2) are associated particularly with an increased susceptibility to immune complex disease, a consequence of the failure of immune complex solubilization. The fre-quency and severity of disease is greatest with deficiencies of one of the subunits of C1 (C1q, C1r, C1s), closely followed by total C4 deficiency, each giving rise to a severe immune complex disease which closely resem-bles systemic lupus erythematosus (SLE). Subtotal deficiency of C4 is com-mon, due to the extremely high frequency of null alleles at both the C4A and C4B loci, but total C4 deficiency is very rare. Deficiency of C2 is the commonest homozygous C deficiency in Caucasoids but causes much less severe disease; it appears that deposition of the early components, C1 and C4, provides some solubilization of immune complexes.

C3 is an essential component of both activation pathways and is vital for efficient opsonization of bacteria. Deficiency thus causes a marked susceptibility to bacterial infections. Immune complex disease is not a common finding in C3 deficiency, although solubilization of preformed immune complexes is severely compromised in the absence of C3. Indi-viduals with C3 deficiency run a stormy course through childhood but with prompt therapy of infections can survive. In adulthood, the number and severity of infections is much reduced as other arms of the immune system take up the challenge. Deficiencies of the regulators fI and fH cause

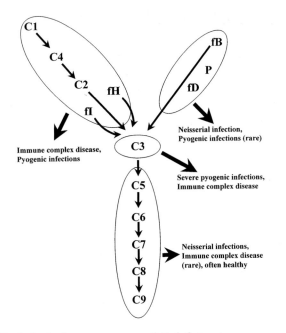

**Figure 1.10. Pathological consequences of C deficiencies.**
Deficiencies of every component of the C system have been described. The consequences of the deficiency are dependent on the portion of the system disrupted.
Deficiencies of CP components (C1, C4, C2) present with immune complex diseases; deficiency of C3 presents with severe bacterial infections; deficiencies of
components of the AP and terminal pathway present with neisserial infections,
although the nature and course of the infections differs in the two groups.

a secondary deficiency of C3 and present with similar symptoms. These
deficiencies will be detailed in Chapter 5.

   Deficiencies of components of the AP are rare and do not predispose to
immune complex disease or pyogenic infections. A few individuals
deficient in fD have been described all of whom have presented with
recurrent *Neisseria* infections, usually meningococcal meningitis. Very
recently, two cases of fH deficiency have been described, again presenting
with meningococcal disease. Deficiency of the positive regulator, properdin, is the commonest disorder of the AP and will be discussed in
Chapter 5.

   Deficiencies of terminal pathway components (C5, C6, C7, C8 or C9)
also cause susceptibility to infection with organisms of the genus *Neisseria*.
Deficiency of C6 is the second most common C deficiency among
Caucasians and is frequently associated with meningococcal meningitis
and systemic infection with the meningococcus. C9 deficiency, rare in

Caucasians, is by far the most frequent C deficiency in Japan with an incidence approaching 1 in 1000 [123].

## C in inflammatory disease

The C system contributes to tissue damage in a large number of auto-immune diseases. Autoantibodies and immune complexes deposit in the affected organs where they trigger activation of C and exacerbate inflammation. In many autoimmune diseases, from the organ-specific (e.g. autoimmune thyroid disease) to the disseminated (e.g. systemic lupus erythematosus), C deposition can be detected in the affected tissues and products of C activation are found in the plasma [124–126]. In all these auto-immune diseases, C is just one of several factors which contribute to pathogenesis. Nevertheless, modulation of C activation may be of therapeutic benefit in these conditions. In systemic lupus erythematosus, autoantibodies are present which recognise DNA and other components of normal cells. Following any stimulus to cell death, these components will be released and immune complexes will form. Immune complexes deposit in capillary beds, particularly in skin and kidney, where they activate C to cause inflammation and further tissue destruction. A vicious cycle is triggered in which more cell killing drives the production of more immune complexes which in turn exacerbates C activation and tissue destruction.

Treatment-precipitated or 'iatrogenic' C activation is a common consequence of the many therapies which involve contact of blood with a foreign surface. Many materials used in the tubing and membrane components of extracorporeal circuits are 'bioincompatible' and hence potentially C-activating. Exposure of blood to these materials in renal haemodialysis, cardiopulmonary bypass or plasmapheresis is a frequent and hazardous event [127]. In haemodialysis, blood is in contact with a large surface area of dialysis membrane, commonly made of cuprophane, a complex polysaccharide material which efficiently activates C via the AP. Release of active fragments, particularly C3a and C5a, causes activation of neutrophils and monocytes which aggregate and become deposited in the lung. Release of toxic molecules from these cells causes the lung damage which is a common complication of long-term dialysis [128]. Numerous modifications to the membrane surface have been investigated to reduce C activation. The most promising strategy involves coating the membrane with heparin a simple procedure which renders the circuit much less C-activating [129–132].

Activation of C may also occur in transplanted organs. When a poorly matched organ is used for transplant it is rapidly rejected due, in part, to C activation in the graft, the phenomenon of hyperacute rejection. C-mediated hyperacute rejection is an almost universal outcome of cross-species transplants and is thus the major barrier to the use of animal organs for human recipients (xenotransplantation)[133]. The attempts, now underway, to overcome this problem by producing transgenic animals (pigs) expressing human C regulatory molecules are described in Chapter 9.

## WHY A BOOK ON C REGULATORS?

The C system was first described over a century ago and all of the components were known by the mid-1970s, causing many immunologists to assume at that time that everything in C was solved and further study of the system was not necessary. Despite this assumption, the C field remains active and productive. In the last decade the sequences of all the components have been ascertained, revealing interesting homologies between C components and with non-C proteins. The locations and structures of many of the C genes have been reported and have provided important clues to the evolution of the C system; the mutations causing C deficiency have been characterized in many individuals and new and improved methods for measuring C have been developed and used to demonstrate C activation in many diseases. Perhaps the most consistently productive area over the last two decades of C research has been the study of the molecules which function as regulators and receptors for C. An area which was born in the late 1970s has grown to dominate the field and to offer real prospects for therapy. For this reason we contend that a comprehensive overview of the field will be of benefit to many researchers and clinicians. For the remainder of this volume we will focus on the C regulators, seeking to present in an easily digestible manner the history, current knowledge and future directions of work on the various regulators.

*Chapter 2*

# REGULATION IN THE COMPLEMENT SYSTEM

## INTRODUCTION

The importance of control in the C system is clearly indicated by the fact that the system includes almost as many regulators as it does pathway components. At least 11 proteins, present in the plasma and on cell membranes, function to downregulate activation of C (Table 2.1). While some of these proteins also serve other functions, regulation of C appears to be the sole physiological role of several. Why has nature evolved such an abundance of C regulators? The answer lies in two aspects of the C system which make it a potent threat to self-cells: first, C is continually activated at a low, tickover rate on all cells in the body; second, the activation pathways of C are proteolytic cascades which endow an enormous propensity for amplification, a small initiating stimulus becoming rapidly magnified to a large and potentially harmful response in the absence of regulation. Not only would uncontrolled activation cause local damage, it would also rapidly consume C, rendering the individual C-deficient, as is graphically illustrated in individuals deficient in the fluid-phase regulator factor I (fI) [134]. In order to prevent such a catastrophe, each part of the C system is tightly regulated. Control is provided in part by the inherent instability of the activation pathway enzymes and fluid-phase precursors of the MAC. This instability limits the effects of C to the vicinity of the activation site. Nevertheless, in the absence of efficient regulation, host damage will occur.

The purpose of this book is to review the current state of knowledge regarding regulation of the C system. In this brief chapter we will introduce the characters, each of which will be described in much greater detail in later chapters, and overview their interactions with C and with each other.

## HISTORY

The history of C regulation is almost as old as that of C itself. Although Ehrlich, Bordet and others at the turn of the century had noted some

**Table 2.1. C *regulatory proteins*.**

RCA, regulators of complement activation gene cluster; tm, transmembrane; GPI, glycosyl phosphatidylinositol; conv., convertase.
C1 inh, C1 inhibitor; CPN, carboxypeptidase N; MCP, membrane cofactor protein; DAF, decay accelerating factor; CR1, C receptor 1; HRF, homologous restriction factor.

| Molecule | Structure | Serum concentration (mg/l) | Gene location | Target |
|---|---|---|---|---|
| **Plasma** | | | | |
| **C1inh** | Single chain, 76 kDa | 200 | 11 | C1 |
| **fH** | Single chain, 150 kDa | 450 | 1 (RCA) | C3/C5 conv. |
| **fI** | 2 chains: 50 kDa; 38 kDa | 35 | 4 | C3/C5 conv. |
| **C4bp** | 6 or 7 α chains (70 kDa), | 250 | 1 (RCA) | CP C3 conv. |
| | 1 or 0 β chain (45 kDa) | | 1 (RCA) | |
| **S protein** | single chain, 83 kDa | 500 | 17q | C5b-7 |
| **Clusterin** | 2 chains: α, 35 kDa; β, 38 kDa | 50 | 8p | C5b-7 |
| **CPN** | dimeric heterodimer, 290 kDa | 30 | ? | C3a, C4a, C5a |
| **(anaphylatoxin inactivator)** | Individual chains of 85 kDa, 50 kDa | | | |
| **Membrane** | | | | |
| **MCP** | Single chain, 60 kDa; tm | — | 1 (RCA) | C3/C5 conv. |
| **DAF** | Single chain, 65 kDa; GPI | — | 1 (RCA) | C3/C5 conv. |
| **CR1** | Single chain, 200 kDa; tm | — | 1 (RCA) | C3/C5 conv. |
| **HRF** | Single chain, 65 kDa; GPI | — | ? | C5b-8/C5b-9 |
| **CD59** | Single chain, 20 kDa; GPI | — | 11 | C5b-8/C5b-9 |

aspects of the behaviour of C which were suggestive of regulation[135], the credit for the first formal studies in this area must go to the Edinburgh pathologist Robert Muir[136, 137]. He noted that human erythrocytes were much more difficult to lyse when human serum was used as a source of C than when sera from a variety of species was substituted. The same held for erythrocytes from other species: rat erythrocytes were least sensitive to rat C and so on. This phenomenon implied the existence of factors in human serum or on human cells which provided resistance against lysis in a homologous combination but not in a heterologous combination. Several studies in the early 1980s re-addressed this issue and confirmed Muir's observation, C from diverse species lysed antibody-sensitized or unsensitized heterologous erythrocytes much more effectively than homologous erythrocytes [138, 139]. The phenomenon persisted even when a reactive lysis system was used to directly deposit MACs on the membrane of the target erythrocyte, eliminating the activation steps [140]. Hansch and co-workers coined the term 'homologous restriction', a term which has since been widely used to describe the apparent species selectivity in C lysis.

The evolution of the phenomenon into a process understandable at the molecular level had begun with the discovery of C inhibitory activities in extracts of erythrocyte stroma [141–143]. Hoffmann described the partial purification of components from human (or non-human) erythrocyte membranes which, when added to appropriate target erythrocytes, rendered the target resistant to lysis by human (or the matching non-human) serum. Nicholson-Weller and colleagues refined the fractionation procedures used by Hoffmann and succeeded in purifying single components from the membranes of guinea pig and human erythrocytes which, when incubated with non-human erythrocytes, protected against lysis by guinea pig and human serum respectively [144, 145]. The protective factor was shown to be a protein with a molecular weight of approximately 60 kDa which caused the degradation of the C3-cleaving enzymes and was given the descriptive name 'decay-accelerating factor' (DAF). The biochemistry and biology of DAF is described in Chapter 3. There followed a busy decade during which several more membrane proteins protecting against C were defined: C receptor 1 (CR1) by Fearon in 1979 [146]; membrane cofactor protein (MCP) by Atkinson and co-workers in 1985 [147, 148]; homologous restriction factor (HRF) by Zalman and by Hansch in 1986 [149, 150] and CD59 by several groups in 1988 and 1989.

The first plasma regulator of C to be well characterized was C1 inhibitor (C1inh). Lepow and co-workers partially purified C1inh from normal plasma in 1961 [151]. The discovery a year or so later that the enigmatic disease hereditary angioneurotic edema (now termed hereditary angiedema; HAE) was caused by a deficiency of C1inh [152] caused an explosion of interest in this molecule and the development of treatment regimens for HAE involving the restoration of serum levels of C1inh by infusing fresh plasma or concentrates of partially purified C1inh, the first example of therapy using a C regulator. The serine protease now known as factor I (fI) was first described by Lachmann and Müller-Eberhard in 1968 (then given the descriptive name C3 inactivator) [153, 154]; the plasma cofactor factor H (fH) was described by Whaley and Ruddy in 1976 [155]; the C4b-binding protein

(C4bp) was isolated from mouse serum by Ferreira *et al.* in 1977 [156]. Table 2.1 provides a summary of the large family of C inhibitors present on cell membranes and in the fluid phase and Figure 2.1 illustrates their points of action in the C system.

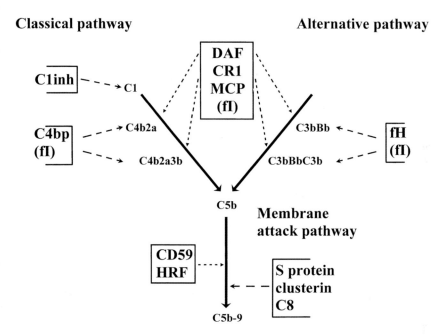

**Classical pathway**                                  **Alternative pathway**

**Figure 2.1. Regulation in the C system.**
The C system is tightly controlled by a number of fluid-phase and membrane-associated proteins acting either to inhibit the enzymes of the activation pathways (activated C1, C3 convertases, C5 convertases) or to restrict assembly of the MAC. This figure indicates where in the system each of the regulators exerts its effect. Membrane-bound regulators are boxed, fluid-phase inhibitors are bracketed. C1inh, C1 inhibitor; C4bp, C4b binding protein; DAF, decay-accelerating factor; MCP, membrane cofactor protein; CR1, C receptor 1; HRF, homologous restriction factor; fH, factor H; fI, factor I.

## CONTROL IN THE ACTIVATION PATHWAYS

The first step of the classical pathway (CP) is regulated by C1-inhibitor (C1inh), a 76-kDa plasma protein, a member of the serine protease inhibitor (serpin) family, which binds activated C1, physically removing C1r and C1s and thus causing disruption of the multimolecular C1 complex [157, 158]. C1inh is the only serum protease inhibitor capable of inhibiting activated C1 and even partial deficiency can result in uncontrolled activation in peripheral sites with resultant inflammation.

Control of the later steps of the CP and alternative pathway (AP) primarily involves fI, a two-chain (50 kDa and 38 kDa) plasma serine protease which, in the presence of specific cofactors, cleaves C3b and C4b in the activation pathway convertases, rendering the convertase inactive. The cofactors are proteins present on the membrane and in plasma which are essential for regulation. In the plasma two proteins act as cofactors for fI: fH is a large (approx. 150-kDa) single-chain glycoprotein which binds C3b in the AP convertases (C3bBb and C3bBbP) and catalyses the cleavage of C3b by fI [159]; C4bp is a large, heptameric plasma protein (six or seven α chains, each of approx. 70 kDa, and zero or one β chain) which binds C4b in the CP convertase (C4b2a) and catalyses the cleavage of C4b by fI. When present in the complex, the β chain confers upon C4bp the capacity to bind protein S, a cofactor involved in the coagulation system [160].

Both fH and C4bp also inhibit in a second way, by acting to break up (decay) the multi-component convertases, a property termed *decay acceleration* [22, 161]. The convertases undergo spontaneous decay with half-lives of several minutes. Decay accelerators enhance this decay by binding tightly to the convertase, thereby displacing the enzymatic component of the complex (C2a in the CP convertases, Bb in the AP convertases) and preventing the association of more C2 or fB.

On the membrane, at least three proteins help regulate the activation pathway convertases. Decay-accelerating factor (DAF) is a 65-kDa single-chain protein tethered to the membrane by a glycosyl phosphatidylinositol (GPI) anchor. DAF binds to and causes dissociation of the enzymatic component of the complex on the membrane. In the CP convertases C4b2a and C4b2a3b, C2a is displaced; in the AP convertases C3bBb and C3bBbC3b, Bb is displaced. Once the convertase is disrupted, DAF is released and can bind and inhibit other convertases on the membrane.

Membrane cofactor protein (MCP) is a 60-kDa transmembrane protein which, like DAF, binds to the activation pathway convertases but instead of causing dissociation, acts as a cofactor for the cleavage of C4b and C3b in the complexes by fI, thus irreversibly inactivating the enzyme.

C receptor 1 (CR1) is a large (approx. 200-kDa) transmembrane protein which inactivates the activation pathway convertases both by causing dissociation/decay (like DAF) and by acting as a cofactor for cleavage by fI (like MCP).

Factor I (fI), in the presence of an appropriate cofactor, can enzymatically cleave both C4b (C4bp, MCP or CR1 as cofactor) and C3b (fH,

MCP or CR1 as cofactor) and hence inactivate the CP and AP C3 and C5 convertases. The cleavage events involved are similar for C4b and C3b. In the case of C3b, fI sequentially cleaves at two sites in the α chain, releasing a small fragment (C3f, 3 kDa) and yielding iC3b (i = inactive). FH (fH), MCP and CR1 all act as cofactors for this first cleavage of C3b. FI can cause further cleavage of iC3b, cutting at a single site in the molecule, releasing the large fragment C3c and leaving C3dg attached to the membrane; however, only CR1 has cofactor activity for this second cleavage (Figure 2.2). Further cleavage of C3dg by serum proteases may subsequently occur, producing a number of breakdown products. C4bp, MCP and CR1 all act as cofactors for fI cleavage of C4b. Cleavage occurs at two sites in the α′ chain to yield the large fragment C4c which is released and the small fragment C4d which remains surface-associated (Figure 2.2)

DAF, MCP, CR1, fH and C4bp all share important structural features which indicate that they have arisen by gene reduplication of an ancestral gene encoding the primal C regulatory molecule. All are tightly linked in a region known as the regulators of C activation (RCA) cluster on the short arm of chromosome 1. All are described in detail in Chapter 3.

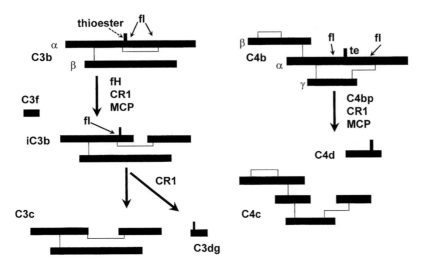

**Figure 2.2. Cofactors for fI cleavage of C3b and C4b.**

Inactivation of C3b and C4b involves enzymatic cleavage of the molecules by serum fI. Cleavage requires the presence of a specific cofactor. The figure illustrates the multiple cleavages made by fI, the products of cleavage and the cofactors operating to facilitate each of the cleavage steps.

A unique feature of the AP is the existence of a protein which stabilizes the C3 and C5 cleaving enzymes. Properdin (P) was the first of the components specific to the AP to be discovered [43]. It is a large, oligomeric (two, three, four or more identical 53-kDa subunits) plasma protein which binds C3b in the convertase and inhibits the spontaneous and accelerated (fH, DAF, CR1) decay [56, 162].

## CONTROL IN THE MEMBRANE ATTACK PATHWAY

The membrane attack pathway, like the activation pathways, is tightly regulated by inhibitors present in the fluid phase and on membranes (Figure 2.3). The hydrophobic membrane binding site in the fluid-phase C5b-7 complex is intrinsically unstable, being susceptible to inactivation by hydrolysis; the bulk of the C5b-7 generated during activation thus spontaneously decays. Binding of C5b-7 to membranes is further restricted by several serum proteins. The most studied of these is S-protein (vitronectin), an 80-kDa single-chain protein, a single molecule of which binds tightly to the C5b-7 complex and prevents binding to membranes [163]. C8 and C9 incorporate into the complex but the resultant SC5b-9 complex is inactive. S-protein also has important roles in cell adhesion as its

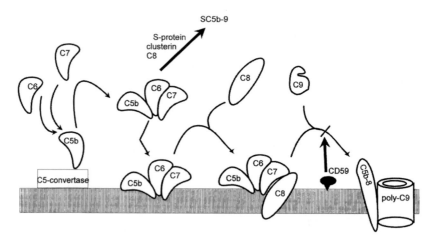

**Figure 2.3. Regulation of assembly of the MAC.**

The membrane attack pathway is inhibited from the fluid phase by serum proteins (S-protein, clusterin, C8) which bind the fluid-phase C5b67 complex and prevent its association with the membrane. On the membrane, control is provided by CD59 which prevents the incorporation of C9 into the forming MAC, thereby preventing formation of the lytic lesion.

alternative names (vitronectin, cell-spreading factor) imply. The fluid-phase C5b-7 complex can also be inactivated by binding to a recently described serum protein, clusterin, which consists of two disulphide-bonded 40-kDa chains [164]. Clusterin, a multi-functional protein which is a serum lipoprotein, is present in high concentration in seminal plasma and also appears to be a marker for cell death. C5b-7 is also inactivated by binding other lipoproteins and, perhaps most important of all, by binding C8 in the fluid phase [165, 166]. With the exception of C8, all these fluid-phase inhibitors have functions outside the C system and it is likely that their C5b-7 binding properties are by-products of their non-specific affinity for hydrophobic molecules.

On the membrane, two inhibitors have been described, homologous restriction factor (HRF), a 70-kDa GPI-anchored protein the significance of which is unproven [167, 168], and CD59 antigen (CD59), a 20-kDa GPI-anchored molecule which is the major MAC inhibitor on most cells. CD59 binds to C8 in the C5b-8 complex and blocks incorporation of C9, thereby preventing formation of the lytic MAC [169, 170]. A detailed description of each of the fluid-phase and membrane regulators of MAC assembly is given in Chapter 4.

Nucleated cells have another defence against MAC lysis: active recovery processes which physically remove the MAC from the membrane and restore the integrity of the bilayer [171]. In different cell types, removal may involve either shedding of MAC on membrane vesicles or internalization and degradation of MAC. The relative importance of MAC inhibitory proteins and recovery processes in nucleated cell survival has not been addressed, but some cell types have an enormous capacity for rapidly removing MAC. This neglected aspect of cell defence is also described in Chapter 4.

## CONTROL OF THE ANAPHYLACTIC PEPTIDES

The anaphylatoxins C3a, C4a and C5a are 74–77-amino acid peptides released from the amino termini of the $\alpha$ chains of C3, C4 and C5 during C activation. C3a and C5a express powerful biological activities, acting through specific membrane receptors on target cells, notably phagocytes [108]. C4a in humans is probably without function. Their effects are restricted to the site of C activation by the presence of inhibitors in plasma. The principal inhibitor for all the anaphylatoxins is the serum carboxypeptidase-N (anaphylatoxin inactivator), a plasma enzyme which cleaves the C-terminal Arg residue from each of the anaphylatoxins. The des-Arg metabolites thus formed have much reduced or absent biological activity, the exception being C5adesArg which retains much of the neutrophil-stimulating and basophil-activating activities of C5a [172]. Carboxypeptidase-N is a large, heterodimeric molecule composed of two subunits of molecular weights 83 kDa and 50 kDa which is itself dimerized to give a four-chain molecule with a total molecular weight in excess of 200 kDa. In addition to its effects on the C-derived anaphylatoxins, carboxypeptidase-N also cleaves and inactivates several other biologically important kinins which are generated at inflammatory sites.

## SUMMARY

The C system is rigidly controlled by a battery of regulators, present in the fluid phase and on membranes, which interfere at almost every step in the pathway. The majority of the regulators have been discovered within the last 20 years and over the past decade interest in the regulators has intensified, fuelled by the realization that regulators designed by nature might be utilized to control C activation in disease. Inherited deficiencies of individual C regulators exist and give rise to predictable clinical syndromes (Chapter 5); the expression of C regulators in tissues and biological fluids is altered in many diseases, either as a consequence of C activation or as a protective strategy against C damage; C regulators have unique roles in the reproductive system (Chapter 6) and micro-organisms have pirated C regulators for their own protection in the host (Chapter 7). Perhaps most exciting of all in this fast-moving field, C regulators are forming the basis of designer drugs for treatment of disease (Chapter 9). We hope that this volume will not only provide a comprehensive account of the information currently available on the C regulators but also convey the air of excitement which currently pervades the field.

# Chapter 3

# REGULATION IN THE ACTIVATION PATHWAYS

## REGULATION OF C1

*Introduction*

Unlike the alternative pathway (AP), which 'ticks over' constantly in plasma thereby priming the C system for immediate action when confronted by foreign micro-organisms, the classical pathway (CP) forms part of the acquired immune system – in most cases antibody is required for its initiation. The CP is activated following cross-linking of surface- or particle-bound IgG Fc or IgM Fc by the C1 complex[173]. This results in conformational changes within C1q that transmit to the rest of the complex resulting in autoactivation of C1r and subsequent activation of C1s. C1s is the serine protease responsible for cleavage of both C4 and C2 and initiation of CP convertase formation. Several inhibitors are present either on the cell membrane or in plasma to control inappropriate convertase activity; these are discussed in the later sections of this chapter. In this section we focus on the regulation of C1. Binding of C1 to antibody is the first step in the CP and consequently the first opportunity for control of C activation. A single C regulator, termed C1 inhibitor (C1inh), has evolved to fulfil this function.

*History*

In 1957 Ratnoff & Lepow described an activity present in fresh plasma with the ability to inactivate C1[174]. C1 esterase inhibitor, as it was then termed, was further characterized during the following few years and subsequently isolated in 1961 by Pensky and colleagues using Dowex-2 anion exchange chromatography of the supernatant from a 40% ammonium sulphate cut of human plasma[175]. These workers demonstrated that the partially purified inhibitor was distinct from other protease inhibitors known at that time, was heat-labile (being inactivated by heating at 56°C for 30 minutes) and unstable at acidic pH. The purified protein demonstrated an instantaneous, 1:1 stoichiometric inhibition of the C1-mediated hydrolysis of the artificial substrate N-acetyl-L-tyrosine ethyl ester. In 1962, a protein of unknown function, $\alpha_2$-neuraminoglycoprotein, was independently isolated from serum [176]. The identity of this protein with the previously described C1inh was recognised in 1969 by comparisons using immunoelectrophoresis, Ouchterlony double diffusion analysis and SDS-PAGE [177]. However, $\alpha_2$-neuraminoglycoprotein did not inhibit the esterolytic activity of C1. This was later shown to be due to the inclusion of acidic conditions in the purification protocol which are now known to inactivate C1inh.

*Function*

C1inh is a member of the *serine protease inhibitor* (serpin) family which now has more than 40 recognised members [178, 179]. Serpins are involved in control of proteases from a diverse range of physiological systems, such as coagulation, fibrinolysis and complement. In vertebrates, they form a major component of plasma, constituting as much as 10% of the total protein. Predictably, serpins share many characteristics. They are single-chain proteins containing a conserved region of about 380 amino acids in which resides the protease-inhibitory activity. They are often heavily glycosylated and, depending on their role, frequently possess functional domains in addition to the serpin domain [179]. Inhibitory serpins fish for appropriate targets by presenting a reactive site loop as a potential substrate to the target protease. If the protease takes this 'bait' and cleaves the serpin in the reactive site loop, it becomes irreversibly trapped and inactivated by formation of a stable and tight protease–serpin complex (discussed in detail below). C1inh is thus a suicide inhibitor which is consumed by interaction with the protease. Whilst C1inh is the only plasma inhibitor of activated C1, it also serves as an inhibitor of various other plasma proteases such as Factor XIIa, Factor XIa, plasmin and kallikrein; in some cases (for example in the case of Factor XIIa) it is the major inhibitor of the protease [180–186].

C1 is converted from the circulating inactive complex to an activated state following binding of the complex to an antibody-coated surface, and ligation of two or more of the globular 'heads' of C1q. The requirement of cross-linking two or more Fc in order to activate C1q ensures that fluid-phase circulating antibody does not activate the CP and that activation of C1 occurs only on appropriate particles or surfaces. Activation of C1 is characterized by the cleavage of the single-chain zymogens, C1r and C1s, resulting in both cases in formation of the active two-chain serine protease. Cleavage occurs through an intramolecular autocatalytic mechanism whereby a conformational change in C1q results in autoactivation of C1r which in turn cleaves and activates C1s [187, 188]. The two chains of activated C1r and C1s remain joined by a single disulphide bond and are termed 'heavy' (N-terminal part of C1s; 56 kDa; also known as A chain) and 'light' (27 kDa; also known as B chain). The serine protease domain resides in the light chain.

The role of C1inh in C1 regulation is two-fold. Firstly, C1inh binds reversibly to native C1 under physiological conditions and acts as a fluid-phase 'chaperone' to the assembled C1 complex which can otherwise spontaneously autoactivate at low levels in the absence of antibody [189, 190]. In the absence of C1inh, fluid-phase auto-activation results in consumption of C1, C2 and C4. The role of C1inh as a 'stabilizing' factor in plasma is immediately evident when C1inh is depleted from serum or when the isolated components of C1 are reassembled *in vitro*; in either situation C1 activity is rapidly lost upon incubation at 37°C [188, 190–192]. The second, and perhaps more obvious role of C1inh is in control of the activated C1 complex. C1inh limits activation of the classical pathway by rapidly inactivating C1; indeed the half-life of activated C1 is reported to be only 13 seconds in serum, although C1 bound to activating particles is less sensitive to inactivation by C1inh than is fluid-phase C1 [193, 194]. Inactivation proceeds through formation of a tight complex between the inhibitor and the light chains of activated C1s and C1r in a manner typical of the known activities of serpins (discussed below). Interaction between C1inh, C1r and C1s results in dissociation of these C1 subcomponents from C1q in the form of fluid-phase complexes composed of one molecule each of C1s and C1r, and two molecules of C1inh (C1r$_1$:C1s$_1$:C1inh$_2$) (Figure 3.1) [195–198]. The stable fluid-phase complexes thus formed can be found in plasma following C1 activation and can be used as an index of CP activation [199, 200].

*Distribution/biosynthesis*
C1inh is present in plasma at between 200 and 250 μg/ml and is synthesized primarily in the liver [201–207]. C1inh can also be synthesized by cells other than hepatocytes, the list including fibroblasts [207, 208], endothelial cells [207, 209], monocytes [207, 210–212], platelets [213, 214], chondrocytes [207] and glial cells [215, 216].

A deficiency in C1inh results in a condition known as hereditary angioedema (HAE) [217, 218]. Clinical disease occurs in those heterozygous for the deficiency and is characterized by recurrent, acute oedema of the skin or mucosa, usually involving the extremities, face, larynx and gastrointestinal tract. There are two types of HAE characterized either by the presence in plasma of low amounts of a normal C1inh (Type I), or by the presence of a dysfunctional protein, the levels of which may be normal or elevated (Type II). In Type I HAE, plasma levels of C1inh are typically less than 25% of normal, rather than the 50% which might be predicted for a

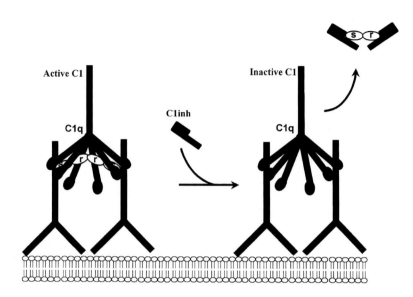

**Figure 3.1. Inhibition of the classical pathway (CP) by C1inh.**

Cross-linking two or more globular 'heads' of C1q following binding to an antibody-coated surface results in a conformational change in C1q. This change 'transmits' to the C1 subcomponents resulting in cleavage and activation of the serine protease C1r through an intramolecular autocatalytic mechanism. Activated C1r in turn activates C1s by proteolytic cleavage. This active form of C1s cleaves both C4 and C2 during formation of the CP C3 convertase. Inactivation of C1 is brought about by irreversible binding of C1inh to the active serine proteases, C1r and C1s, which results in their dissociation from C1q.

heterozygous deficiency, due to a normal or increased catabolic rate compounded by a low synthesis rate [219, 220]. Many mutations of C1inh have been characterized and frequently involve changes in the reactive site loop, particularly of the P1 residue (Arg-444, discussed below) [221, 222]. Another form of angiedema, acquired angiedema, is occasionally identified in patients with malignancy, particularly lymphoma or leukaemia [223-225]. Low level activation of C on tumour cells and consumption of C1inh may be responsible for the depressed levels of the inhibitor in this syndrome. Deficiency of C1inh and HAE is discussed in depth in Chapter 5.

*Molecular cloning and structural analysis*
Initial studies of purified C1inh demonstrated that it had a sedimentation coefficient of 3.7–4.5 S giving an estimate of protein molecular weight of about 105 kDa, a figure which was supported by SDS-PAGE analysis [201, 226, 227]. Detailed analysis of the amino acid composition and carbohydrate

content indicated that carbohydrate contributed as much as 35% by weight to the mass of C1inh, indicating that the polypeptide backbone was probably about 70 kDa [201, 226, 227]. A smaller (96-kDa), functional form of C1inh had also been reported and was thought to represent a cleavage product of the native molecule or a form with altered carbohydrate content [227]. These initial structural studies of C1inh were hard to reconcile with structural data that emerged following publication of the cDNA sequence of C1inh in 1986 [228–232]. From the cDNA sequence it was demonstrated that the mature protein consisted of 478 amino acids and the molecular weight of the polypeptide backbone was predicted to be only 52.8 kDa (Figure 3.2). Further study of carbohydrate content by NMR spectroscopy and chemical analysis indicated that C1inh contained seven O-linked carbohydrate groups in addition to six N-linked carbohydrate moieties, which contributed a total of 18.3 kDa to the molecular weight of the protein [233, 234]. The calculated predicted molecular weight of the glycosylated molecule was therefore only 71 kDa. In support of this value, a figure of 76 kDa was obtained from neutron-scattering studies [234]. The reason for the discrepancy between the molecular weight estimated from analysis of the polypeptide backbone and carbohydrate and that evident by sedimentation analysis and SDS-PAGE is still unresolved, although it is possible that high levels of carbohydrate or the high proportion of Pro residues in C1inh is responsible for the unusual sedimentation characteristics and electrophoretic mobility, resulting in an over-estimation of the molecular weight.

Of the 478 amino acids in C1inh, residues 120–478 constitute the serpin domain showing 26% amino-acid identity with α1-antitrypsin, 39% if conserved residues are taken into account [228]. The amino-terminal

**Figure 3.2. Derived amino acid sequence of C1inh.**

```
MASRLTLLTLLLLLLAGDRASSNPNATSSSSQDPESLQDRGEGKVATTVI    28
SKMLFVEPILEVSSLPTTNSTTNSATKITANTTDEPTTQPTTEPTTQPTI    78
QPTQPTTQLPTDSPTQPTTGSFCPGPVTLCSDLESHSTEAVLGDALVDFS   128
LKLYHAFSAMKKVETNMAFSPFSIASLLTQVLLGAGQNTKTNLESILSYP   178
KDFTCVHQALKGFTTKGVTSVSQIFHSPDLAIRDTFVNASRTLYSSSPRV   228
LSNNSDANLELINTWVAKNTNNKISRLLDSLPSDTRLVLLNAIYLSAKWK   278
TTFDPKKTRMEPFHFKNSVIKVPMMNSKKYPVAHFIDQTLKAKVGQLQLS   328
HNLSLVILVPQNLKHRLEDMEQALSPSVFKAIMEKLEMSKFQPTLLTLPR   378
IKVTTSQDMLSIMEKLEFFDFSYDLNLCGLTEDPDLQVSAMQHQTVLELT   428
ETGVEAAAASAISVARTLLVFEVQQPFLFVLWDQQHKFPVFMGRVYDPRA   478
```

The amino-terminal signal peptide is underlined. The first residue of the mature protein is **N** in bold. Numbering is for the mature protein.

**GenBank Accession No:** M13656

domain has no homology to any other serpin and contains a multitude of potential sites for O-glycosylation and three of the six N-linked oligo-saccharides in C1inh. Two disulphide bridges link the serpin domain to the carboxy-terminal end of the glycosylated 'tail' (Cys-101 to Cys-406, Cys-108 to Cys-183) (Figure 3.3)[228]. The amino-terminal region of C1inh also contains an unusual repeat unit -Glx-Pro-Thr-Thr- (Glx represents either Glu or Gln) that is repeated seven times within a span of 35 amino acids between Glu-63 and Thr-97. The function of this repeating unit is still unclear although six out of the seven O-linked carbohydrate moieties are located in these repeats. The function of the heavily glycosylated amino-terminal domain characteristic only of C1inh has not been resolved, although it is clear that the carbohydrate residues are not essential for inhibition of C1 and control of the C cascade. Removal of the O- and N-linked carbohydrates with O-glycanase or N-glycanase has no effect on C1 inhibition, and unglycosylated C1inh, synthesized and secreted by a human hepatoma cell line in the presence of tunicamycin, is fully active[206, 235]. More recently, recombinant truncated forms of the molecule have been synthesized that lack either the amino-terminal 76 or 98 residues[236]. Both forms bound C1r, C1s, kallikrein and factor XIIa and effectively inhibited C1 activity, indicating that the unusual amino-terminal domain is not required for protease inactivation.

(a)      (b)

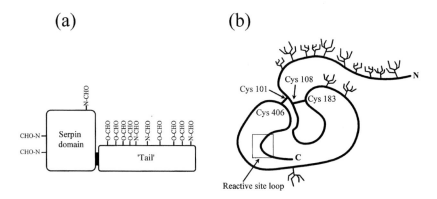

**Figure 3.3. Structure of C1inh.**

(a) C1inh is comprised of two domains: the serpin domain which contains the C-regulatory activity, and a heavily glycosylated amino-terminal domain, the function of which is not yet clear.

(b) Amino acids 120–478 constitute the serpin domain of C1inh. The reactive site loop is located towards the carboxy terminus of the protein and the P1–P1′ residues are located as Arg-444-Thr-445. Two disulphide bridges link the amino-terminal and serpin domains. Figure modified from[228].

The first electron microscopy images of C1inh revealed a two-domain molecule consisting of a globular head with a very long tail[237]. The overall length of the protein was estimated to be about 36 nm, with the long tail contributing approximately 33 nm to the length. A later electron microscopy study confirmed the basic two-domain structure, but the thin rod-like 'tail' of C1inh was estimated from this analysis to have an average length of only 13 nm and the globular 'head' to be 5 nm × 6 nm in size[238]. This latter data was in good agreement with neutron scattering studies in which the scattering data fitted a model of C1inh in which the total length was 18 nm[234]. It has been suggested that the elongated amino-terminal domain evident in the early microscopy studies was an artefact generated by 'shearing' of the C1inh tail. The predicted C1inh structure derived from the neutron-scattering studies utilized as a model the crystal structure of α1-antitrypsin, in which the serpin domain has the dimensions 7 nm × 3 nm × 3 nm[239]. The heavily glycosylated tail was modelled as an elongated structure extending for 15 nm, and the predicted sedimentation coefficient of C1inh based on this model (using a molecular weight of 71 kDa) was calculated as 3.7 S, corresponding closely to figures produced experimentally. The 120 amino acids in the glycosylated amino-terminal domain do not assume a random conformation, but rather are folded into a compact conformation giving rise to a domain of diameter 2 nm[234, 240].

The gene for human C1inh maps to chromosome 11 and spans approximately 1700 bp (Figure 3.4)[241–243]. There are eight exons in total, although surprisingly the intron/exon structure bears no obvious similarity to other serpin genes. Sequence encoding the first 17 amino acids of the signal peptide is in exon 2; the remaining six amino acids of the signal are encoded on exon 3, along with the first 161 amino acids of the mature protein, including a portion of the serpin domain. The remaining exons

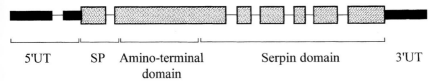

| 5'UT | SP | Amino-terminal domain | Serpin domain | 3'UT |

**Figure 3.4. Organization of the C1inh gene.**

The *C1inh* gene spans approximately 1700 bp on chromosome 11 and contains eight exons. The contribution of each exon to the structure of the mature protein is indicated in the figure and discussed in the text. The 5′ and 3′ untranslated regions (5′UT and 3′UT respectively) are indicated by black boxes. SP, signal peptide.

encode the rest of the serpin domain, with the intron/exon boundaries showing no obvious segregation into structural elements. The reactive centre loop is encoded on exon 8. One characteristic of the C1inh gene is the presence of a high number of Alu repeats: 17 in the intron sequences and one in each of the 5′ and 3′ untranslated regions. A region of alternating purine/pyrimidine repeats is also present in the second intron of the gene. Whilst the functional significance of these repeats is unclear, they may contribute to instability in the gene. It has been suggested that the Alu repeats may be involved in generating deletions and duplications within the C1inh gene, and that the purine/pyrimidine repeating sequence may predispose to mutation and gene rearrangement [244, 245].

*Binding site/serpin activity*

The serpin domain in C1inh is typical of this family of proteins, showing 39% homology (conserved/identical residues) with α1-antitrypsin, the first serpin to be crystallized [178, 239, 246, 247]. The crystal structures of many different serpins have now been elucidated and compared. In all cases, the serpin fold is formed from three β-sheets, termed A, B and C, which form the core of the protein and are flanked by nine α-helices. The reactive site of the native serpin is on an exposed loop of the protein near to the carboxy terminus; in ovalbumin this loop is an α-helix and this is thought to be the case for most serpins [248, 249]. Serpins are 'suicide inhibitors', they act as false substrates for the target protease which binds to the serpin and cleaves it at its reactive centre loop [250]. This results in formation of a protease–serpin complex in which the serpin is no longer active. Until recently, the only crystal structures available were of cleaved inhibitory serpins. However, in 1994 the crystal structure of an uncleaved inhibitory serpin (a variant of human antichymotrypsin) was published, providing the first clear structure for the uncleaved reactive centre loop, a distorted helical conformation independent of any β-sheet structure [251]. Fourier transform infrared spectroscopy (FT-IR) studies indicate that the α-helices and β-sheets in native serpins are not fully stabilized and that reactive site cleavage results in a strengthening of the hydrogen bonds and an increase in thermal stabilization [252, 253]. Hence, the uncleaved form of the protein is often termed the 'stressed' form, and the form that has bound its target protease is termed the 'relaxed' form. This change has been demonstrated with various inhibitory serpins but does not occur in non-inhibitors of the serpin family such as ovalbumin and angiotensin [253–257].

The residues at the serpin active site between which cleavage occurs are termed P1 and P1′ and the specificity of any serpin appears to be determined by the structure of the P1 residue. In C1inh, P1 and P1′ are located as Arg-444 and Thr-445 [258]. However, cleavage at four other sites has been demonstrated in C1inh, at peptide bonds 439–440, 440–441, 441–442, 442–443 [259]. In the cleaved inhibitory serpins that have been crystallized, the residues amino-terminal to the cleavage site (P1–P14 in C1inh), that were on the exposed loop in the uncleaved molecule, form strand A4 in the β-sheet A of the cleaved molecule (Figure 3.5). With one exception, latent plasminogen activator inhibitor-1 (PAI-1), this β-sheet is formed from five strands in the native, uncleaved form and six strands in the reactive centre cleaved form. Latent PAI-1 is an inactive, but not cleaved, form of the serpin which appears to maintain its active conformation in plasma by interaction with S-protein. Once latent PAI-1 has formed, its activity can only be restored by denaturation and subsequent refolding of the protein [260]. The crystal structure of this latent form of a serpin reveals that the amino acids amino-terminal to the reactive site are incorporated into the β-sheet in a manner similar to that of a cleaved serpin [261]. In cleaved serpins, the free carboxy and amino termini formed during the

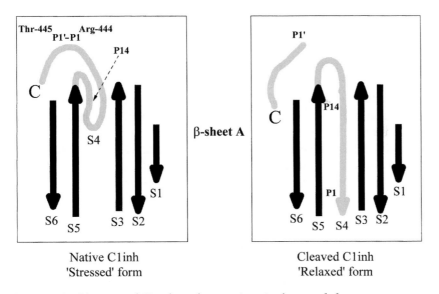

Native C1inh                     Cleaved C1inh
'Stressed' form                  'Relaxed' form

**Figure 3.5. Cleavage of C1inh at the reactive site loop and the consequent conformational transition.**

If activated C1r or C1s takes the 'bait' and cleaves C1inh at the exposed reactive site loop between Arg-444 and Thr-445 (P1–P1′ residues), formation of a stable complex between the protease and the inhibitor ensues. Formation of this complex results in protease inactivation. A conformational transition accompanies cleavage of C1inh in which strand 4 (S4), which previously formed an exposed loop on the protein surface, becomes incorporated into the β sheet A. Following cleavage, the P1 and P1′ residues are located at opposite ends of the serpin fold.

cleavage reaction are 7 nm apart at opposite ends of the serpin fold. While this implies a dramatic change in conformation, H[1]NMR studies indicate only small changes following cleavage of the native form corresponding to small rearrangements of 0.01 to 0.05 nm within the hydrophobic protein core. The tertiary structure associations are altered very little between native and cleaved forms of the inhibitor[253, 262].

C1inh interacts with its target proteases, C1r and C1s, in a manner typical of serpins. The resulting complex is stable to heat and denaturation, and analysis by reducing SDS-PAGE reveals protein bands characteristic of a C1inh–C1s or C1inh–C1r light-chain complex and free heavy chain. The complex can only be dissociated under alkaline conditions and is particularly sensitive to hydroxylamine, indicating that the covalent bond is an ester[196, 198, 227, 263, 264]. The exact mechanism of inhibition is still a matter of discussion, some suggesting that the covalent bond may be an artefact generated through denaturation in preparation for SDS-PAGE. However, recent evidence indicates that, following the initial binding between serpin and enzyme, an acyl-enzyme covalent intermediate containing cleaved inhibitor is rapidly formed[265–267]. Initial complex formation, P1-P1' cleavage and insertion of strand A4 into β-sheet A proceeds very rapidly. The complex probably remains as this stable acyl-enzyme intermediate with the conformational change locking the inhibitor–protease complex. It has been speculated that release of the protease might occur if a water molecule was to enter the active site and promote deacylation. Details of the precise mechanism of protease inhibition by serpins such as C1inh are still emerging. Definition of crystal structures of uncleaved inhibitory serpins and protease-complexed serpins may further aid elucidation of the structures formed during the inhibitory process.

## REGULATORS ENCODED IN THE RCA GENE CLUSTER

### The RCA gene cluster

*Background*

As discussed in Chapter 2, a multitude of protective molecules exist in order to defend host cells from inappropriate lysis by homologous C. Of these a subset of proteins, including both fluid-phase and membrane-bound molecules, are related through evolution as indicated by structural similarity and clustering of their genes on the same chromosome. The first evidence for a gene cluster encoding complement regulatory proteins (CRP) was obtained in 1984 when the loci for C4b binding protein

(C4bp) and complement receptor 1 (CR1) were shown to be closely linked [268]. In 1985, Atkinson and co-workers reviewed the structural, functional and genetic data available at that time on CRP which acted on the components C3 and C4 and the concept of a new superfamily was proposed [269]. Many other articles and reviews which further corroborated the existence of the multi-gene family were published within the following few years [23, 270–272]. To date, there are six members of the RCA (regulators of C activation) cluster identified in humans. These proteins are complement receptor 1 (CR1; CD35), complement receptor 2 (CR2; CD21), factor H (fH), decay accelerating factor (DAF; CD55), membrane cofactor protein (MCP; CD46) and C4bp. CR1, CR2, DAF and MCP are primarily located at the plasma membrane, although fluid-phase counterparts of all these proteins have been identified. Factor H and C4bp are fluid-phase C regulators, although an fH-like molecule has been reported on the surface of certain cells.

*Function of RCA proteins*

Without exception, all the RCA proteins interact with activation products of C3 and/or C4. CR1 binds to C3b leading to destabilization of the C3 convertase and, by acting as a cofactor for fI, triggers inactivation of C3b through formation of iC3b. CR1 retains a weak affinity for the cleavage product, iC3b, and promotes further cleavage by fI, resulting in formation of the smaller fragments C3c and C3dg. CR1 also binds C4b, leading to decay of the convertase and formation of the inactive cleavage products, C4c and C4d. As its name implies, DAF acts to accelerate decay of the C3 convertases C3bBb and C4b2a and also the C5 convertases. Unlike CR1 it does not have any cofactor activity for fI. Comparison of the domain structure of DAF with another RCA protein, MCP, reveals a remarkable structural similarity between the two proteins. However, the activity of MCP is quite different from that of DAF. MCP has cofactor activity for fI leading to inactivation of C3b through formation of iC3b; it possesses no decay-accelerating activity. Whilst the distribution of CR1 is rather limited (mainly on circulating cells), MCP and DAF are widely expressed and their activities are complementary. The plasma proteins fH and C4bp are both cofactors for fI: fH promotes cleavage of C3b whilst C4bp promotes cleavage of C4b. Both proteins also prevent amplification of either activation pathway by preventing binding of fB or C2 and hence formation of the convertase proenzyme, or by promoting dissociation of pre-formed convertase. FH plays a crucial role in the homeostasis of the AP by keeping 'tick over' activation of the AP in check.

Unlike the other RCA proteins, CR2 has no direct influence on C activation although it does bind the C3 activation product, C3dg. Recent research indicates that CR2 plays an important role in the humoral immune response. Opsonized antigen which has bound to a specific B cell effectively cross-links surface IgM (bound to antigen) with CR2 (bound to C3dg) [118, 273, 274]. Cross-linking of these proteins, which are constitutively associated with CD19 and intracellular protein tyrosine kinases, leads to synergistic interaction and cell activation. Other RCA proteins also have roles that are not directly associated with convertase decay or C3b inactivation. For example, CR1 plays a crucial role in immune complex clearance [111, 275]. Most circulating CR1 is carried on erythrocytes where it is clustered on the cell surface enabling binding to fluid-phase immune complexes which have been opsonized by C3b. This allows transportation of the large insoluble complexes to the liver where they can be degraded. Similarly, opsonization of particulate antigen by C3b leads to ligation of C receptors (CR1 and CR3) as well as Fc receptors on macrophages and polymorphonuclear cells, resulting in enhanced phagocytosis of bacteria and other particles. The GPI-anchored RCA protein, DAF, has been implicated in cell signalling leading ultimately to activation of intracellular protein tyrosine kinases [276–278]. It is not yet clear how a signal is transmitted across the plasma membrane, but presumably the DAF protein itself, or its GPI anchor, interacts with a transmembrane protein that in turn interacts with an intracellular kinase. MCP may have a role in fertilization, where it is suggested that C3b forms a bridge between the oocyte and sperm; C3b which is bound to MCP on acrosome-reacted sperm may in turn bind to C receptors (such as CR1 or CR3) on the oocyte, facilitating fertilization [279, 280].

Interestingly, a number of the membrane-bound RCA proteins may have a detrimental effect on the host cell as they can act as receptors for several viruses, such as Epstein–Barr virus (EBV) (CR2) [281], echovirus 7 and other echoviruses (DAF) [282, 283], enterovirus 70 (DAF) [284], coxsackieviruses A21, B1, B3 and B5 (DAF) [285–288] and measles virus (MCP) [289–291]. RCA proteins are also implicated in HIV infection. The envelope proteins of HIV, gp41 and gp120, bind to fH, bestowing the virus with some protection from C attack [292–295]. Similarly HIV 'hijacks' DAF and MCP during the budding process with the result that protective molecules are 'expressed' on the viral surface [295–297]. To add insult to injury, it seems that HIV also uses C activation and subsequent opsonization with C3 activation fragments to gain entry to cells via C receptors [298, 299]. Several RCA proteins also bind the M protein of *Streptococcus pyogenes*. Factor H and C4bp are held on the bacterial surface where they probably protect the bacterium from host C [300, 301]. *S. pyogenes* also binds MCP on keratinocytes, possibly aiding bacterial invasion through the epidermis [302].

*Organization of the RCA cluster*

The genes for several groups of structurally related C proteins, both those involved in activation and in regulation, are closely linked. For example the genes for C2, fB and C4 are found on chromosome 6 in the major histocompatibility complex [303–305]. The genes for C6, C7 and C9 are also linked on chromosome 5 [306, 307], and those for C1r and C1s on chromosome 12 [308]. These groups of proteins share with each other many structural features and show high amino acid identities, e.g. 29% for C6 and C7 [309], 39% for C2 and factor B [310] and 40% for C1r and C1s [308], indicating that the genes probably arose through duplication and subsequent divergence. Polymorphisms within the regulatory proteins have enabled analysis of gene linkage. The first indication of the existence of the RCA cluster came from the work of Rodriguez de Cordoba and colleagues in 1984, who used segregation analysis in families informative for the CR1 and C4bp loci, to show that CR1 and C4bp were closely linked [268]. In 1985 the gene for fH was included in the RCA cluster [311] and, using recently identified partial cDNA clones for CR1 and CR2 as probes, CR2 was included and the cluster was localized to band 1q32 on chromosome 1 by fluorescence *in situ* hybridization [312–314]. In 1987 the gene encoding DAF was also localized to band 1q32 on chromosome 1 by *in situ* hybridization, establishing the close linkage of this protein to CR1, CR2, fH and C4bp [315]. The same conclusion was drawn by Rey-Campos *et al.* following segregation analysis of polymorphic variants of these proteins in informative families [316]. A close linkage of the last member of this cluster, MCP, was soon established, its location being upstream but within 100 kb of the CR1 gene [147, 317]. The gene encoding for the β chain of C4bp was shown to be tightly linked to the gene encoding the α chain in 1990, with a gap of only 3.5–5 kb between them [318]. Current knowledge of the organization of the RCA cluster links the RCA genes and other gene-like elements in the order (5′ to 3′) C4bpβ, C4bpα, C4bpα-like, DAF, CR2, CR1, MCP-like, CR1-like and MCP on a 900-kb segment of DNA (Figure 3.6) [317–322]. The gene for fH is unusual in that it is located a relatively long distance from the main

**Figure 3.6. Linkage of the regulation of complement activation (RCA) genes on chromosome 1.**

The genes encoding the RCA proteins are clustered within a 900-kb segment of DNA on chromosome 1, band 1q32. The gene encoding fH is located approximately 7 Mb distant from the main cluster, this gene is more closely linked to that for hfHR-2 and factor XIII B.

cluster. Indeed, identification of further alleles of C4bp and fH has permitted the localization of this gene at about 7 Mb distant from the main cluster [323]. Interestingly, the genes for fH and factor XIII B are linked, and are located closer to each other than fH is to the rest of the RCA cluster [324]. Recently the gene for the human fH-related 2 (fHR-2) protein (see below) has also been located within 165 kb of the factor XIII B gene, demonstrating linkage between fH and hfHR-2 [325].

*The short consensus repeat (SCR)*

As well as close gene linkage, the RCA proteins also share structural similarity. The elucidation of the amino acid sequences of the RCA proteins demonstrated that all contained stretches of amino acid sequence, consisting of 60–70 amino acids, which showed marked homology or conservation with one another and with similar sequences in other RCA proteins (Table 3.1; Figure 3.7). The protein domain present in all these proteins is known as the short consensus repeat (SCR); also called the complement

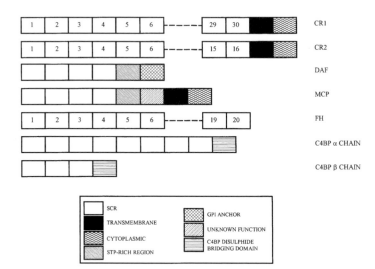

**Figure 3.7. Organization of protein domains in complement-regulatory proteins (CRPs) encoded in the RCA gene cluster.**

All proteins encoded in the RCA gene cluster contain short consensus repeat (SCR) domains, from as few as three SCRs (in the C4bp β chain) to as many as 37 (in the CR1 B allele). These repeats are arranged contiguously within the protein. Other characteristics are shared between the proteins, such as the Ser/Thr/Pro (STP)-rich region in membrane cofactor protein (MCP) and decay-accelerating factor (DAF), and the disulphide bridging domain encoded by the C4bp genes.

Table 3.1. *Characteristics of complement-regulatory proteins (CRPs) located in the regulators of complement activation (RCA) cluster.*

Abbreviations: SCR, short consensus repeat; MCP, membrane cofactor protein; DAF, decay-accelerating factor; CR, complement receptor.

| Protein | Primary structure (reference number) | Number of SCRs | Interacts with ... | Activity |
|---|---|---|---|---|
| Factor H | (390, 406, 407) | 20 | C3b | • Decay acceleration<br>• Cofactor |
| C4bp α chain | (457, 486, 487, 875) | 8 | C4b | • Decay acceleration<br>• Cofactor |
| C4bp β chain | (330) | 3 | Protein S | • Renders protein S inactive<br>• May localize C4bp to damaged membranes |
| MCP | (147, 358, 741, 742) | 4 | C3b, C4b | • Cofactor |
| DAF | (678, 679) | 4 | C3bBb, C4b2a | • Decay acceleration |
| CR1 | (344–346) | 30 (F allele) | C3b, C4b (iC3b, iC4b) | • Decay acceleration<br>• Cofactor |
| CR2 | (353) | 15 (16) | C3dg | • Role in B-cell response to antigen |

control protein repeat (CCP), Sushi repeat, or even 'B-type module' as it was originally identified in factor B. Some RCA proteins contain a small number of these domains, for example DAF and MCP contain only four SCR repeats, as well as other unrelated domains. In contrast, other members of the family are composed almost entirely from multiple copies of this domain, these include fH (20 repeats), CR1 (30 repeats in the most common form) and CR2 (15/16 repeats). The extracellular domains of both CR1 and CR2 are composed entirely from SCRs, although both have non-SCR transmembrane and cytoplasmic domains. Individual SCRs from fH have been expressed in yeast, and their three-dimensional structures determined by nuclear magnetic resonance[326–328]. Each SCR is a compact, globular domain consisting of a hydrophobic core wrapped in β-sheet, with the carboxy and amino termini at opposite ends of the domain. The consensus sequence includes four invariant Cys and several other residues which are conserved in most SCRs including a Trp residue, and several Pro and Gly residues. Other residues are also conserved, or are substituted with similar amino acids such as Tyr/Phe and Ile/Leu/Val (Figure 3.8). The four invari-

**(a)**

$\cdot \underline{C} \cdot \cdot P \cdot \cdot I/L/V \cdot \cdot \cdot NG \cdot \cdot \cdot \cdot \cdot \cdot \cdot F/Y \cdot \cdot \cdot G \cdot \cdot I/L/V \cdot \cdot F/Y \cdot \underline{C} \cdot \cdot G \cdot F/Y \cdot \cdot \cdot G \cdot \cdot \cdot I/L/V \cdot \cdot \underline{C} \cdot \cdot \cdot \cdot \underline{W} \cdot \cdot \cdot P \cdot \cdot \underline{C} \cdot$

**(b)**

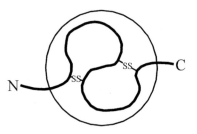

**Figure 3.8. The SCR domain.**
**(a)** All proteins encoded in the RCA gene cluster contain the SCR domain. These domains comprise about 60 amino acids, a number of which are highly conserved. The four Cys residues and the Trp residue (underlined) are conserved in at least 95% of known SCR domains. Other amino acids including certain Gly, Phe, Tyr, Pro, Ile, Leu, Val are also conserved.
**(b)** Each SCR is a compact, globular domain consisting of a hydrophobic core wrapped in β sheet which is held together by two disulphide bonds arranged in a 1–3, 2–4 pattern. The carboxy and amino termini are located at opposite ends of the domain.

ant Cys residues form two intradomain disulphide bonds that help maintain the structure of the domain and are always linked in a 1–3, 2–4 pattern. SCRs fold as independent protein domains and are arranged within all RCA proteins contiguously, resembling 'beads on a string'. This arrangement can be visualized using electron microscopy, and has been demonstrated for various RCA proteins, including C4bp which has a remarkable structure. The most common isoform of this regulator consists of seven α chains and one β chain. The α chain consists of eight SCRs and the β chain of three SCRs. Each of the chains has a non-SCR carboxy-terminal domain through which they are covalently linked by disulphide bonding[329–331]. The molecule resembles a spider, in that there is a central body with eight elongated, flexible 'arms' consisting of SCRs stretching away from the central core[332, 333]. All RCA proteins contain stretches of contiguously organized SCR domains, giving a typical elongated structure.

The SCR repeat is not confined to RCA proteins. It is also found in other C proteins such as factor B and C2 (three repeats each)[304, 310, 334, 335], C6 and C7 (two repeats each)[309, 336, 337] and also C1r and C1s (two repeats each)[308, 338]. There are indications that fI also contains this repeat although in this case the consensus is rather weak[339]. Therefore the C proteins containing this repeat include DAF, MCP, CR1, CR2, C4bp, fH, fB, C2, C6, C7, C1r and C1s. The majority of these proteins (which are located on several different chromosomes) interact with C3b and/or homologous proteins such as C4b and C5b (interacting with C6 and C7), suggesting that this particular arrangement of amino acids in these SCR domains generates C3b/C4b-binding sites. These binding sites have been defined in many of the RCA proteins and have been shown to require at least two SCRs. It is important to note that the SCR domain is also present in non-C proteins such as the β subunit of clotting factor XIII[340], haptoglobin[341], β2 glycoprotein[342], IL2 receptor[343] and several other proteins, implying that the SCR domain forms a 'structural building block', rather like the immunoglobulin fold in that it is used in proteins of diverse function.

*Evolutionary relationship between RCA proteins*
Proteins of the RCA family have a strong evolutionary relationship, the cluster having arisen through duplication of an SCR encoding sequence and subsequent divergence. There is also evidence of intragenic duplication in at least two members of the family, CR1 and CR2. The CR1 duplication unit consists of a 7-SCR (about 450-amino acid) stretch of sequence, and the two most common isoforms of CR1 contain either four sets of this repeat unit (A allele) or five sets (B-allele)[344–346]. These 7-SCR units are termed long homologous repeats (LHR) and show between 65

and 90% sequence homology with each other. When the component SCRs in each LHR of CR1 are analysed and compared separately an even more compelling pattern becomes apparent with the emergence of repeating units with as much as 99% sequence homology (Figure 3.9)[345]. For example, SCRs 3–9 show 99% homology to SCRs 10–16 (the equivalent SCRs in LHRs 1 and 2), SCRs 10–11 are 99% homologous to SCRs 17–18, and SCRs 19–21 are 91% homologous to SCRs 26–28. In each case the homology resides within equivalent SCRs in each LHR, conserving the fundamental 7-SCR repeating unit structure[345, 347]. In order to explain

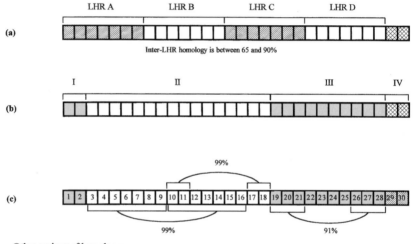

**Other regions of homology:**

| Region | Homology with region.. | |
|---|---|---|
| SCR 12-18 | SCR19-25 | 67% |
| SCR 1-2 | SCR8-9 | 61% |
| SCR1-2 | SCR15-16 | 61% |
| SCR1-2 | SCR22-23 | 59% |

**Figure 3.9. Organization of internal regions of homology in complement receptor 1 (CR1) (A/F allele).**

**(a)** CR1 is arranged into repeating units consisting of seven SCR domains; each of these is termed a 'long homologous repeat' (LHR). The carboxy-terminal two SCRs do not conform to this pattern.
**(b)** The SCRs in CR1 can be arranged into regions of even higher identity, showing in some cases as much as 99% sequence homology. (Figure modified from [345].)
**(c)** Percent identity between SCRs in homology regions represented in (b) (according to[345]). The transmembrane and cytoplasmic domains are not represented in this figure.

this amazing degree of symmetry and homology it has been proposed that duplication of the 7-SCR ancestral unit has been accompanied by unequal cross-over and/or gene conversion in the *cr1* gene [348–350]. This provides a striking example of concerted evolution within a gene where mutations in individual repeats are propagated to other corresponding sites in homology regions.

The CR2 repeating unit consists of four SCR domains, repeated four times [351, 352]. Analysis of the homology between CR1 LHRs and the CR2 repeating units suggests that both genes may have arisen through duplication of the same primordial group of SCRs [351, 353]. Most SCR domains are encoded entirely by a single exon, although a few are split over two exons and some exons encode two SCRs. Examples of each of these patterns can be found in the CR1 and CR2 genes. For example, within the LHRs of CR1, the first, fifth and seventh SCRs are encoded on a single exon, the second and sixth on two exons and the third and fourth are encoded together on a single exon. This pattern is repeated throughout each LHR, once more illustrating the duplication events that occurred during evolution of the RCA cluster [354, 355]. Similarly, analysis of the genomic organization of CR2 reveals that within each repeating unit the first exon encodes two SCRs, the second exon encodes the third SCR and the fourth SCR is encoded by two separate exons [352]. CR1 and CR2 are not the only RCA proteins which contain an SCR encoded on two exons; the second SCR of the C4bp $\alpha$ chain is encoded on two separate exons [329], as is the third SCR of the C4bp $\beta$ chain [356], the third SCR of DAF [357] and the second of MCP [358]. The structure of the human fH gene has not yet been determined but the gene for murine fH contains an SCR that is encoded on two separate exons [359]. The location of the intron that splits the SCR is identical in all RCA proteins and is found in the codon encoding a Gly residue three amino acids carboxy terminal to the second Cys. The only exception is the C4bp $\beta$ chain gene where the split codon encodes the amino acid Arg, although the location in the SCR remains the same. This provides strong evidence for the existence of a common ancestral SCR from which all the RCA proteins have evolved.


# Fluid phase RCA proteins

## Factor H

*History*
Factor H (fH) was first described in 1965 as a protein of unknown function contaminating preparations of C3 [360]. It was then termed $\beta_1 H$ due to its

electrophoretic mobility in agarose gels. Its role as a decay-accelerating factor and cofactor for fI was recognised in 1976 when several groups described a plasma protein that potentiated the inactivation of C3b by fI; this was termed C3b inactivator accelerator (A·C3bINA) [49, 155, 361–363]. These groups demonstrated its identity with β1H by double diffusion analysis which showed that antisera to A·C3bINA and β1H produced lines of identity. It was also demonstrated that A·C3bINA, henceforth termed fH, could physically bind to cell-bound C3b, limit binding of fB and formation of the AP convertase, and increase the rate of loss of pre-formed enzymatic sites from both cell-bound C3bB and C3bBP.

*Function*
FH prevents amplification of the AP by accelerating the dissociation of fB or Bb from both the C3 and C5 convertases (Figure 3.10). The primary interaction of fH with the convertase is with the non-catalytic subunit C3b, resulting in displacement of fB or Bb [364]. It also binds to C3b (or C3i) preventing the initial binding of fB and thus formation of the proenzyme, C3bB. The second regulatory role of fH in the AP is as an essential cofactor for the plasma serine protease fI (Figure 3.10). In the presence of fH, fI cleaves C3b at two sites in the α' chain, resulting in the release of C3f (3 kDa) and formation of the enzymatically inactive molecule iC3b [365]. Under physiological conditions, fH can only act as cofactor for the first two cleavages of C3b. However, it has been reported that under conditions

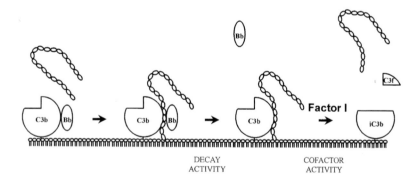

**Figure 3.10. Inhibition of the alternative pathway (AP) by Factor H (fH).**

Factor H has both decay accelerating activity and cofactor activity for fI. It thus acts to dissociate the components of the C3 convertase and also promotes cleavage and irreversible inactivation of C3b through formation of iC3b. The affinity of fH for C3b is dependent on the nature of the surface to which C3b has bound; fH binds to C3b with high affinity if the surface is 'non-activating' (see text and Figure 3.11).

of low ionic strength, fH can also act as cofactor for the third fl-mediated cleavage (for which CR1 normally acts as cofactor) resulting in the formation of C3dg and C3c [366]. Initial reports indicated that fH also bound C4b and acted as a cofactor for degradation by fl under physiological conditions but this was not corroborated by others and contamination of fH preparations with C4bp in initial studies seems a likely explanation [361, 362, 367, 368]. Factor H only binds C4b and acts as cofactor for cleavage of C4b to C4c and C4d under non-physiological conditions of low ionic strength, or at supraphysiological concentrations of fH [362, 364, 369].

The AP ticks over constantly at a very low level in plasma, priming the C system for immediate response to a foreign organism and subjecting all tissues to C attack [370]. Functional C3 convertases are formed in plasma either by continuous low-rate hydrolysis of the thioester in C3 producing C3i (which has C3b-like properties) or by cleavage of native C3 by plasma proteases, particularly at a site of injury and inflammation. Both C3i and protease-cleaved C3b can bind fB, which is in turn cleaved by factor D (fD) to form C3 convertases (see Chapter 1). FH plays a crucial role in maintaining the homeostasis of the AP. Amplification is dependent on the affinity of binding of either fH or fB to surface-bound C3b [371-373]. If C3b is fluid-phase, or bound to a non-activating surface, fH binds preferentially and further amplification of the pathway is prevented (Figure 3.11). On the other hand, C3b bound to an activating surface preferentially binds fB and the AP amplifies through activation of the convertase by fD and generation of further nascent C3b. In the absence of fH, the AP cycles out of control as evidenced by individuals deficient in fH (or fl) who have severe hypocomplementaemia with reduced levels of C3 and fB. Factor H deficiency is very rare and, other than hypocomplementaemia, symptoms are variable; diseases associated with this deficiency have included glomerulonephritis, haemolytic uraemic syndrome and meningococcal disease [374-379].

It is not yet clear what features make a surface activating and whether there are structures on the cell surface that prevent fH binding or aid fB binding. However, it is clear that surface charge and sialic acid content have a role to play. This was evident during the early characterization of fH when it was demonstrated that sialic acid on a particle surface made it non-activating, and its removal using sialidase converted cells from non-activators to activators of the AP [380-383]. Reduction of sialic acid content diminishes the number of high affinity binding sites for fH whilst leaving the affinity of fB for surface-bound C3b unaffected. The consensus is that fH, C3b and surface structures form a ternary complex on non-activating surfaces, and that polyanions on such a surface bind to a specific site on fH promoting C3b inactivation probably by increasing the affinity of the

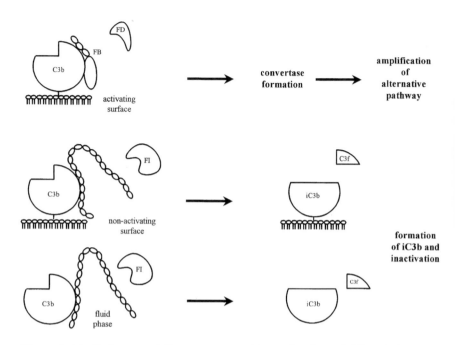

**Figure 3.11. Influence of C-activating surface on binding of fH to C3b.**

The nature of the surface to which C3b has bound determines whether fH or fB binds to C3b. If the surface is 'activating' then fB binds preferentially, the C3 convertase is formed and the AP amplifies. If C3b is bound to a 'non-activating' surface, fH binds preferentially to C3b. This results in displacement of any bound fB or Bb from C3b, prevention of further fB binding, and as a result of cofactor activity of fH for fI, irreversible inactivation of C3b through formation of iC3b. Fluid phase C3b is not 'protected' by an activating surface, hence it binds fH and iC3b formation is promoted.

C3b–fH interaction [384–386]. However, the precise nature of non-activating surfaces is likely to be complex. It appears that C3b also contains a binding site for 'activators' of the AP as fH binding to C3b can be inhibited by incubation of C3b with certain polysaccharides [387, 388]. Other molecules, such as immunoglobulin, can also protect covalently attached C3b from inactivation by fH and fI; C3b covalently bound to such a 'protective' surface has increased resistance to inactivation compared to free C3b [389]. Protection by antibody may have several consequences, including increased C activation on bacteria or other foreign surfaces, C activation on non-activating surfaces, and increased and prolonged C3b deposition on immune complexes.

*Distribution/biosynthesis*
Factor H is present in plasma at about 400 μg/ml. It is synthesized predominantly in the liver[390, 391] but also at many extrahepatic sites, notably fibroblasts[392], endothelial cells[393, 394], monocytes[395, 396], platelets (in the α granules)[397], myoblasts[398] and glial cells[399]. There are reports that fH, or a fH-like molecule, is also expressed on the surface of certain cells (B-cell lines, U937-cell line) where it may contribute to C regulation on the cell surface or control binding of C3b-coated immune complexes to the cells[400–403]. Characterization of the membrane-associated molecule has yielded variable results. Three reports describe a single-chain surface protein with a molecular weight characteristic of fH[400–402], whereas a fourth describes a two-chain molecule on the surface of B cells (68 kDa and 75 kDa)[403]. Two of these reports also describe a cell-associated cofactor activity that can be blocked by anti-fH antibodies[401, 403]. No secretion of fH from the cells into the supernatant was detected, indicating that the membrane-associated form was not secreted fH that had absorbed onto the cell surface, and that fluid-phase fH was not responsible for the cofactor activity.

*Molecular cloning and structural analysis*
The *fH* gene is located on Chromosome 1q32, about 7 Mb distant from the main RCA cluster; it is closely linked to the gene encoding factor XIII B and an fH-related protein, hfHR-2 (see below)[311, 323–325]. Factor H is a single-chain molecule of 150 kDa containing about 18% by weight of carbohydrate, although the oligosaccharides are not essential for function as their removal does not abrogate regulatory activity[404, 405]. Partial protein sequencing of fH indicated that it contained SCR domains characteristic of an RCA protein[406, 407]. The full sequence was derived from the cDNA and revealed that the molecule consisted of 1213 amino acids (Figure 3.12), and was composed entirely of SCRs, 20 in all, arranged as 'beads on a string' (Figure 3.13(a))[390]. The assembly of fH entirely from SCR domains is reflected in the high β-sheet content evident by Fourier transform infrared spectroscopy[408]. Transmission electron microscopy analysis of fH shows an extended, flexible molecule that can fold back on itself, with only one end appearing to interact with C3b[409]. Although by electron microscopy the molecule appears monomeric, X-ray and neutron solution scattering studies suggest that the protein exists predominantly in a dimeric form[410].

There are two common variants of fH in the Caucasian population, FH*1 (HF*A) and FH*2 (HF*B), and three rare variants[323, 411]. All are encoded by codominant alleles at a single autosomal locus and the gene frequencies in Caucasoids are 0.685, 0.301, 0.006, 0.002 and 0.006 (FH*1,

**Figure 3. 12. Derived amino acid sequence of fH.**

```
MRLLAKIICLMLWAICVAEDCNELPPRRNTEILTGSWSDQTYPEGTQAIY     32
KCRPGYRSLGNVIMVCRKGEWVALNPLRKCQKRPCGHPGDTPFGTFTLTG     82
GNVFEYGVKAVYTCNEGYQLLGEINYRECDTDGWTNDIPICEVVKCLPVT    132
APENGKIVSSAMEPDREYHFGQAVRFVCNSGYKIEGDEEMHCSDDGFWSK    182
EKPKCVEISCKSPDVINGSPISQKIIYKENERFQYKCNMGYEYSERGDAV    232
CTESGWRPLPSCEEKSCDNPYIPNGDYSPLRIKHRTGDEITYQCRNGFYP    282
ATRGNTAKCTSTGWIPAPRCTLKPCDYPDIKHGGLYHENMRRPYFPVAVG    332
KYYSYYCDEHFETPSGSYWDHIHCTQDGWSPAVPCLRKCYFPYLENGYNQ    382
NHGRKFVQGKSIDVACHPGYALPKAQTTVTCMENGWSPTPRCIRVKTCSK    432
SSIDIENGFISESQYTYALKEKAKYQCKLGYVTADGETSGSIRCGKDGWS    482
AQPTCIKSCDIPVFMNARTKNDFTWFKLNDTLDYECHDGYESNTGSTTGS    532
IVCGYNGWSDLPICYERECELPKIDVHLVPDRKKDQYKVGEVLKFSCKPG    582
FTIVGPNSVQCYHFGLSPDLPICKEQVQSCGPPPELLNGNVKEKTKEEYG    632
HSEVVEYYCNPRFLMKGPNKIQCVDGEWTTLPVCIVEESTCGDIPELEHG    682
WAQLSSPPYYYGDSVEFNCSESFTMIGHRSITCIHGVWTQLPQCVAIDKL    732
KKCKSSNLIILEEHLKNKKEFDHNSNIRYRCRGKEGWIHTVCINGRWDPE    782
VNCSMAQIQLCPPPPQIPNSHNMTTTLNYRDGEKVSVLCQENYLIQEGEE    832
ITCKDGRWQSIPLCVEKIPCSQPPQIEHGTINSSRSSQESYAHGTKLSYT    882
CEGGFRISEENETTCYMGKWSSPPQCEGLPCKSPPEISHGVVAHMSDSYQ    932
YGEEVTYKCFEGFGIDGPAIAKCLGEKWSHPPSCIKTDCLSLPSFENAIP    982
MGEKKDVYKAGEQVTYTCATYYKMDGASNVTCINSRWTGRPTCRDTSCVN   1032
PPTVQNAYIVSRQMSKYPSGERVRYQCRSPYEMFGDEEVMCLNGNWTEPP   1082
QCKDSTGKCGPPPPIDNGDITSFPLSVYAPASSVEYQCQNLYQLEGNKRI   1132
TCRNGQWSEPPKCLHPCVISREIMENYNIALRWTAKQKLYSRTGESVEFV   1182
CKRGYRLSSRSHTLRTTCWDGKLEYPTCAKR                      1213
```

The amino-terminal signal peptide is underlined. The first residue of the mature protein is **E** in bold. Numbering is for the mature protein.

**GenBank Accession No:**   Y00716; M32093; X07525

FH*2, FH*3, FH*4, FH*5 respectively). The two major allelic variants differ due to a T/C change in the cDNA which results in a Tyr to His substitution at position 384 in the translated protein; the polymorphism has been confirmed by protein sequencing[390, 412]. Similar polymorphisms have also been described in the Chinese Han population with the common FH*1 and FH*2 genes at frequencies of 0.4828 and 0.5172 respectively [413]. In the Japanese population, FH*1 and FH*2 have frequencies of 0.4261 and 0.4895, rare alleles include HF*A1, HF*M (FH*3) and a null allele, HF*Q0[414].

*Binding/active sites*
The cofactor activity of fH resides in the amino terminal portion of the molecule (Figure 3.14). This has been demonstrated by various means. Initial studies established that limited tryptic digestion of fH resulted in

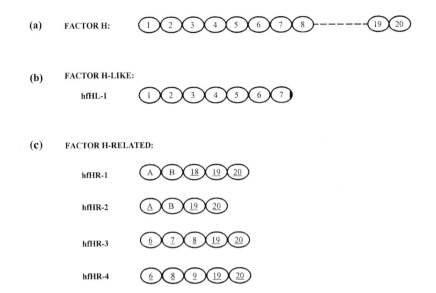

**Figure 3.13. Structure of fH, fH-like proteins, and fH-related proteins.**

**(a)** FH is comprised from 20 SCR domains arranged contiguously in the protein, resembling 'beads on a string'; this confers an elongated, flexible structure on fH.
**(b)** Alternative splicing of the *fH* gene results in a truncated form of the protein (42 kDa) termed human fH-like-1. This truncated form of fH has been identified in plasma and consists of seven SCR domains which are identical to the first seven SCR of full-length fH other than the carboxy-terminal four amino acids which are unique to this protein.
**(c)** Four human fH-related (hfHR) proteins have been identified to date. Whilst these proteins are encoded on separate genes, they bear a high degree of homology to fH and consist only of SCR domains. The carboxy-terminal two SCR domains in all four fHR proteins are homologous to SCRs 19 and 20 of fH. The amino-terminal two SCR of hfHR-1 are distinct from those in fH but are conserved in hfHR-2 (100% identical in the case of the SCR termed 'B' in the figure). Human fHR-1 also contains an SCR homologous to SCR18 in fH. Human fHR-3 and fHR-4 contain three SCRs at their amino terminus which are homologous to other SCRs found in fH. SCRs which are highly homologous to (but not identical with) those in fH or hfHR-1 are underlined.

the generation of two disulphide-linked fragments. The smaller 38-kDa fragment, when purified, could bind to C3b and act as a cofactor for fI [415, 416]. However, this fragment could not act efficiently as cofactor for cleavage of surface-bound C3b, indicating that trypsin cleavage abrogated recognition of sialic acid or other similar surface structure [417, 418]. This small cleavage product contained the first five amino-terminal SCRs and part of the sixth. Similarly, a truncated form of fH (human fH-like protein-1; hfHL-1; see below) consisting of only the first seven SCRs has

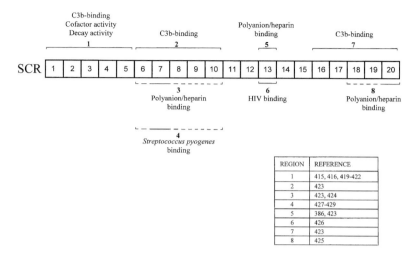

**Figure 3.14. Binding sites located in fH.**

Many binding sites have been located on fH, including putative binding sites for its main ligand, C3b and binding sites for polyanions, HIV and the bacterium *S. pyogenes*. Cofactor activity resides in the amino-terminal portion of the molecule. In regions 3, 4 and 8, the major binding site has been pinpointed to a single SCR (SCR7 or SCR20; indicated by a solid line). Dashed lines indicate other SCRs which may contribute to ligand binding.

cofactor activity. In order to analyse further the role of the SCRs in cofactor activity, various deletion mutants and truncated forms of fH have been expressed in CHO cells [419]. This work confirmed that full functional activity required four SCRs: either SCR 1–4 or SCR 1–3 plus SCR 5. Others have also demonstrated the importance of the amino-terminal SCRs in cofactor activity and decay-accelerating activity [420, 421]. Systematic deletion of SCRs in the hfHL-1 protein demonstrated that all recombinant proteins containing SCRs 1 through 4 had cofactor and decay-accelerating activity. The presence of both SCR1 and SCR4 was essential (although replacement of SCR4 with SCR5 was not tested). The spacing between these two SCR domains was also critical for maintaining activity; insertion of just a few amino acids between SCRs abolished activity. A C3b-binding domain also resides in this portion of the molecule, possibly within SCR 3–5 [422]. Whilst the C3b-binding site in these amino-terminal SCRs is essential for cofactor activity, other C3b-binding sites have been described in SCRs 6–10 and SCRs 16–20 [423]. It is likely that all three C3b-binding sites contribute to binding affinity for C3b and hence are involved in regulation of the AP. Other putative functional domains within fH include the polyanion/heparin binding site(s), which has been variously located in SCR13 [386, 423], a region between SCRs 6 to 10, probably in SCR7 [423, 424], or to SCR20 [425]. It is worth noting that whilst single SCRs may give ligand specificity, such as SCR7 or SCR20, it is likely that adjacent SCRs are required for a functional binding site [425]. Similar sites have also been implicated in binding both HIV (SCR13 [426]) and *Streptococcus pyogenes* (SCRs 6–10, most likely SCR7 [427–429]). SCRs 19–20 are highly conserved in fH and in human fH-related proteins (see below) and are therefore likely to have a functional role, although the latter is as yet unclear.

*Alternative forms of fH*

While the 150-kDa fH is the most abundant form in plasma, it has recently become apparent that several other fH-like molecules also exist [430]. There are two distinct groups of 'alternative' fH molecules: those derived from the fH gene by alternative splicing and those derived from distinct genes which are highly homologous to fH, probably having arisen through gene duplication (Figure 3.13). Human fH (150 kDa) is encoded by a 4.4 kb mRNA transcript; however, there are many other mRNA species demonstrated by northern blot analysis of liver RNA including species of 5.5, 5.0, 2.2, 2.0, 1.8 and 1.4 kb [431, 432]. The 1.8-kb transcript is a splicing variant of the fH gene which encodes a 42-kDa protein, consisting of seven SCRs with identical sequence to the amino-terminal portion of fH with the exception of four unique amino acids at the carboxy terminus [390, 391, 433, 434]. This truncated form of fH, human fH-like-1 (hfHL-1), is present in plasma at between 10 and 50 µg/ml and has been demonstrated by western blot [391, 430, 433, 435]. In common with fH, hfHL-1 can regulate complement [420, 435], bind to heparin and also to the pathogen *Streptococcus pyogenes* [428]. It also demonstrates cell adhesion activity, probably mediated through the RGD (Arg-Gly-Asp) motif in SCR4 [436]. This cell attachment activity was initially considered unique to hfHL-1, as full length fH was inactive in the assays used in this study [436]. However, recent data suggests that fH may fulfil a role, albeit minor, in neutrophil adhesion to endothelia [437].

The 1.4 kb band on northern blots is actually composed of three different mRNA species. Two encode proteins consisting of five SCRs (recently termed human fH-related (hfHR)-1 and hfHR-3) and the third a protein of four SCRs (hfHR-2) [430]. All three forms are derived from distinct genes as the nucleotide sequences vary substantially from that of fH. Human fHR-1 and hfHR-2 each contain at their amino terminus two SCRs quite distinct from those in fH [432, 438, 439]. The first SCR of hfHR-1 shows 95.5% amino acid identity to the first SCR of hfHR-2, and the second SCR of hfHR-1 is 100% identical to that of hfHR-2. The third

SCR in hfHR-1 is homologous to SCR18 in fH (absent in hfHR-2), and the two carboxy terminal SCRs in both these proteins are highly homologous to SCRs 19 and 20 of fH. Differential use of glycosylation sites results in a wide spread of molecular weights. Glycosylated hfHR-1 has an apparent molecular weight of either 37 or 43 kDa [440] and hfHR-2 of either 24 or 29 kDa [439]. The hfHR-3 clone encodes a protein of 37.5 kDa [441]. The protein is predicted to contain five SCRs with homology to the fH SCRs 6, 7, 8, 19 and 20. A further member of the fH-related family, hfHR-4, has recently been cloned and identified in plasma [442, 443]. It is also organized into five SCRs with the first three being similar to SCRs 6, 8 and 9 in fH and, in common with all the other characterized hfHR proteins, the two carboxy terminal SCRs are similar to SCRs 19 and 20 in fH. Human fHR-4 is very closely related to hfHR-3, with SCRs 2, 3 and 5 having over 93% homology.

Human fHR-1, hfHR-2 and hfHR-4 have all been demonstrated in human plasma, whereas hfHR-3 has yet to be detected [420, 430, 433, 438, 439]. Human fHR-4 exists as a dimer with a molecular weight of 86 kDa (106 kDa when reduced) and is either free, or associated with lipoprotein particles. Interestingly, hfHR-1 and -2 have also been found associated with a complex of phospholipid and other proteins [444, 445]. Whilst these fH-related proteins have been demonstrated in human plasma they do not have C-regulatory activity, probably because they lack the amino-terminal SCRs of fH implicated in these functions (see previous section). However, all four forms contain SCRs homologous to the carboxy-terminal SCRs of fH. This region has been implicated in both C3b and heparin binding. Furthermore, hfHR-3 and hfHR-4 are also comprised of SCRs homologous to SCRs 6–9 of fH, an additional region which may be involved in C3b and heparin binding (see above). Indeed, it has recently been demonstrated that hfHR-3 can bind to heparin [425]. These fH-related proteins may therefore fulfil a function distinct from cofactor activity or decay acceleration, but related to other as yet poorly defined roles of fH in plasma, such as cell activation (see below). No other fH-related molecules have been demonstrated as yet. However, there remain several abundant fH-related mRNA transcripts in liver which are uncloned and several uncharacterized proteins present in plasma that cross-react in a western blot with anti-fH antiserum [430]. It is thus likely that other fH-like and fH-related proteins will emerge to further complicate the picture.

### Other roles

As well as its role as a decay accelerator and cofactor for fI, fH has also been implicated in cell activation. There are reports that fH can bind to certain cells, possibly to a specific receptor [437, 446–448]. Factor H triggers a variety of responses from different cell types such as release of fI from lymphocytes [449], and prostaglandin E and thromboxane from macrophages [450]. Factor H can also cause activation of monocytes [451] and inhibit differentiation of B cells [452]. A recent study implicates fH as an adhesion molecule with the potential to form a 'bridge' between activated endothelial cells and neutrophils [437]. This report suggests that fH might bind to glycosaminoglycans exposed at sites of tissue damage and inflammation, possibly through one of its polyanion binding sites, and to CR3 on neutrophils. Furthermore, binding of fH to neutrophils enhances the cellular response to inflammatory mediators such as C5a and TNFα. A similar, but seemingly more potent, cell adhesion activity has also been demonstrated with hfHL-1 [436].

*Interaction with micro-organisms*
It has been suggested that various pathogens acquire fH in order to protect themselves from attack by host C (see Chapter 7). Factor H binds to envelope proteins of HIV (gp41 and gp120) and confers resistance to C attack upon the virus [293–295]. Not only does fH protect HIV from lysis in plasma but, as a consequence of C fixation on the virus and formation of iC3b on the virus surface through the activity of fH, HIV infection of cells may be enhanced through binding to C receptors [299, 431]. Similarly, another pathogen, *Streptococcus pyogenes*, has taken advantage of fH as well as other RCA proteins, such as C4bp and MCP. The M protein of group A streptococci contains a binding site for fH, enabling the bacterium to localize fH at the bacterial surface, resulting in a decreased deposition of C3b and protection from phagocytosis [300, 453].

# C4b-binding protein

*History*
In 1977 a protein was described in mouse plasma that bound the murine C4b, and also human and guinea pig C4b; it was thus termed C4b-binding protein (C4bp) [156]. As the mouse C system was poorly characterized at this time, efforts were made to characterize the C4bp analogue in human plasma [454]. Complexes of C4b and its binding proteins were precipitated from serum in which C had been activated. Specific precipitation was achieved by incubating the serum with increasing amounts of goat anti-human C4 until antigen–antibody equivalence was reached. The aggregates were harvested and used to raise a polyclonal antiserum which was absorbed with Sepharose-C4b to render it monospecific for C4bp. This antiserum was then used for crossed immunoelectrophoresis to demonstrate the formation of complexes between C4b and C4bp, and to monitor

fractionation of C4bp through the purification procedures of PEG precipitation and ion exchange chromatography. The purified protein was a glycoprotein which ran as a doublet on non-reducing SDS-PAGE with bands of apparent molecular mass 540 kDa and 590 kDa. On reduction, the molecule ran as a single band with a molecular weight of 70 kDa, implying that the large C4bp molecule was composed of multiple (seven or eight) disulphide-linked subunits. Its ability to form complexes with C4b was demonstrated by sucrose gradient ultracentrifugation. Interestingly, C4bp was first isolated from human plasma in 1976 but in the guise of 'proline rich protein' (PRP). PRP was recognised as an acute phase reactant, and also as a constituent of chylomicrons; indeed it was originally isolated by absorption to lecithin-stabilized triglyceride emulsion [455, 456]. It is unclear why C4bp should associate with lipid, although it is possible that this property is conferred by association with protein S (see below). It is intriguing that fH-related molecules (hfHR-1, -2 and -4) have also been found associated with lipoprotein particles (see above). The identity of PRP with C4bp was finally realised in 1989 when the cDNA sequence of PRP was deduced and, other than three nucleotide differences, was found to be identical to the published sequence of the C4bp α chain [457].

*Function*
C4bp has a function in the CP analogous to that of fH in the AP. It regulates the CP convertases (C4b2a and C4b2a3b) on cell surfaces by accelerating decay of the preformed convertase and preventing association of the convertase subunits [368] (Figure 3.15). It is also a cofactor for fI, pro-

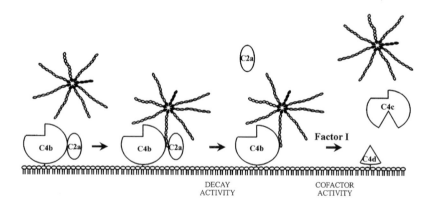

**Figure 3.15. Inhibition of the CP by C4bp.**
C4bp has both decay-accelerating activity and cofactor activity for fI. It thus acts to dissociate the components of the CP C3 convertase and also promotes cleavage and irreversible inactivation of C4b through formation of C4c and C4d.

moting cleavage and inactivation of C4b through formation of iC4b and subsequently C4c and C4d (Figure 3.15). It performs the same function in the fluid phase, binding to C4b and providing cofactor activity for fl. Indeed, the first evidence for cofactor activity of C4bp was obtained in 1975 when Shiraishi & Stroud demonstrated that a high molecular weight cofactor was required for the fl-mediated cleavage of fluid phase C4b into C4c and C4d [458]. A preparation of the same cofactor (then termed C3b-C4bINA cofactor) was also demonstrated to function as a cofactor for cleavage and inactivation of C3b [459]. Shortly after the purification and characterization of human C4bp in 1978 [454] and demonstration of its function as a cofactor for fl [367], it was suggested, and subsequently confirmed, that C4bp was identical to the high molecular weight cofactor described a few years earlier by Shiraishi & Stroud [460, 461]. There is no functional difference between the two forms of C4bp (540- and 590-kDa) evident on SDS-PAGE. The two forms have been separated and both shown to serve as a cofactor for fl [460]. C4bp can also function as a cofactor for fl-mediated inactivation of C3b, although with a much lower efficiency, and under physiological conditions it is unlikely that C4bp will regulate activity of C3b [369, 459, 460, 462].

An unusual feature of C4bp is its association in plasma with another protein unrelated to the C system, a single-chain 70 kDa protein, called protein S, which is involved in the coagulation cascade. As much as 50% of C4bp in plasma circulates as a complex with protein S. To avoid confusion we stress that this protein is unrelated to the S protein which plays a role in the regulation of the terminal C pathway (Chapter 4). Protein S has vitamin K-dependent anticoagulant activity; it functions as a cofactor for activated protein C in the inactivation of coagulation factors Va and VIIIa [463, 464]. Binding of protein S to C4bp renders protein S inactive, but has no effect on the function of C4bp as an inhibitor of the C cascade [465, 466]. The association of C4bp with protein S is non-covalent and was first demonstrated in 1981 when it was shown that protein S in plasma existed in two forms: in a 1:1 complex with C4bp, or free [467]. The binding sites for C4b and protein S on C4bp were shown to be distinct and independent with the presence of protein S having no effect on the interaction of C4b with C4bp [468, 469]. The binding site for C4b is on the 70-kDa subunit of C4bp (the $\alpha$ chain), whereas protein S binds a second chain, the $\beta$ chain (45-kDa). The physiological purpose of the association between C4bp and protein S is unclear. The reaction is specific, of high affinity, and is enhanced by $Ca^{2+}$ [470, 471]. Protein S has a high affinity for negatively charged phospholipid membranes, and it has been suggested that this may localize C4bp at the surface of injured or activated cells and thereby enhance C regulation on these surfaces.

C4bp also interacts with serum amyloid P component (SAP), a member of the pentraxin protein family related in structure to C-reactive protein[472]. The physiological function of SAP is unclear although it is found associated with amyloid deposits in a variety of diseases, including Alzheimer's disease. In contrast to the C4bp–protein S interaction which has been thoroughly analysed, the interaction of C4bp with SAP is poorly defined. Only in recent years have the molecular interactions between SAP and C4bp been analysed. SAP forms a $Ca^{2+}$-dependent 1:1 complex with C4bp in plasma, the interaction having apparently no effect on association of C4bp with either protein S or C4b, indicating that the binding sites are different[473]. The physiological role for this association is unknown, although it has been recently reported that association of C4bp with SAP can interfere with its cofactor activity for fI in a fluid-phase system[474].

*Distribution/biosynthesis*
The main site of C4bp synthesis is the liver; indeed, current evidence indicates that this is the sole site of synthesis. The molecule is present in plasma at about 160 μg/ml and deficiency is extremely rare. A subtotal C4bp deficiency has been identified in one family where the father and daughter had 20–25% of normal C4bp plasma levels[475–477]. The daughter presented with angiedema and symptoms of Behçet's disease, characterized by oral ulceration, genital ulcers and uveitis. Total protein S concentrations (free and bound to C4bp) were normal, although levels of free protein S were elevated; it is interesting that no coagulation defect was noted in this individual despite the role of protein S in the coagulation system[476]. There are reports that levels of C4bp are affected by autoimmune diseases such as systemic lupus erythematosus (SLE). However, elevated levels[478, 479], depressed levels[480], and normal levels of C4bp[481] have all been demonstrated in this disease. Levels of C4bp are also elevated during the acute-phase response[478, 482], and it appears that in the acute phase the genes for the α and β chains may be differentially regulated (see below)[483, 484].

*Molecular cloning and structural analysis*
C4bp has an oligomeric structure with several isoforms consisting of different combinations of α and β chains. The most common isoform consists of seven α chains and a single β chain (α7β1) (Figure 3.16). The α chain is the 70-kDa band originally demonstrated on reducing SDS-PAGE of C4bp[454], whereas the β chain has a molecular weight of 45 kDa. Both chains have the repeating SCR domain structure common to all RCA proteins, and are linked together in their non-SCR carboxy-terminal domains

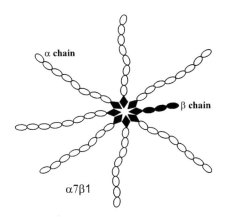

**Figure 3.16.  Structure of C4bp.**

The most common isoform of C4bp consists of seven α chains and one β chain. These are covalently linked by disulphide bonds through Cys residues located in the carboxy-terminal 'bridging' domain of each chain. The remainder of the C4bp chains are comprised entirely of SCR domains (eight in each α chain, three in the β chain).

through disulphide bonds. The amino acid sequence of C4bp α chain was deduced in 1985 through a combination of amino acid and cDNA sequencing (Figure 3.17) [485-487]. The first indication that the protein contained the SCR domain came from sequencing of multiple C4bp peptides produced by treatment with CNBr or chymotrypsin [486]. This yielded 55% of the protein sequence and enabled comparison to the factor B sequence obtained by Morley & Campbell (1984) which had first identified the SCR domain [334]. The cloning and sequencing of a C4bp cDNA clone provided the sequence from the 32nd amino acid through to the 3′ untranslated region [487]. This, when combined with the amino acid sequencing data previously obtained, gave the complete protein sequence of the mature protein. A cDNA clone was later obtained which included the signal peptide and confirmed the protein sequencing data [488]. The α chain consists of eight SCR domains (491 amino acids) and one non-SCR carboxy-terminal domain (58 amino acids). Four potential N-linked glycosylation sites are present, of which three are presumed to carry carbohydrate as these specific residues are not identified by Edman degradation.

The β chain of C4bp was first discovered in 1988 [489]. Close scrutiny of reducing SDS-PAGE gels of C4bp preparations demonstrated the presence of another band at 45 kDa, as well as the expected band at 70 kDa. The 45-kDa band stained poorly with Coomassie Blue and was only visible when gels were overloaded with protein. Isolation of the reduced,

### Figure 3.17. Derived amino acid sequence of C4bp.

α chain:

```
GALHRKRKMAAWPFSRLWKVSDPILFQMTLIAALLPAVLGNCGPPPTLSF      10
AAPMDITLTETRFKTGTTLKYTCLPGYVRSHSTQTLTCNSDGEWVYNTFC      60
IYKRCRHPGELRNGQVEIKTDLSFGSQIEFSCSEGFFLIGSTTSRCEVQD     110
RGVGWSHPLPQCEIVKCKPPPDIRNGRHSGEENFYAYGFSVTYSCDPRFS     160
LLGHASISCTVENETIGVWRPSPPTCEKITCRKPDVSHGEMVSGFGPIYN     210
YKDTIVFKCQKGFVLRGSSVIHCDADSKWNPSPPACEPNSCINLPDIPHA     260
SWETYPRPTKEDVYVVGTVLRYRCHPGYKPTTDEPTTVICQKNLRWTPYQ     310
GCEALCCPEPKLNNGEITQHRKSRPANHCVYFYGDEISFSCHETSRFSAI     360
CQGDGTWSPRTPSCGDICNFPPPKIAHGHYKQSSSYSFFKEEIIYECDKGY    410
ILVGQAKLSCSYSHWSAPAPQCKALCRKPELVNGRLSVDKDQYVEPENVT     460
IQCDSGYGVVGPQSITCSGNRTWYPEVPKCEWETPEGCEQVLTGKRLMQC     510
LPNPEDVKMALEVYKLSLEIEQLELQRDSARQSTLDKEL                549
```

β chain:

```
MFFWCACCLMVAWRVSASDAEHCPELPPVDNSIFVAKEVEGQILGTYVCI      33
KGYHLVGKKTLFCNASKEWDNTTTECRLGHCPDPVLVNGEFSSSGPVNVS      83
DKITFMCNDHYILKGSNRSQCLEDHTWAPPFPICKSRDCDPPGNPVHGYF     133
EGNNFTLGSTISYYCEDRYYLVGVQEQQCVDGEWSSALPVCKLIQEAPKP     183
ECEKALLAFQESKNLCEAMENFMQQLKESGMTMEELKYSLELKKAELKAK     233
LL                                                     235
```

The amino-terminal signal peptides are underlined. The first residue of the mature protein chains are **N** (α chain) or **S** (β chain) in bold. Numbering is for the mature protein.

**GenBank Accession No:**    α chain: X02865, X07853; β chain: M29964

carboxymethylated 45-kDa band enabled amino-terminal sequence analysis which produced a hitherto unknown sequence, suggesting the isolation of a novel subunit. The cDNA encoding the β chain was isolated in 1990 and revealed that the mature β chain consisted of 235 amino acids, and in common with the α chain, contained SCR domains (three in total)[330]. It also contained a 60-amino acid carboxy-terminal domain with similarities to the non-SCR carboxy-terminal domain in the α chain, indicating that the two chains probably arose through gene duplication; indeed the genes for the two chains of C4bp are very closely linked in the RCA gene cluster. Whilst the SCR domains were predicted to consist mainly of β sheet, the carboxy-terminal domains of both chains consisted mainly of α helix and contained two Cys likely to be involved in cross-linking to other subunits. There are five potential N-glycosylation sites predicted from the sequence of the β chain and, as shown by its reduction in apparent molecular weight from 45 kDa to 29 kDa when treated with endoglycosidase F, the β chain is heavily glycosylated[489].

As well as the $\alpha7\beta1$ isoform, there are two other isoforms of C4bp in plasma: $\alpha7\beta0$ and $\alpha6\beta1$ [331, 490]. These oligomers are also disulphide-bonded together at their carboxy-terminal ends. The $\alpha7\beta1$ isoform has also been termed 'C4bp-high' due to its reduced mobility on SDS-PAGE and is responsible for the 590-kDa band on non-reduced gels. The other two isoforms ($\alpha7\beta0$ and $\alpha6\beta1$) run with a similar mobility on SDS-PAGE and probably together form the 'C4bp-low' (540-kDa) band originally identified on gels (Figure 3.18) [490]. The plasma concentration of C4bp and the isoform pattern vary enormously between individuals. In one study of normal individuals, the $\alpha7\beta0$ isoform was shown to range from 1 to 40% of the total C4bp [331]. It has been suggested that the relative expression levels of the individual chains of C4bp play a major role in determining the proportions of the different isoforms found in plasma. Genetic factors are thought to control the expression of different genes, and initial linkage studies indicate that these genetic factors are closely linked to the RCA gene cluster [331]. Differential regulation of the two genes may reflect a mechanism for changing the plasma content of one isoform relative to the others, and indeed differential regulation has been demonstrated, particularly during the acute-phase response. The physiological consequences of an alteration in the levels of individual isoforms are most likely to be found in the coagulation system. Isoforms containing the $\beta$ chain bind the anticoagulatory factor protein S and render it inactive in the coagulation cascade. Hence, high levels of $\beta$ chain-containing isoforms may decrease levels of free (active) protein S, and increase the risk of thrombotic disease [465, 479]. It is intriguing that bovine and murine C4bp cannot bind protein S (the murine $\beta$ chain is a pseudogene, and the bovine $\beta$ chain has lost the protein S binding site; see below). The absence of protein S binding activity in at least two species calls into doubt the physiological importance of this association. Interestingly, free $\beta$ chain has been found in the supernatant of the HepG2 cell line [331]. These cells express relatively high levels of $\beta$ chain compared to the $\alpha$ chain, resulting in secretion of C4bp isoforms containing the $\beta$ chain ($\alpha6\beta1$, $\alpha7\beta1$) and presumably 'excess' $\beta$ chain. Free $\beta$ chain has not been reported in normal human plasma.

C4bp shows genetic polymorphism with at least three codominant alleles coding for the protein, with gene frequencies of 0.986 (C4BP*1), 0.010 (C4BP*2) and 0.004 (C4BP*3) in the Caucasian population [323, 491]. These polymorphisms are not linked to the 'high' and 'low' forms of C4bp demonstrated by SDS-PAGE. Analyses of families informative for the C4bp and CR1 loci demonstrated that these two loci were linked, and provided the first evidence for the existence of the RCA gene cluster [268, 269]. A heterogeneity in the C4bp $\beta$ chain has been identified consisting of the

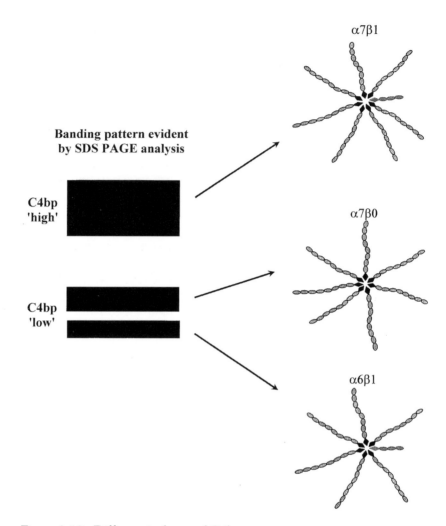

**Figure 3.18.  Different isoforms of C4bp.**
There are three isoforms of C4bp which vary in the number of $\alpha$ and $\beta$ chains which constitute the protein. The form of C4bp termed 'C4bp high' in early studies, due to its electrophoretic mobility, is comprised of seven $\alpha$ chains and one $\beta$ chain. The form of C4bp originally termed 'C4bp low' is comprised from two different isoforms with slightly different mobilities. One of these isoforms consists of seven $\alpha$ chains and no $\beta$ chains, and the second isoform comprises six $\alpha$ chains and one $\beta$ chain.

presence or absence of an Ala residue at position 3 of the mature protein which defines a second allele of the C4bp $\beta$ chain [330]. The genes for the $\alpha$ and $\beta$ chains of C4bp are closely linked in the RCA cluster with the 3' end of the C4bp β gene lying within 3.5–5 kb of the 5′ end of the C4bp α gene [318]. There is also a 'C4bp-like' region that maps between the DAF gene and the C4bp α gene which is predicted to be a pseudogene [318, 322]. The α chain gene spans 40 kb of DNA and is composed of 12 exons (Figure 3.19) [329, 492]. Each SCR is encoded by a single exon, apart from SCR2 which is split into two exons. The intron that splits the SCR is located in the codon encoding a Gly residue which is three amino acids carboxy-terminal to the second Cys. This splice site is identical in split SCRs of all other RCA proteins. The first two exons encode the 5′ untranslated sequence and the signal peptide, and the last exon encodes the C-terminal non-SCR domain and the 3′ untranslated sequence. The β chain gene spans more than 10 kb of sequence (Figure 3.19) [356]. The first two SCRs are encoded on separate exons and the third SCR on two exons. The location of the split occurs in the same position as in other RCA gene products, although the amino acid is an Arg rather than a Gly. One unusual feature is that the exon encoding the second half of the third SCR also contains sequence for part of the carboxy-terminal domain. The

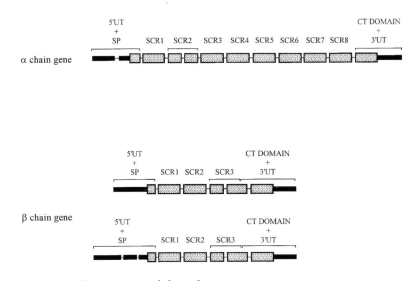

**Figure 3.19. Organization of the *c4bp* genes.**

Both *c4bp* genes are closely linked in the RCA cluster separated by only 3.5–5 kb of DNA. The α-chain gene spans 40 kb and is comprised of 12 exons and the β-chain gene spans at least 10 kb of sequence. Two classes of β-chain transcripts have been identified which vary in their 5′ untranslated region. The contribution of each exon to the structure of the mature protein is indicated in the figure and discussed in the text. The 5′ and 3′ untranslated regions (5′UT and 3′UT respectively) are indicated by black boxes. SP, signal peptide; CT, carboxy-terminal bridging domain.

remainder of the carboxy-terminal domain and the 3′ untranslated region are encoded on the last exon. Two classes of β chain gene transcript have been isolated with different 5′ untranslated regions, one encoded on a single exon, and the other on three exons. The reason for this is still unclear. It is interesting to note that the gene encoding C4bp β chain has become a pseudogene in the mouse but not the rat [493, 494]. No mRNA coding for the murine β chain could be detected by northern blot of liver RNA using a human C4bp β-cDNA probe, and cloning of the gene indicated that it contained only two exons which shared homology with the human C4bp β gene. The first exon was homologous with the exon encoding SCR1 of the human β–chain (72.4% homology) and the second was homologous with the exon encoding SCR2 (84.2% homology). No other exons coding for C4bp-like sequences were found. The predicted cDNA sequence contained two in-phase stop codons that would result in the production of a truncated (60 amino acids) form of the protein, unlikely to be functional.

During the acute-phase response, total levels of C4bp (isoforms with and without the β chain) are elevated compared to normal levels [478, 482–484]. One study reported an increase of total C4bp concentration in acute phase sera to 162% of that found in controls, whereas levels of C4bp containing the β chain were only slightly elevated (122%) and were comparable to the elevation seen with protein S (124%) [483]. This differential regulation of the α and β chains of C4bp would cause the increase in protein S to be masked by its binding to the increased levels of C4bp isoforms containing the β chain, resulting in a stable concentration of free protein S in plasma even though the total concentrations of C4bp and protein S had altered. A subsequent study also described differential regulation of the two genes, but demonstrated that whilst total levels of C4bp were elevated, some individuals demonstrated an increase in isoforms without the β chain, and others showed an increase in isoforms with the β chain [484]. This suggests that both C4bp α and C4bp β are acute-phase reactants. Acute-phase cytokines used *in vitro* have different effects on the levels of α and β chain expression in Hep3B cells. IL-1β, IL-6 and IFN-γ increased mRNA levels for both α and β, whereas TNF-α had the opposite effect and decreased the expression of both chains. However, synergistic effects of combinations of cytokines resulted in large increases in C4bp α expression with minimal increases in C4bp β. The combined effects of cytokines during the acute-phase response may therefore play a crucial role in maintaining stable levels of the β chain and free protein S in plasma [484]. The picture still needs clarification, particularly as differing effects on C4bp expression have been reported for the same cytokine in different studies [482, 484, 495].

*Binding/active sites*

The individual chains of C4bp are elongated structures due to the organization of their SCR domains. Imaging of C4bp by electron microscopy clearly reveals its unusual structure[332]. The $\alpha$ chains stretch away from the central 'core' of the protein, resulting in a spider-like structure. The $\alpha$ chain arms are each about 330 Å in length and are flexible, with the ability to bend at multiple points along their length. The $\beta$ chain appears as a much shorter arm. The two ligands, C4b and protein S, can be visualized binding either to the central core (protein S) or at the amino-terminal ends of the $\alpha$ chains (C4b). If C4bp is subjected to limited digestion with chymotrypsin, fragments of 48 kDa and 160 kDa are produced[496, 497]. The 48-kDa fragment retains C4b-binding and cofactor activity and represents the amino-terminal portion of the $\alpha$ chain, as confirmed by amino-terminal sequencing. The 160-kDa fragment is the 'core' of the protein consisting of the carboxy-terminal portions of the chains disulphide bonded into a ring structure. The protein S binding site is present on the central core (on the short $\beta$ chain), although this binding site is also sensitive to treatment with chymotrypsin and needs to be 'protected' by the presence of protein S during the digestion. Whilst interaction with SAP is only partially characterized, it appears that it also involves a site on the 160-kDa fragment, close to the central core of the molecule; in this case the carbohydrate moieties on the $\alpha$ chains are required for binding[498]. When studied by electron microscopy, the 48-kDa fragment is an elongated 290-Å × 30-Å structure, as expected[499]. The 160 kDa fragment is visualised as a ring structure with an outer diameter of 60 Å and an inner diameter of 13 Å. The truncated arms extending from the ring are about 40 Å in length. Analysis of the C4bp structure by synchrotron X-ray scattering and hydrodynamic analysis support the structure delineated by the electron microscopy studies, although a more compact structure is predicted where the angle between the $\alpha$ chains and the central axis is smaller[333]. This implies that the molecule is not planar and that the $\alpha$ chains project out as a cluster from one face of the central core.

Each $\alpha$ chain has a binding site for C4b indicating that the interaction between C4bp and C4b is likely to be multivalent. Initial studies showed that C4bp could bind at least four C4b molecules[454, 462, 469]. In one of these studies analytical ultracentrifugation was used to study the C4bp–C4b interaction and showed that, at physiological ionic strength, four molecules of C4b bound to C4bp, and at low ionic strength six C4b molecules bound[462]. C4bp did not bind native C4 or C4d and demonstrated only a weak binding affinity for C4c. The sites were independent and binding was not co-operative. Hydrodynamic analysis demonstrated a significant increase in the observed sedimentation rates of the C4bp–C4b complex after binding of three C4b, in accordance with the generation of a more compact three-dimensional structure [462]. It was hypothesized that at physiological ionic strength, the structure of C4bp partially collapsed following binding of three or four C4b, sterically hindering any further binding of C4b. However, at low ionic strength the structure of C4bp was more stable, permitting simultaneous binding of up to six C4b. A low affinity interaction of C4bp with C3b was also demonstrated by analytical ultracentrifugation under low ionic strength conditions [462].

Contradictory results have generated controversy in the characterization of the binding site for C4b on the C4bp α chain (Figure 3.20). Two studies have located the binding site towards the carboxy-terminal end of the α chain, in SCRs 6 and 7 [485, 500]. In the first, C4bp was subjected to limited digestion by several different enzymes and the cofactor and C4b-

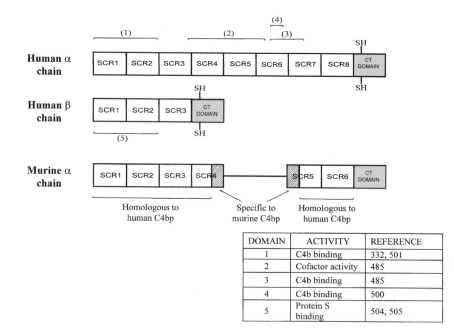

**Figure 3.20. Binding sites located in C4bp.**

The binding site(s) for C4b in the α chain have been located variously either at the amino-terminal end of the α chain, or towards the carboxy-terminal end of the chain in SCRs 6 and 7. Murine C4bp can also bind human C4b and function as a fI cofactor. Comparison of the human α-chain structure with that of the murine C4bp α chain, indicates that SCRs 5 and 6 in human C4bp α chain are not involved in C4b binding as these SCR are 'missing' in the murine form of the protein. The binding site for protein S has been firmly located in SCR1 of the β chain.

binding activities were compared before and after digestion[485]. The cofactor activity site was tentatively assigned to the region between residues 177 and 322, and the binding activity to residues 332–395. In the second study, a monoclonal antibody was generated with specificity for the α chain which prevented binding of C4b to C4bp. Its epitope was mapped by submitting C4bp to trypsin digestion and analysing binding of the resulting peptides to the antibody[500]. This located the epitope in the region between residues Ser-333 and Arg-356 (SCR6). In contrast, electron microscopy studies have implicated the amino-terminal end of the α chain in binding[332], as does a recent study using recombinant proteins[501]. The recombinant proteins were chimeras based on the α chain sequence in which one, two or three of the amino-terminal SCRs were replaced with the corresponding SCRs from the β chain. None of the mutant proteins bound to C4b, indicating that the amino-terminal SCRs were essential for binding. In support of this, monoclonal antibodies specific for either SCR1 or SCR2 inhibited C4b binding, and in some cases blocked cofactor activity, providing good evidence that the C4b binding site is located in the three amino-terminal SCRs of the α chain. In the mouse, C4bp consists of non-covalently associated α chains of only six SCRs; the 'missing' SCRs correspond to SCRs 5 and 6 in the human[502]. Murine C4bp is 51% identical at the amino acid level to the human protein (61% if conserved residues are taken into account), can bind human C4b and provide cofactor activity for cleavage by murine fI[156, 503]. This strongly implies that the carboxy-terminal SCRs are not involved in C4b binding and cofactor activity (Figure 3.20).

The binding site for protein S has been located on SCR1 of the β chain by synthesizing overlapping peptides that spanned the whole of the β chain amino acid sequence and searching for those that inhibited binding of protein S to C4bp[504]. A peptide spanning residues 31–45 not only inhibited C4bp–protein S interaction, but also inhibited the cofactor activity of protein S in coagulation assays. A monoclonal antibody against the peptide could also prevent the protein association, and immobilized peptide could bind protein S. These results convincingly demonstrate that this region in SCR1 contains the binding site for protein S, a surprising finding as protein S bears no homology to C3 or C4. The location of the site has been further corroborated by the expression of recombinant forms of the protein (Figure 3.20)[505]. Deletion mutants of the β chain containing only SCRs 1 and 2 could bind to protein S, whereas mutants comprising SCRs 2 and 3 could not. The importance of SCR1 was further investigated by generating chimeric proteins based on the α-chain sequence[505]. These recombinant proteins had either one, two or three of the amino-terminal SCRs of the α chain replaced with the corresponding

SCRs of the β chain, and were also used to investigate the C4b binding site on the α chain (see above). The recombinant proteins acquired the capacity to bind protein S if SCR1 from the α chain was replaced by SCR1 of the β chain, demonstrating the key involvement of this SCR in protein S binding. Molecular cloning of bovine C4bp α and β chains has revealed that the β chain contains only two SCR and comparison with the human β chain sequence indicates that the 'missing' SCR is SCR1 [506]. The lack of association of bovine C4bp with protein S lends further support to the above data implicating SCR1 in protein S binding.

*Interaction with micro-organisms*
C4bp not only binds C4b, protein S and SAP, but in common with other members of the RCA family (fH and MCP), it is itself a ligand for the pathogen *Streptococcus pyogenes*. *S. pyogenes* can bind to human C4bp through its surface proteins Arp or Sir [301, 507]. These proteins are members of the M protein family implicated in virulence [508]. It has been suggested that surface molecules on *S. pyogenes* mimic C4b epitopes in order that C4bp becomes bound to the bacterial surface [509]. This provides the bacterium with the means to protect itself from the host C system.

## Membrane RCA proteins

### Complement receptor 1 (CR1; C3b receptor; CD35)

*History*
The term 'immune adherence' describes the binding to primate erythrocytes of micro-organisms or immune complexes that have been opsonized with antibody and C [510, 511]. Immune adherence is mediated by specific membrane receptors (immune adherence receptors) on the erythrocyte surface which bind C3b or its breakdown products on the surface of the micro-organism. Whilst the phenomenon of parasite binding to leukocytes in the presence of immune serum was demonstrated many years earlier, the binding of parasites to human and primate erythrocytes was first demonstrated by Duke & Wallace in 1930 [512]. Red cell adhesion was demonstrated by mixing trypanosomes with the blood from immune patients and showing that the parasite became covered in the patients' erythrocytes. C was implicated in the immune adherence reaction when it was demonstrated that a heat-labile factor from normal serum was required to mediate the adherence in conjunction with a heat-stable factor in immune serum. The heat-labile factor was also inactivated by ammonia

and cobra venom factor providing further evidence of the involvement of C in the reaction [513, 514]. The introduction of the rosetting technique enabled the specificity of the binding reactions to be analysed further [515]. The adherence of any cell type to erythrocytes coated with different C components could be studied, demonstrating that either C3b or C4b fixed on the surface of cells mediated the immune adherence reaction and that various cell types other than erythrocytes also bore one or more immune adherence receptor(s) on their cell surface. The receptor on the erythro-cyte surface involved in the 'classical' immune adherence binding reaction is complement receptor 1 (CR1). This receptor binds to bacteria and other particles opsonized with the C3 activation products C3b and iC3b, and the C4 product C4b.

CR1 was isolated in 1979 by virtue of its activity as a decay-accelerating factor for the AP C3 convertase [516]. Membranes of human erythrocytes were solubilized with detergent and the inhibitory protein was purified using conventional column chromatographic techniques (ion exchange and gel filtration), affinity chromatography on Sepharose-bound C3 and Sepharose-lentil lectin. CR1, initially termed gp205 due to its elec-trophoretic mobility, was purified to homogeneity by monitoring the activity of fractions for inhibition of the C3bBbP convertase on sheep erythrocytes. Its ability to act as cofactor for the factor I-mediated inacti-vation of C3 was also demonstrated. The identity of gp205 with the ery-throcyte C receptor mediating immune adherence was reported the following year, when it was demonstrated that treatment of a variety of tar-get cells including erythrocytes, B lymphocytes, polymorphonuclear cells and monocytes, with a polyclonal antibody to purified gp205 inhibited rosette formation between sheep EC3b and the target cells [517].

*Function*
CR1 is involved in the regulation of both the CP and AP, and encom-passes the functions of both DAF and MCP (see next two sections) by act-ing as a decay-accelerating factor and a cofactor for fI-mediated breakdown of C3b and C4b. When CR1 was initially purified, it was shown to be a cofactor for the fI-mediated inactivation of C3b through formation of iC3b [516]. However, subsequent studies demonstrated that it was also a cofactor for the third fI-mediated cleavage, generating C3c and C3dg from iC3b [366, 518, 519]. C3b bound to solid-phase reagents, such as ATS (activated thiol sepharose), or sheep E, could be degraded to C3c and C3dg by the addition of fluid-phase CR1 in isotonic serum [366]. FH would only act as cofactor for this third cleavage to C3c under conditions of low ionic strength, indicating that, under physiological conditions, CR1 was the only cofactor for this cleavage. At about the same time others also

demonstrated that C3b, bound to immune complexes or on sheep E, was cleaved to C3c and C3dg by purified CR1 or in the presence of human erythrocytes, but that cleavage was inhibited if antibody against CR1 was included in the incubation [518, 519]. Although the affinity of CR1 for C4b is less than for C3b, CR1 also binds C4b in the C4b2a CP convertase leading to decay of the convertase and formation of the inactive cleavage products, C4c and C4d [520–525].

CR1 is multifunctional and has several roles other than inhibition of the C cascade. It is a receptor, and its ligands are the C3 and C4 activation products, C3b and C4b, and the inactive cleaved form, iC3b, although binding to the latter is with a much reduced affinity [519, 526]. As a C3b receptor, erythrocyte CR1 plays a crucial role in the clearance of fluid-phase immune complexes from the circulation, by binding to fragments of C3 and C4 that become attached to immune complexes during C activation (Figure 3.21) [111, 275, 527, 528]. Due to the low affinity binding between monomeric C3b and CR1, multivalent interactions between CR1 and C3b-bearing surfaces or complexes are essential for efficient attachment to CR1. Only C3b multimers, such as dimeric C3b (or even C3b/C4b heterodimers) generated at sites of C activation through clustering of C3b, bind to CR1 with a high affinity [529, 530]. Despite the relatively low number of CR1 on erythrocytes, efficient binding of immune complexes is achieved by organization of CR1 into clusters on the surface of these cells, favouring CR1/immune complex interactions [531–533]. Erythrocyte binding permits the safe transportation of immune complexes to the reticuloendothelial system where they are degraded. Binding of immune complexes to erythrocytes is a dynamic event with binding and release occurring rapidly as C3b on the complex is degraded to iC3b through the interaction with CR1 and fI. The low affinity of CR1 for iC3b causes release of the complex, which is then rapidly re-bound as further C3b on the complex interacts with CR1. This 'on-off' mechanism of binding of immune complexes to CR1 facilitates transfer from erythrocytes to tissue macrophages in the spleen and liver.

A recent study reports the intriguing finding that CR1 is also a receptor for the collagen-like domains of cell bound C1q [534]. It has long been known that cellular receptors for C1q exist on leukocytes, but their functional relevance is still unclear. Shortly following C1-mediated activation of the CP, the C1r and C1s subunits are dissociated through binding to C1inh. This leaves the collagenous 'stalk' of C1q projecting away from cell membranes, or immune complexes, and available for receptor binding. It has been suggested that opsonization of immune complexes with C1q is of equal or greater importance than C3b/C4b opsonization for immune complex clearance [534]. The coating of immune complexes with

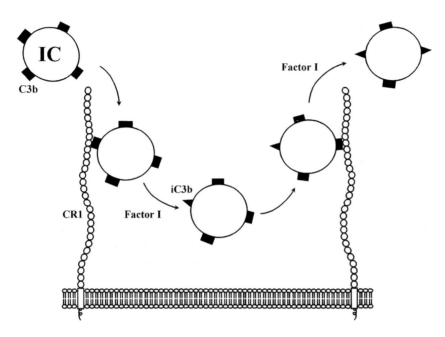

**Figure 3.21. Processing of immune complexes by CR1.**
CR1 binds C3b/C4b (and to a lesser extent, iC3b) and acts as a decay-accelerating factor promoting dissociation of components of the C3 and C5 convertases. It is also a cofactor for all three fI-mediated cleavages of C3b, the only human cofactor which supports formation of C3dg and C3c under physiological conditions. CR1 is an extrinsic C regulator, and its major physiological role is processing of immune complexes (ICs). C3b-opsonized IC are bound to the surface of E through interaction with CR1; subsequent cleavage of C3b and formation of iC3b results in release of the IC due to the low-affinity interaction between iC3b and CR1. The IC rebinds to the E surface via another C3b–CR1 interaction. Binding of IC to E is therefore a 'dynamic' event with the rapid 'on-off' binding mechanism facilitating transfer to tissue macrophages in the spleen and liver.

any one of three opsonins, C3b, C4b or C1q, facilitates binding to erythrocyte CR1 and aids their clearance. It follows that deficiency of C1q results in a total lack of opsonization, whereas a deficiency of C2, or C4, results in lack of C3b-opsonization, or both C3b- and C4b-opsonization respectively. This 'progressive' lack of opsonization parallels the severity of immune complex disease seen in individuals deficient in components of the CP; individuals deficient in C1q are particularly prone to severe immune complex disease.

Leukocyte CR1 also facilitates phagocytosis of bacteria and other particles which have been opsonized by C3b, or even by C1q. By binding to particles thus opsonized, CR1 increases the number of sites of attachment between the particle and leukocyte and enhances phagocytosis mediated by other receptors such as CR3 and Fc receptors [535, 536].

*Distribution*
In contrast to the wide distribution of MCP and DAF, expression of CR1 is restricted. CR1 is found mainly on circulating cells such as erythrocytes, monocytes, macrophages, lymphocytes (excluding some T cells), neutrophils and eosinophils [515, 517, 537–539]. Whilst erythrocytes have on average only 500 copies of CR1 on the cell surface, due to the sheer number of these cells compared to leukocytes they carry most of the CR1 in the bloodstream. Leukocytes have far greater numbers of CR1 on the surface per cell than do erythrocytes. 'Resting' neutrophils and monocytes have about 5000 CR1 per cell, although following activation of neutrophils, levels of CR1 are rapidly increased (within a few minutes) by as much as ten-fold [540, 541]. This enhanced expression occurred in the absence of degranulation and in the presence of protein synthesis inhibitors, implying a non-granule intracellular pool, and was triggered by nanomolar quantities of inflammatory mediators. Immuno-electron microscopy localized the intracellular CR1 pool to small vesicles, and further analysis indicated that these were the secretory vesicles that also contained alkaline phosphatase, cytochrome $b_{558}$, FMLP (formyl-methionyl-leucyl-phenylalanine) receptors, DAF and one of the intracellular pools of CR3 [542, 543]. It is thought that these vesicles are formed through an endocytic process as they contain albumin, a protein not synthesized by neutrophils [544]. B cells express very high levels of CR1, between 20,000 and 40,000 CR1 per cell [540]. These high levels may be necessary in order to capture immune complexes on the B cell surface. CR1 cofactor activity will then trigger fI cleavage to generate C3dg-coated antigen; this in turn binds CR2 and cross-links the CD19/CR2 signalling complex with sIgM, resulting in B cell proliferation and antibody production. CR1 is also found at several other sites including podocytes in the kidney [545, 546], follicular dendritic cells in lymphoid organs [547] and astrocytes in the brain [548].

A soluble form of CR1 with a molecular weight comparable to erythrocyte CR1 has also been described in plasma [549]. In normal individuals the level of circulating soluble CR1 is about 30 ng/ml; this is increased in disease such as liver failure or renal failure or in certain haematological malignancies [550]. The circulating molecule is probably released from the surface of leukocytes by enzymatic cleavage at a site within the carboxy-terminal end of the transmembrane domain [551]. Interestingly, neutrophils release large amounts of soluble CR1, particularly when stimulated with various agents such as FMLP and TNF-$\alpha$ [552]. Soluble CR1 is also found in the synovial fluid of patients with inflammatory joint disease, where high levels of CR1 in the joint are associated with severe disease [553]. It is not yet clear whether soluble CR1 has a physiological role, but experiments using soluble recombinant CR1, demonstrate that it can be a very powerful inhibitor of C with a molar activity some 20-fold greater than fH (see Chapter 9).

*Molecular cloning and structural analysis*
Initial studies of CR1 demonstrated that the protein had an apparent molecular weight of about 205 kDa on non-reducing SDS-PAGE or 235 kDa under reducing conditions. However, isolation of CR1 from individuals, rather than from pooled plasma, demonstrated that the protein was polymorphic. Two independent reports demonstrated the presence of two major co-dominant alleles [554, 555]. These CR1 isoforms had molecular weights under non-reducing conditions of 190 kDa and 220 kDa respectively, the reduced forms both had an apparent molecular weight some 30 kDa greater. Purification and functional characterization of the two forms of CR1 demonstrated that each possessed similar cofactor and decay-accelerating activity [556]. In one study of 33 individuals it was found that 70% expressed only the small isoform (termed A or F), 3% expressed only the larger isoform (termed B or S), and the remaining 27% expressed both isoforms [554]. A second study of a larger population (111 Caucasian individuals) put the distribution at 64.9%, 1.8% and 33.3% [555]. Two much rarer isoforms were subsequently identified with molecular weights of 160 kDa (C allele) and 250 kDa (D allele) respectively [557, 558]. In large population studies in Caucasians the gene frequencies of the different alleles were 0.83 (A), 0.16 (B), 0.01 (C) and ~0.0025 (D). mRNA size polymorphisms were identified and these correlated with the protein molecular weight polymorphisms, suggesting that the differences in the allelic variants was present at the genomic level and that the addition or subtraction of a genomic sequence was responsible for the molecular weight differences [559, 560]. Indeed, analysis of the carbohydrate on the different isoforms of CR1 [561], cloning of the CR1 cDNAs (Figure 3.22) [344–346, 355], and characterization of the CR1 genes [354, 355] demonstrated that the differences in molecular weight between the various forms were not due to altered glycosylation, but rather to differences in the polypeptide backbone. There is, however, a difference between the molecular weights of CR1 on erythrocytes and PMN of any given individual, which is due to altered N-glycosylation and can be eliminated by treatment with endoglycosidase F [557, 561, 562].

The gene for CR1 is found in the RCA cluster on chromosome 1 [23, 269–272]. Early indications that CR1 belonged to a gene cluster regulating

**Figure 3.22. Derived amino acid sequence of CR1 (A allele).**

```
MGASSPRSPEPVGPPAPGLPFCCGGSLLAVVVLLALPVAWGQCNAPEWLP      9
FARPTNLTDEFEFPIGTYLNYECRPGYSGRPFSIICLKNSVWTGAKDRCR     59
RKSCRNPPDPVNGMVHVIKGIQFGSQIKYSCTKGYRLIGSSSATCIISGD    109
TVIWDNETPICDRIPCGLPPTITNGDFISTNRENFHYGSVVTYRCNPGSG    159
GRKVFELVGEPSIYCTSNDDQVGIWSGPAPQCIIPNKCTPPNVENGILVS    209
DNRSLFSLNEVVEFRCQPGFVMKGPRRVKCQALNKWEPELPSCSRVCQPP    259
PDVLHAERTQRDKDNFSPGQEVFYSCEPGYDLRGAASMRCTPQGDWSPAA    309
PTCEVKSCDDFMGQLLNGRVLFPVNLQLGAKVDFVCDEGFQLKGSSASYC    359
VLAGMESLWNSSVPVCEQIFCPSPPVIPNGRHTGKPLEVFPFGKAVNYTC    409
DPHPDRGTSFDLIGESTIRCTSDPQGNGVWSSPAPRCGILGHCQAPDHFL    459
FAKLKTQTNASDFPIGTSLKYECRPEYYGRPFSITCLDNLVWSSPKDVCK    509
RKSCKTPPDPVNGMVHVITDIQVGSRINYSCTTGHRLIGHSSAECILSGN    559
AAHWSTKPPICQRIPCGLPPTIANGDFISTNRENFHYGSVVTYRCNPGSG    609
GRKVFELVGEPSIYCTSNDDQVGIWSGPAPQCIIPNKCTPPNVENGILVS    659
DNRSLFSLNEVVEFRCQPGFVMKGPRRVKCQALNKWEPELPSCSRVCQPP    709
PDVLHAERTQRDKDNFSPGQEVFYSCEPGYDLRGAASMRCTPQGDWSPAA    759
PTCEVKSCDDFMGQLLNGRVLFPVNLQLGAKVDFVCDEGFQLKGSSASYC    809
VLAGMESLWNSSVPVCEQIFCPSPPVIPNGRHTGKPLEVFPFGKAVNYTC    859
DPHPDRGTSFDLIGESTIRCTSDPQGNGVWSSPAPRCGILGHCQAPDHFL    909
FAKLKTQTNASDFPIGTSLKYECRPEYYGRPFSITCLDNLVWSSPKDVCK    959
RKSCKTPPDPVNGMVHVITDIQVGSRINYSCTTGHRLIGHSSAECILSGN   1009
TAHWSTKPPICQRIPCGLPPTIANGDFISTNRENFHYGSVVTYRCNLGSR   1059
GRKVFELVGEPSIYCTSNDDQVGIWSGPAPQCIIPNKCTPPNVENGILVS   1109
DNRSLFSLNEVVEFRCQPGFVMKGPRRVKCQALNKWEPELPSCSRVCQPP   1159
PEILHGEHTPSHQDNFSPGQEVFYSCEPGYDLRGAASLHCTPQGDWSPEA   1209
PRCAVKSCDDFLGQLPHGRVLFPLNLQLGAKVSFVCDEGFRLKGSSVSHC   1259
VLVGMRSLWNNSVPVCEHIFCPNPPAILNGRHTGTPSGDIPYGKEISYTC   1309
DPHPDRGMTFNLIGESTIRCTSDPHGNGVWSSPAPRCELSVRAGHCKTPE   1359
QFPFASPTIPINDFEFPVGTSLNYECRPGYFGKMFSISCLENLVWSSVED   1409
NCRRKSCGPPPEPFNGMVHINTDTQFGSTVNYSCNEGFRLIGSPSTTCLV   1459
SGNNVTWDKKAPICEIISCEPPPTISNGDFYSNNRTSFHNGTVVTYQCHT   1509
GPDGEQLFELVGERSIYCTSKDDQVGVWSSPPPRCISTNKCTAPEVENAI   1559
RVPGNRSFFSLTEIIRFRCQPGFVMVGSHTVQCQTNGRWGPKLPHCSRVC   1609
QPPPEILHGEHTLSHQDNFSPGQEVFYSCEPSYDLRGAASLHCTPQGDWS   1659
PEAPRCTVKSCDDFLGQLPHGRVLLPLNLQLGAKVSFVCDEGFRLKGRSA   1709
SHCVLAGMKALWNSSVPVCEQIFCPNPPAILNGRHTGTPFGDIPYGKEIS   1759
YACDTHPDRGMTFNLIGESSIRCTSDPQGNGVWSSPAPRCELSVPAACPH   1809
PPKIQNGHYIGGHVSLYLPGMTISYTCDPGYLLVGKGFIFCTDQGIWSQL   1859
DHYCKEVNCSFPLFMNGISKELEMKKVYHYGDYVTLKCEDGYTLEGSPWS   1909
QCQADDRWDPPLAKCTSRAHDALIVGTLSGTIFFILLIIFLSWIILKHRK   1959
GNNAHENPKEVAIHLHSQGGSSVHPRTLQTNEENSRVLP              1998
```

The amino-terminal signal peptide is underlined. The first residue of the mature protein is **Q** in bold. Numbering is for the mature protein.

**GenBank Accession No:**    X05309, X14358, Y00812, Y00816

C activation came from segregation analysis of the genes for CR1 and C4bp [268]. It was demonstrated that these two genes were closely linked and it was proposed that the genes encoding regulators of the C cascade might be clustered on the same chromosome. Several years later the isolation of partial clones for CR1 and CR2 enabled localization of these two genes to band 1q32 on chromosome 1 [312–314]. The gene for CR1 is located between the CR2 and 'MCP-like' genes [272, 319, 320]. In common with other members of the RCA protein family, CR1 contains SCR domains within its structure. The extracellular domain of CR1 consists solely of SCRs and, while there is usually a high degree of amino acid homology between individual SCRs in RCA proteins (typically 25–30%), their arrangement in CR1 has an unusual feature. In the most common isoform of CR1, the A or F allotype, there are 30 SCR repeats. These can be arranged into four sets of seven SCRs (SCRs 1–28) which are termed 'long homologous repeats' (LHRs) [344–346]. Homology between LHRs is between 65 and 90%, indicating that intragenic duplication played a part in the evolution of the protein. The remaining two SCRs (29 and 30) do not show the same high levels of homology and are not within the same duplication unit. The form of CR1 encoded at the B (or S) allele, consists of 37 SCRs organized into five LHRs [354]. Once again the last two SCRs do not form part of an LHR (Figure 3.23).

Close examination of the CR1 sequence revealed that higher levels of homology between SCRs were present in another pattern which still incorporated the seven-SCR repetition unit (Figure 3.24). For example, SCRs 3–9 showed 99% homology to SCRs 10–16, and SCRs 19–21 were 91% homologous to SCRs 26–28 [345, 347]. These high levels of homology may have been generated initially by gene duplication in the seven-SCR unit and then maintained by unequal cross-over and/or gene conversion [348–350]. Unequal cross-over between sister chromatids is a reciprocal recombination event which results in duplication of a sequence in one chromatid and deletion of that sequence from the other. Gene conversion, on the other hand, is a non-reciprocal recombination which results in 'conversion' of one sequence on a chromatid to that found on the other. One sequence of daughter chromosome remains unaltered, whereas the second daughter chromosome has had part of its sequence 'converted' to that found in the sister chromatid. Both these types of recombination events are important mechanisms for concerted evolution, promoting spread of repeats through members of a gene family which evolve together through genetic interactions amongst members. In one study, analysis of homology in intron sequences of the CR1 gene has identified a candidate site for one unequal cross-over event [355]. A breakpoint in homology was discovered around exon four in LHR-B. Sequence 5' to the exon was

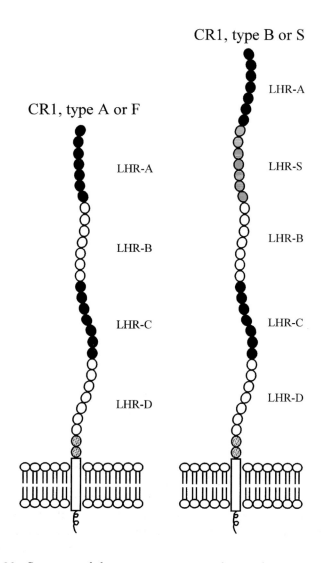

**Figure 3.23. Structure of the two most common forms of CR1.**

The extracellular portion of CR1 is comprised entirely of SCR domains which can be arranged into repeating units consisting of seven SCRs; these are termed 'long homologous repeats' (LHRs) and are indicated in the figure by shading. The carboxy-terminal two SCRs do not form part of an LHR. The form of CR1 encoded at the A allele consists of 30 SCRs, and that encoded at the B allele consists of 37 SCRs (this form of CR1 contains an 'extra' LHR termed LHR-S).

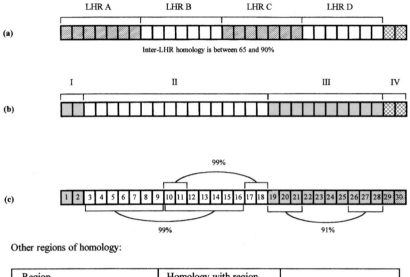

**Figure 3.24. Organization of internal regions of homology in CR1 (A/F allele).**

**(a)** CR1 is arranged into repeating units termed LHRs, which consist of seven SCR domains. The carboxy-terminal two SCR do not conform to this pattern.

**(b)** The SCRs in CR1 can be arranged into regions of even higher identity, showing in some cases as much as 99% sequence homology (figure modified from [345]).

**(c)** Percent identity between SCRs in homology regions represented in (b) (according to [345]). The transmembrane and cytoplasmic domains are not represented in this figure.

found to be homologus (95%) to LHR-C, and sequence 3′ to the exon was homologous (97%) to LHR-A, suggesting that a cross-over event occurred within this exon (Figure 3.25). These unequal cross-over events could also give rise to the large form of CR1 (addition of sequence; D allotype) and the small isoform (deletion of sequence; C allotype).

The allotypic variants of CR1 described above all possess transmembrane domains and a cytoplasmic tail. However, an alternative form of CR1 has been predicted from a cDNA clone that would result in a secreted form of the protein consisting of LHR-A and the first one-and-a-half SCRs of LHR-B [345]. This form of the protein is too small to be that identified by

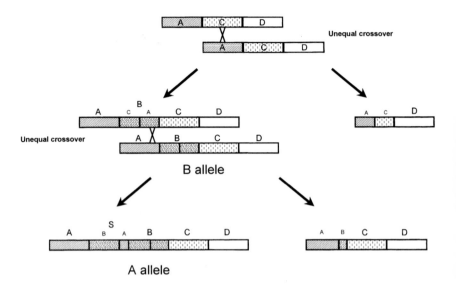

**Figure 3.25. Proposed mechanism for generation of CR1 alleles.**
Unequal cross-over between sister chromatids results in duplication of a sequence
in one chromatid, and deletion of that sequence from the other. Unequal cross-
over of a primordial CR1 allele, indicated in this figure as consisting of three LHRs
(A, C and D) may have given rise to alleles encoding longer forms of CR1 with
additional LHRs (A and B alleles), and truncated forms of the molecule. Figure
modified from [355].

Yoon & Fearon in 1985 and has yet to be identified in plasma. The truncated form would result from the use of an alternative polyadenylation site
located within the intron that splits the two exons encoding SCR9. One
other alternative form of CR1 has been proposed; a 'CR1-like' allele has
been identified in the RCA cluster with 95% nucleotide homology to CR1
and 91% amino acid identity [345, 347, 354]. The clone consists of the signal peptide, and six SCRs comparable to the first five SCRs of CR1 LHR-A and
SCR2 of LHR-B [347]. Whilst no stop codon has been identified, any translated protein is likely to be truncated, as CR1 probes could not identify
further homologous coding sequences. However, the expressed 'CR1-like'
protein has not yet been identified, indicating that the CR1-like allele
may be a pseudogene.

The duplication events that occurred during evolution of the RCA
cluster are clearly demonstrated by examining the exon arrangement in
CR1 [355]. The gene for the A (F) allotype of CR1 contains 39 exons and
spans 133 kb (Figure 3.26). The 5′ untranslated region and signal peptide
are encoded together on the first exon, and the two SCRs that do not fit

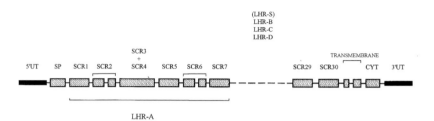

**Figure 3.26. Organization of the *cr1* genes.**

The *cr1* gene is located in the RCA cluster on chromosome 1. The gene encoding the A allotype spans 133 kb of DNA and is comprised of 39 exons. The exon organization of each LHR is identical. The contribution of each exon to the structure of the mature protein is indicated in the figure and discussed in the text. The 5′ and 3′ untranslated regions (5′UT and 3′UT respectively) are indicated by black boxes. SP, signal peptide; CYT, cytoplasmic tail.

into a homology region (29 and 30) are each encoded on a single exon. The transmembrane domain is encoded on two exons, the cytoplasmic tail on another and the 3′ untranslated region on the last exon. The exon organization of each LHR is the same. Within each LHR, the first, fifth and seventh SCRs are encoded on a single exon, whilst the second and sixth are on two exons and the third and fourth are encoded together on a single exon [354, 355]. The B (S) allotype has an extra eight exons corresponding to the fifth LHR (LHR-S) and the organization is identical to the other LHRs.

Quantitation of the different CR1 allelic variants on the surface of erythrocytes from heterozygous individuals demonstrated that the expression of the two alleles was not equal [554, 562]. For example, in two unrelated individuals heterozygous for the A and B forms of CR1, one may express 90% of the A type on the erythrocyte surface, whereas the other may express 90% of the B type. The ratios of one type of CR1 to another vary enormously. In one study, the ratio of the 220-kDa to the 190-kDa form, on erythrocytes from different individuals, varied from 0.1 to 3.1 [562]. The pattern of expression was stable in any one individual and analysis of expression of the different isoforms among heterozygous related individuals suggested that the ratio of the two isoforms was genetically determined [557]. Although the average number of CR1 on erythrocytes is about 500, there is enormous inter-individual variation, from as few as 150 to as many as 3000 copies per cell [517, 563–565]. The level of CR1 expressed on erythrocytes in an individual is a stable characteristic and is not linked to any particular isoform; high or low expression is an independent inherited trait and is associated with a Hind III restriction fragment length polymorphism

(RFLP) [354, 564, 566, 567]. Individuals homozygous for an allelic HindIII fragment of 7.4 kb were shown to express high levels of CR1, whereas those homozygous for a 6.9-kb fragment expressed low levels and heterozygotes expressed intermediate levels. Gene frequencies of the high (H) and low (L) alleles were calculated as 0.73 and 0.27 [568]. This RFLP cosegregated with the different structural allotypes of CR1, indicating that the regulatory element identified by the RFLP was linked to the CR1 gene [567]. The association of either HindIII fragment with quantitative expression of different allotypes suggested that the regulatory element was *cis*-acting. However, the localization of this polymorphism in the A (F) allele to a region between the two exons encoding the second SCR in LHRD is confusing, as this lies some 70 kb distant from the 5′ promoter region [354]. The link between this polymorphism and expression of CR1 is still unclear. The broad range of CR1 expression present on erythrocytes of individuals from any one particular group (HH, HL, or LL) means that there is an enormous overlap in expression between the groups making it impossible to categorize individuals solely from quantification of CR1 expression.

Individuals suffering from SLE often show reduced levels of CR1 [563, 569]. SLE is an autoimmune disease associated with the production of autoantibodies to native DNA and other nuclear antigens. Although SLE is a prominent feature in individuals with deficiencies in components of the CP (e.g. C1q, C2 and C4), the vast majority of cases of SLE have no primary deficiency of C. SLE is characterized by immune complex disease which may be contributed to by a number of different factors such as inefficient C activation on the complexes and/or decreased numbers of CR1 on erythrocytes [570]. There was early controversy as to whether the reduced number of CR1 on erythrocytes in SLE was inherited [564, 569] or acquired [565, 568, 571–575]. It is now broadly accepted that the low level of CR1 does not itself induce disease, but is actually a consequence of disease. The decreased expression of CR1 in SLE does not correlate with the Hind III polymorphism described above, and individuals who have low CR1 levels due to this polymorphism do not have an increased propensity to develop SLE. Removal of immune complexes from the surface of erythrocytes by mononuclear phagocytic cells in the liver and spleen may proceed in part by proteolytic cleavage of CR1 from the surface of the carrier erythrocyte. It has been suggested that this mechanism is enhanced in SLE patients due to an increase in immune complex load [574, 576, 577]. It is also possible that release of vesicles from the surface of erythrocytes results in a decrease in levels of expression of surface CR1 [550]. CR1 is also lost from the surface of erythrocytes as they age, through an as yet unidentified process, although the mechanisms mentioned above may also be responsible for the decreasing number of CR1 on aged erythrocytes [576, 578, 579]. There are conflicting reports as to whether erythrocytes from different groups of individuals (HH, HL or LL) and from SLE patients show different rates of peripheral catabolism of CR1 [576, 578, 579].

*Binding/active sites*
Binding sites for C3b or C4b are located at the amino-terminal ends of LHRs A, B, and C. LHR A binds C4b, whilst LHRs B and C bind both C3b and C4b; in all cases the first two SCRs (1 and 2, 8 and 9, 15 and 16) are required for ligand binding (Figure 3.27) [346]. Amino acid substitution studies demonstrate that there is one amino acid in SCR1, and three in SCR2, essential for C4b binding [580]. However, it is possible to alter the binding specificity of LHR-A such that it will also bind C3b by substituting a few key amino acids in SCRs1 and 2 with the equivalent amino acids from SCR9 [580–583]. Whilst the amino acids in the two amino-terminal SCRs of each LHR define binding specificity, three SCRs are required before even a low-affinity C3b binding is achieved, and four SCRs are required to achieve a C3b-binding affinity and cofactor activity comparable to the native CR1 protein [584].

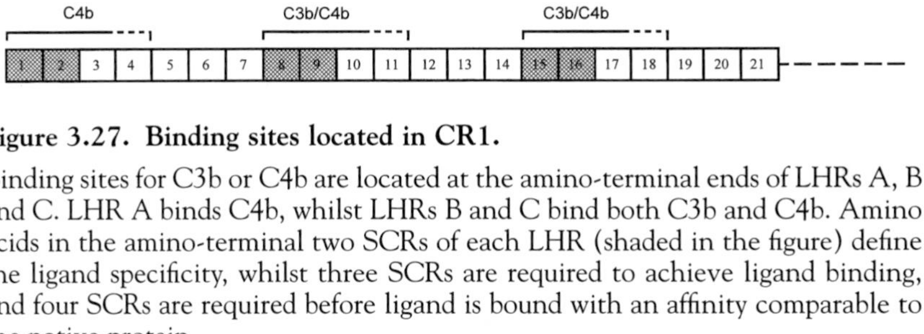

**Figure 3.27. Binding sites located in CR1.**
Binding sites for C3b or C4b are located at the amino-terminal ends of LHRs A, B and C. LHR A binds C4b, whilst LHRs B and C bind both C3b and C4b. Amino acids in the amino-terminal two SCRs of each LHR (shaded in the figure) define the ligand specificity, whilst three SCRs are required to achieve ligand binding, and four SCRs are required before ligand is bound with an affinity comparable to the native protein.

*Interaction with micro-organisms*
Recognition of a foreign surface, such as a bacterial membrane, by the C system results in opsonization of the micro-organism with C3 activation fragments, with consequent lysis or enhanced clearance through phagocytosis. A growing number of organisms are recognized which, despite activating human C, fail to be lysed and even use opsonization as a means to further invasion. The prime example of this is HIV, which is capable of infecting cells using a multitude of receptors (discussed in Chapter 7). The 'classical' receptor for HIV is CD4, although complement receptors, including CR1, have been shown to bind opsonized HIV and enhance infection [299, 585, 586]. Follicular dendritic cells (FDCs) located in germinal centres of lymph nodes are a major reservoir of virions in an infected

individual and here too, C receptors (CR1 included) have been implicated in 'tethering' opsonized virions to the cell surface. This accumulation of virus in the germinal centre may be an important mechanism for viral persistence and infection of T cells [587–589]. It has also recently been reported that CR1 may play a role in permitting entry of *Mycobacterium tuberculosis* into macrophages through interaction with complement receptors [590, 591]. Again, opsonization of the micro-organism with C3 is one mechanism whereby internalization can be achieved. It is interesting to note that CR1 may even play a role in malarial infection [592]. Severe disease has been associated with the ability of erythrocytes infected by *Plasmodium falciparum* to rosette with uninfected erythrocytes, thus enhancing transmission. Erythrocytes from individuals with a deficiency of CR1 show a reduced tendency to form rosettes, and soluble recombinant CR1 can inhibit rosette formation, indicating that CR1 may mediate, at least in part, the rosetting interaction between erythrocytes [592]. Individuals with certain polymorphisms in CR1, common in Africa, have a reduced affinity for binding *P. falciparum* through CR1 and it has been proposed that this represents a protective mutation arising under selective pressure from the parasite.

## Complement receptor 2 (CR2; C3d receptor; CD21)

C receptor 2 (CR2) is not an inhibitor of the C cascade. It has a key role in B-cell activation and is expressed primarily on these cells where it is a receptor for C3dg, the end-product of C3 inactivation. The location of the gene encoding CR2 in the RCA gene cluster, and the high homology of the protein to CR1 merits a brief description of its structure and function here.

### History
CR2 was first identified in 1973 when it was demonstrated that lymphocytes carried a receptor for C3 fragments which was distinct from the immune adherence receptor, CR1 [593–595]. However, efforts to purify CR2 bore fruit only in 1981 when a protein with C3d-binding properties was isolated from the supernatant of Raji cell cultures. It had a molecular weight of 72 kDa (hence originally termed gp72) and, due to its soluble nature, was assumed to have been spontaneously shed from the cell surface [596]. Antiserum raised against gp72 was used to purify the C3d receptor from membranes of solubilized lymphocytes and indeed a 72-kDa protein was isolated and tentatively identified as CR2. In an independent study later that year, a 140-kDa glycoprotein was isolated from solubilized Raji lymphoblastoid membranes on the basis of its C3b-binding activity [597].

However, in this case it was unclear whether the isolated receptor also had C3d-binding activity. The subsequent characterization of two antibodies (anti-B2 and HB-5) to a 140-kDa membrane protein (gp140) found on human B cells helped clarify the situation [598–601]. Both antibodies specifically detected a B-cell membrane protein that bound C3d and was involved in rosetting with sheep erythrocytes. These properties indicated that gp140 was identical with the earlier described lymphocyte receptor CR2, whilst the previously isolated protein, gp72, was shown to be a degradation product of CR2 containing the C3d-binding site [602, 603].

*Function*
CR2 forms a functional complex on the B-cell surface with a number of other proteins, including a member of the IgG superfamily, CD19, and a member of the tetraspan membrane protein family, TAPA-1 (CD81) (Figure 3.28). These three proteins can be co-immunoprecipitated from

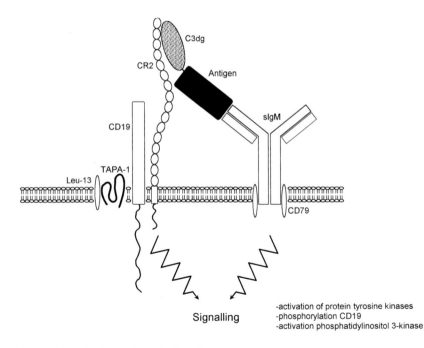

**Figure 3.28. Role of CR2 in B-cell activation.**
CR2 is constitutively associated with CD19 and TAPA-1 (CD81) on the B-cell surface. TAPA-1 is itself associated with Leu-13. Ligation of the CD19–CR2 complex results in cell activation signals, as does ligation of surface IgM. Simultaneous ligation of both these complexes, for example with C3dg coated antigen, markedly lowers the threshold required for signalling through the antigen receptor.

digitonin lysates of B cells indicating that they are physically associated on the B cell surface [604-606]. A fourth molecule, Leu-13, may also be involved in the CD19 complex as it can be co-immunoprecipitated with TAPA1 [605]. The TAPA1–Leu-13 complex is widely expressed on leukocytes and may represent a general signal transduction complex which attains specific signalling functions upon association with other membrane complexes. Whilst it is uncertain whether surface IgM (sIgM) is also incorporated into the CR2–CD19 complex, it is clear that together they play a crucial role in signalling and activation of B cells (reviewed in [274, 607]). Simultaneous ligation of either CR2 or CD19 in the CD19–CR2 complex with sIgM markedly lowers the threshold required for signalling through the antigen receptor [273, 608]. This synergistic interaction of different stimulatory signals, which probably involves several biochemical pathways, is the key to B-cell activation [609]. *In vivo*, crosslinking between sIgM and the CR2 complex is probably provided by C3dg-coated antigen, or by cells bearing antigen and another ligand for either CD19 or CR2, such as CD23 [610-612]. Interestingly, there are also indications that CR2 may associate with other molecules in the plasma membrane of T cells upon ligation of the receptor with C3d [613]. It is not yet clear whether these other complexes have any role in T-cell signalling.

The role of CR2 and C3dg in humoral immunity is evidenced *in vivo* by the impaired immune response in individuals and other animals with a defective C system, or in animals in which CR2 function has been blocked by specific antibody or competed out by soluble CR2 [614-618]. A striking demonstration of the interplay between the C system and the humoral immune response has recently been demonstrated in a study which utilized recombinant C3d-hen egg lysozyme (HEL) fusion proteins to immunize mice [119]. These fusion proteins contained HEL fused to increasing numbers of murine C3d moieties. HEL bearing two copies of C3d was 1000-fold more immunogenic than HEL alone, and HEL bearing three copies of C3d was 10,000-fold more immunogenic. Thus, C3d can profoundly influence the acquired immune response.

*Distribution*

CR2 is expressed primarily on late pre-B cells and mature B cells expressing surface antibody, reflecting its role in the B-cell response to antigen [593, 594, 619]. There are also reports of CR2 on other cell types such as epithelial cells [620-622], follicular dendritic cells [547, 623], thymocytes and some peripheral blood T lymphocytes [299, 624-627], some T-cell lines [628], and astrocytes [548]. A soluble form of CR2 (135-kDa) is also found in human serum which is probably derived by shedding from lymphocytes [629, 630].

Levels of soluble CR2 may be elevated in individuals with EBV-associated disease [631].

*Molecular cloning and structural analysis*

A partial cDNA clone encoding CR2 was originally isolated from a human tonsillar library using a probe made from CR1 cDNA [313]. This partial sequence suggested that CR2 belonged to the family of C3b-binding proteins which contained the 60-amino acid repeating unit, the SCR domain. The full cDNA for CR2 was cloned the following year from a library generated from Raji B lymphoblastoid cells [353]. The cDNA encoded a 20-amino acid signal peptide, 1005 amino acids of SCR sequence (16 SCRs in total), a 28-amino acid transmembrane domain and a 34-amino acid cytoplasmic tail containing potential sites for Ser and Thr phosphorylation (Figure 3.29). The mature protein contains 1067 amino acids in total indicating that the polypeptide backbone should have a molecular weight of approximately 117 kDa. This is close to that determined for the non-glycosylated molecule synthesized by B-cell lines in the presence of tunicamycin [632]. Whilst the original cloning data indicated that CR2 comprised 16 SCRs, cDNA sequences published the following year demonstrated two different forms of CR2 with either 15 or 16 SCRs [351]. The 11th SCR (numbered SCR10a) is 'missing' in the shorter transcript. In contrast to CR1, the alternative form is not due to the presence of another allele, but is the result of alternative splicing of the gene [633]. Alternative alleles have been identified which differ from the original clone by several nucleotide differences resulting in multiple amino acid changes [352, 633].

As in CR1, the SCRs in CR2 are arranged into repeating units. Each of these units is based on four SCRs which includes SCR10a (homologous to SCRs 3, 7, 14) but excludes SCR15 (Figure 3.29). This latter SCR is more homologous to SCRs 6, 10 and 13 [351]. The CR2 gene is located in the RCA cluster between the genes for DAF and CR1 and is about 30 kb in length [313, 314, 319]. The gene has a similar exon/intron organization to CR1 in that SCRs are encoded on single, split or fused exons and the pattern is repeated throughout the gene [352]. The first exon of each repeating unit encodes two SCRs, the second exon encodes the third SCR and the fourth SCR is encoded by two separate exons (with the exception of SCR15) (Figure 3.30).

*Binding/active sites*

The binding sites for both C3d and EBV have been located in the two amino-terminal SCRs of CR2 by analysis of CR2 deletion mutants and chimeric proteins formed with CR1 [634–636]. However, whilst these two

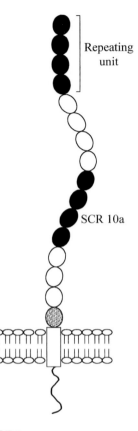

**Figure 3.29. Structure of CR2.**
The entire extracellular portion of CR2 is comprised from SCR domains. These are organized into repeating units consisting of four SCRs (shaded in the figure) which show high levels of homology. The carboxy-terminal SCR is not included in a repeating unit. Alternative splicing of the CR2 gene results in isoforms containing either 15 or 16 SCRs. The alternatively spliced exon encodes SCR10a and is indicated in the figure.

amino-terminal SCRs are sufficient for both C3d and EBV binding, four SCRs are required to attain a binding affinity comparable to the native protein[635]. The amino-terminal two SCRs also contain the epitope for OKB7, a monoclonal antibody capable of blocking both EBV infection and C3dg binding[637, 638]. The epitope for this antibody has been mapped to a region in SCR1 between the first two Cys residues and studies using peptides derived from this region of SCR1 have confirmed that this is the major site of interaction in CR2 for both C3dg and EBV[636]. The construction of human–mouse CR2 chimeras and analysis of their binding

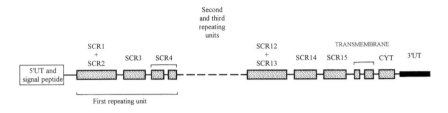

**Figure 3.30. Organization of the *cr2* gene.**

The *cr2* gene is located in the RCA cluster on chromosome 1. It spans approximately 30 kb and comprises at least 30 exons (the intron/exon arrangement of the gene encoding the 5′UT and SP has not yet been defined). The exon organization of each repeating unit is identical, other than that of the fourth repeating unit. In this case the fourth SCR (which is not homologous to the fourth SCRs in other repeating units) is encoded on a single exon. The contribution of each exon to the structure of the mature protein is indicated in the figure and discussed in the text. The 3′ untranslated region (3′UT) is indicated by a narrow black box. CYT, cytoplasmic tail.

properties (mouse CR2 does not bind EBV) also implicates this region as the primary site of interaction, although other binding sites for both ligands have been identified in SCR1 and in SCR2 [636, 639].

*Interaction with micro-organisms*

CR2 is the cellular receptor for Epstein–Barr virus (EBV), and is hence often referred to as the EBV receptor. Identity of the EBV receptor with CR2 was established using antibodies with specificity for either CR2 or the EBV receptor. Multiple studies indicated that the EBV receptor and CR2 were co-expressed to the same extent on cells, that anti-CR2 antibodies could block attachment of EBV to lymphocytes and that the biochemical properties of the antigens recognised by the antibodies were identical [281, 640, 641]. In addition to its role as the EBV receptor, CR2 has also been implicated in HIV infection. In common with other receptors for C3 degradation products, it can mediate cell surface attachment and internalization of HIV virions [299, 585]. This interaction is discussed in detail elsewhere (Chapter 7).

**Decay-accelerating factor (DAF; CD55)**

*History*

A decay-accelerating activity present on human erythrocytes (E) was first described in 1969 by Hoffmann who made a 20% butanol extract of human E membranes and assayed the ability of the membrane extracts to inhibit lysis of antibody-coated sheep E by guinea pig C[642]. Inhibitory activity was detected in the aqueous phase of the extract and Hoffmann further demonstrated that it accelerated the decay of the CP convertase[643]. In the following years C inhibitory activities were also described in E extracts from other species, such as guinea pig and rabbits[644]. The first species from which DAF was purified and characterized was the guinea pig[645]. Guinea pig DAF had a molecular weight of about 60 kDa and accelerated the decay of the CP C3 convertase. Adaptation of the same purification procedure to human E resulted in the isolation of human DAF in the following year[646]. Human E proteins were extracted with butanol and subjected to column chromatography on DEAE-Sephacel, hydroxylapatite, phenyl-Sepharose and trypan blue-Sepharose. Purification was monitored using an assay for the decay of the CP C3 convertase. The purified protein had a molecular weight of 70 kDa, and was an integral membrane protein with decay-accelerating activity for both CP and AP convertases, although it preferentially inactivated the CP convertase, C4b2a. It was distinguished from the recently purified CR1 on the basis of its molecular weight and the failure of neutralizing anti-CR1 antibodies to influence activity. The demonstration that purified E DAF could spontaneously insert into the membranes of foreign E and mediate protection against C indicated that it contained a hydrophobic membrane-interactive domain[647]. This was identified as a GPI anchor when it was demonstrated that DAF could be released from the cell surface by phosphatidylinositol-specific phospholipase C (PIPLC) treatment[648, 649].

*Function*

As its name suggests, DAF functions to accelerate the decay of either C4b2a or C3bBb, and the corresponding C5 convertases (Figure 3.31). It does not have any cofactor activity for fI. DAF is an intrinsic inhibitor, accelerating decay of convertases bound to the membrane of the cell in which it is anchored, and not extrinsically on convertases present on other targets[647]. This indicates that correct orientation and binding between DAF and its substrate can only be achieved if both are present on the same cell surface. There is considerable evidence that the main action of DAF is on the preformed convertase containing Bb or C2a, and that it does not significantly inhibit binding of either fB or C2 to C3b or C4b respectively[650]. Analysis of binding affinities between DAF and its ligands has revealed dramatic differences in affinities for single subunits and the complexed proteins. For example, the apparent association constants (appKa) of DAF for C3b, Bb and C3bBb are reported to be 45 nM$^{-1}$, 67 nM$^{-1}$ and 910 nM$^{-1}$ respectively, an increase of some 15–20-fold for the complex. Likewise, the appKa of DAF for C4b and C4b2a are 0.45 nM$^{-1}$ and

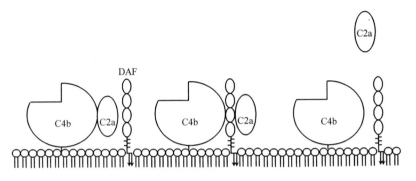

**Figure 3.31. Inhibition of the C activation pathways by decay-accelerating factor (DAF).**

DAF accelerates decay of either the CP or AP C3 and C5 convertases. It is likely that DAF has a low-affinity interaction with C3b or C4b on the cell surface, which is greatly enhanced when a convertase is assembled (see text). Following decay of the convertase, DAF is released enabling it to 'recycle' and interact with further convertases.

530 nM$^{-1}$, a difference of more than 1000-fold[364]. DAF thus has very low affinity for uncomplexed C3b or C4b. In this regard it differs from fH and CR1 which bind strongly to the larger, non-catalytic subunit of the convertases, C3b or C4b. DAF is in part composed of SCR domains which are commonly found in C3b/4b binding proteins and indeed it can be chemically cross-linked to either C3b or C4b on the cell surface, indicating that the inhibitor is in close contact with these subunits[270, 651]. It is likely that DAF has a low affinity interaction with C3b or C4b on the cell surface which is greatly enhanced when a convertase is assembled. The binding site(s) for DAF on the convertases has not yet been defined but there are several possibilities that may explain the data described above. For example, neoantigenic sites may be formed during assembly of the bimolecular enzymes, generating new high-affinity binding site(s) for DAF, or DAF may possess binding sites for both subunits, enhancing binding avidity for the complex. Whatever the explanation, a low-affinity association between C3b/C4b and DAF is probably vital for this inhibitor to work effectively on the cell surface as this enables DAF to dissociate from C3b or C4b following decay of the convertase and to recycle, interacting with other newly formed convertases.

*Distribution*

DAF is widely expressed, being present on most cells including endothelial cells, erythrocytes and leukocytes[652, 653]. Natural killer (NK) cells and a subset of T cells are reported to be deficient in DAF[654, 655]. It is interesting to note here that incorporation of DAF in the plasma membranes of NK cells decreases their capacity for cytotoxicity, and that the presence of DAF in the membranes of target cells reduces their susceptibility to NK cell-mediated cytotoxicity [656]. No explanation for this intriguing effect of DAF on cellular cytotoxicity has yet emerged. DAF is found in many other locations such as the urinary, gastrointestinal and exocrine systems [657–659], placenta [660, 661], spermatozoa [662, 663], skin [664, 665], eye [666], associated with the subendothelial extracellular matrix [667], and at various other sites. Soluble forms of DAF are also present in many biological fluids, including plasma, tears, saliva, synovial fluid, CSF and urine [668].

A deficiency in DAF is associated with the disease, paroxysmal nocturnal haemoglobinuria (PNH), characterized by the unusual sensitivity of E to lysis by homologous C [669–672]. PNH is a clonal disorder, with affected E, leukocytes and platelets arising from an abnormal haematopoietic stem cell clone which lacks the ability to add GPI anchors to proteins [653, 673, 674]. There are also individuals who exhibit a global deficiency in DAF, but not in the other GPI-linked C regulators. These individuals were identified becaused they lacked the Cromer blood group antigens on their E – the 'Inab' phenotype; these antigens have been defined as epitopes on DAF [675–677]. Unlike PNH patients, individuals with the Inab phenotype do not suffer overt E haemolysis, emphasizing the role CD59 plays in the protection of these cells from homologous C attack.

*Molecular cloning and structural analysis*

The cDNA encoding DAF was cloned in 1987, demonstrating that DAF contained multiple SCR modules, in common with other RCA proteins (Figures 3.32, 3.33) [678, 679]. From the amino-terminus, DAF consists of four SCR domains, each of approximately 60 amino acids. The single N-linked sugar moiety is located at Asn-61, between the first and second SCRs. The SCRs are followed by a region rich in Ser, Thr and Pro residues (STP region; approximately 70 amino acids) which contains numerous sites for addition of O-linked oligosaccharides in the mature protein. The terminal 17 amino acids are essential for GPI anchor addition (see below) [678]. Original cloning data demonstrated two alternative species of DAF mRNA [678]. The two predicted proteins differed at the carboxy-terminal ends, where alternative splicing resulted in a mature protein of either 347 or 406 amino acids. Whilst both proteins were identical through the SCRs and STP regions, the 347–amino acid form, highly expressed at the mRNA level, had a hydrophobic sequence at its carboxy terminus with the consensus signal for GPI anchor addition and the 406 amino acid form, expressed at a ten-fold lower level, had a hydrophilic carboxy

**Figure 3.32. Derived amino acid sequence of DAF (glycosyl phosphatidylinositol (GPI)-anchored).**

```
MTVARPSVPAALPLLGELPRLLLLVLLCLPAVWGDCGLPPDVPNAQPALE    16
GRTSFPEDTVITYKCEESFVKIPGEKDSVICLKGSQWSDIEEFCNRSCEV    66
PTRLNSASLKQPYITQNYFPVGTVVEYECRPGYRREPSLSPKLTCLQNLK   116
WSTAVEFCKKKSCPNPGEIRNGQIDVPGGILFGATISFSCNTGYKLFGST   166
SSFCLISGSSVQWSDPLPECREIYCPAPPQIDNGIIQGERDHYGYRQSVT   216
YACNKGFTMIGEHSIYCTVNNDEGEWSGPPPECRGKSLTSKVPPTVQKPT   266
TVNVPTTEVSPTSQKTTTKTTTPNAQATRSTPVSRTTKHFHETTPNKGSG   316
TTSGTTRLLSGHTCFTLTGLLGTLVTMGLLT                      347
```

The amino-terminal signal peptide is underlined, as is the carboxy-terminal signal for GPI anchor attachment. The first residue of the mature protein is **D** in bold. The GPI anchor is attached to Ser-319, indicated in bold. Numbering is for the mature protein.

**GenBank Accession No:** M30142, M15799

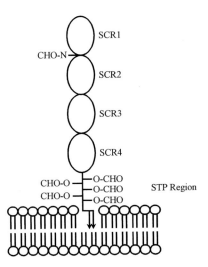

**Figure 3.33. Structure of DAF.**

DAF is comprised from four SCR domains arranged contiguously from the amino terminus of the protein. These are followed by a heavily glycosylated region which is rich in Ser/Thr/Pro residues (STP region); this domain is proximal to the membrane and acts as a 'spacer' region projecting the SCRs the requisite distance away from the plasma membrane. DAF is linked to the membrane by a GPI anchor.

terminus. It has been suggested that this larger form is secreted, although this has yet to be formally demonstrated and the protein has not been detected in plasma.

The amino acids which form the GPI anchor addition signal are predominantly hydrophobic residues such as Leu and Val. However, there are other requirements for GPI anchor attachment including a suitable cleavage/attachment site amino-terminal to the hydrophobic signal[680]. In DAF, the carboxy-terminal 28 amino acids are removed during processing and the GPI anchor is attached to Ser-319 (numbering for the mature protein)[681]. Analysis of GPI anchor attachment by generating synthetic signals for anchoring onto a transmembrane protein (MCP) provides some 'rules' for anchor addition. The attachment site amino acid (ω site) and the residue two amino acids carboxy-terminal to it (ω+2), must be small[682]; in most mammalian GPI-anchored proteins these amino acids are Ala or Gly. DAF is unusual in that the ω+2 residue is a Thr. The spacer region between ω and the hydrophobic domain must be approximately 8–12 amino acids in length and the hydrophobic domain itself must be at least 11 amino acids in length. The GPI anchor is attached in the endoplasmic reticulum by a putative COOH-terminal signal transamidase[683]. The GPI anchor that attaches DAF to the outer leaflet of the membrane contains either two (nucleated cells) or three (E) anchor lipid groups. In the latter case, the inositol ring of the anchor is fatty acid-acylated, providing the third site of attachment and conferring resistance to cleavage by PIPLC[684].

Pulse chase experiments in the HL-60 cell line indicate that DAF is synthesized as a precursor protein of approximately 43 kDa[685]. This species of DAF contains a single N-linked high-mannose unit which rapidly (within minutes) undergoes post-translational modification to yield a 46-kDa species. The addition of multiple O-linked oligosaccharides in the Golgi converts the 46-kDa species to the E (65-kDa) or leukocyte (70-kDa) form, the difference of about 5 kDa being due to differences in the degree of glycosylation. Several atypical forms of DAF have been reported. Smaller forms of the protein have been purified from human erythrocytes and can accelerate decay of fluid-phase, but not cell-bound, convertases[686]. These have been termed DAF-A and DAF-B and have molecular weights of 63 kDa and 55 kDa respectively. Neither form has a GPI anchor and both are thought to result from degradation by endogenous phospholipases and proteases. Two forms of DAF, termed DAF-U1 and DAF-U2, have been purified from human urine[687]. These run as broad bands of 55–65 kDa (DAF-U1) or 60–85 kDa (DAF-U2) on SDS-PAGE analysis, although under physiological conditions they tend to form aggregates of two or three molecules. DAF-U1 has very little activity whereas DAF-U2 has an activity comparable to native DAF. A 100-kDa form of DAF has been reported in tears, along with the typical 70-kDa form [666]. Finally, a dimeric form of DAF (DAF-2) has been purified from human erythrocytes [688, 689]. This form of DAF is functional and appears to result from covalent linkage of two DAF monomers within the SCR domains.

The DAF gene contains 11 exons and spans approximately 40 kb of DNA in the RCA gene cluster (Figure 3.34) [357]. The 5′ untranslated region and signal peptide are encoded on a single exon, and SCRs 1, 2 and 4 are encoded on exons 2 and 3 and 6 respectively. SCR3 is encoded on two exons (4 and 5), with the split occurring in the same position as in most other SCRs (see above). Exons 7, 8 and 9 encode the STP region. Alternative splicing of exon 10 results in the two different classes of mRNA. The major class of mRNA excludes exon 10 and encodes the GPI form of DAF, and the minor form of DAF mRNA includes this exon and encodes the form with the hydrophilic carboxy-terminal sequence. The presence of exon 10 inserts an additional 118 bp in the coding sequence, producing a frameshift which results in translation of the hydrophilic, rather than hydrophobic, carboxy terminus. The last exon is separated from the rest of the gene by a 20-kb intron and encodes the hydrophobic domain (or part of the hydrophilic domain in the case of the minor species) and the 3′ untranslated region.

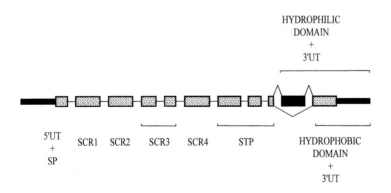

**Figure 3.34. Organization of the *daf* gene.**
The *daf* gene is located in the RCA cluster on chromosome 1. It spans approximately 40 kb and comprises 11 exons. The contribution of each exon to the structure of the mature protein is indicated in the figure and discussed in the text. The 5′ and 3′ untranslated regions (5′UT and 3′UT respectively) are indicated by narrow black boxes, and the alternatively spliced exon is indicated as a thick black box. SP, signal peptide; STP, Ser/Thr/Pro-rich region.

*Binding/active sites*

The SCRs in DAF contain the decay-accelerating activity for both C activation pathways, and the use of recombinant SCR-deletion mutants has enabled elucidation of the 'active' sites [690, 691]. Deletion of SCR1 had no effect on function, whereas removal of any one of SCRs 2–4 abolished regulatory function [690]. Further analysis indicated that regulatory activity for the CP resided in SCRs 2 and 3, and that for the AP in SCRs 2–4 (Figure 3.35) [691]. Molecular modelling of DAF, using the known NMR structures of SCRs from fH, predicts a putative ligand binding site in the groove formed at the SCR2–SCR3 interface, with contributions to binding from a cavity on SCR2 and also surface depressions at the carboxy-terminal ends of SCR3 and SCR4 [692]. Interestingly, the groove formed between SCRs 2 and 3 is lined by three Lys residues, a feature conserved in DAF from other species, implying a role crucial to function. Whilst SCR1 has no role in convertase regulatory activity, a recent study indicates that it may mediate binding to CD97 [693]. CD97 is a leukocyte antigen which is upregulated on activated cells and belongs to the family of seven-

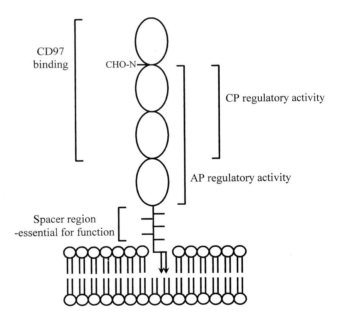

**Figure 3.35. Binding sites located in DAF.**

CP regulatory activity has been located in SCR2-3, and that for the AP in SCR2-4. The amino-terminal SCR is not implicated in C-regulatory activity, but has been implicated in binding to the leukocyte antigen, CD97. The physiological relevance of DAF–CD97 association is not yet apparent.

transmembrane-spanning proteins [694, 695]. The physiological consequences of this interaction are unclear. The STP region of DAF has no convertase regulatory activity. However, deletion of this region was shown to abolish DAF function [690]. Replacement of the STP region by another protein domain of equivalent length, derived from HLA-B44 (residues 66 through to the stop codon, including transmembrane domain and cytoplasmic tail), restored function, indicating that the heavily glycosylated 'stalk' of the DAF molecule acts as an essential spacer region, projecting the SCR domains the requisite distance away from the plasma membrane [690]. The presence of the GPI anchor has been suggested to confer the protein with increased mobility in the plasma membrane. However, recombinant forms of DAF expressed in CHO cells containing either a transmembrane region or GPI anchor have been shown to be equally efficient at protecting cells from C-mediated cytotoxicity [696].

*Other roles*
It has become increasingly clear that GPI-anchored molecules play a role in signal transduction, despite the fact that they are attached only to the outer leaflet of the plasma membrane through two fatty acids linked to the carboxy terminus of the protein. DAF is one such signalling molecule and has been shown to associate with other membrane proteins and to trigger cell activation. The GPI anchor is essential in order for DAF to signal; if it is replaced by the transmembrane and cytoplasmic domains of MCP, the capacity to trigger cell activation is lost [278]. Whilst it is not yet understood how the message is transmitted across the plasma membrane, it is clear that DAF associates with *src* family protein tyrosine kinases in specialized microdomains in the membrane [276–278, 697]. A putative signal-transducing molecule has recently been identified on peripheral blood mononuclear cells [698]. This protein was identified using a monoclonal antibody, 2E12-G7, which co-immunoprecipitated its antigen (43-kDa) and DAF. Cross-linking of either 2E12-G7 Ag or DAF on mononuclear cells resulted in phosphorylation of similarly sized proteins. Molecular cloning and further biochemical characterization of 2E12-G7 Ag will clarify the role of this protein in signal transduction. Cell signalling and activation has also been demonstrated with CD59, another GPI-anchored CRP [699, 700].

*Interaction with micro-organisms*
DAF interacts with a multitude of infectious agents (discussed in detail in Chapter 7). It has recently been identified as a receptor for several viruses, including echovirus7 and other echoviruses [282, 283], enterovirus 70 [284] and

coxsackieviruses A21, B1, B3 and B5 [285-288]. It has also been identified as a ligand for the Dr family of adhesins expressed by *Escherichia coli* [701-703].

## Membrane cofactor protein (MCP; CD46)

*History*

MCP was first identified in 1985 as a C3-binding membrane protein iso-lated from leukocytes [704-706]. Membranes of peripheral blood leukocytes or leukocyte cell lines were iodinated and solubilized with detergent. The cell extracts were subjected to C3 affinity chromatography and those pro-teins binding to the Sepharose were analysed by SDS-PAGE and auto-radiography. Three classes of C3-binding proteins were identified on various cell types; these proteins included the previously identified CR1 and CR2, and a smaller novel protein running as a broad band (45 kDa–70 kDa) or doublet on SDS-PAGE. The protein was termed gp45-70 due to its mobility. Gp45-70 bound to immobilized C3i and C3b, but not C3d and also bound to immobilized C4i, indicating that gp45-70 could bind component proteins from both complement activation pathways. Gp45-70 was present on all leukocyte populations examined, but not on E. Its distribution therefore differed from that of DAF and it could not be immunoprecipitated with anti-DAF antibodies, indicating that it was indeed a unique protein. Gp45-70 was isolated from the membranes of U937 cells (monocyte-like cell line) and HSB-2 cells (T cell line) by four-stage chromatography (chromatofocusing, hydroxylapatite, C3i Sepharose and anion exchange) and also from the membranes of platelets [707, 708]. Analysis of its functional properties demonstrated that, in contrast to DAF, gp45-70 had no decay-accelerating activity, rather it had cofactor activity for the fI mediated cleavage of C3b and thereby complemented the decay-accelerating activity of DAF on the cell membrane. C4b could also be cleaved using gp45-70 as cofactor, although the reaction was less efficient than with C3b. Due to this activity the protein was renamed 'membrane cofactor protein (MCP)'.

Several years prior to its recognition as a regulator of the complement cascade, MCP was discovered independently by several groups using monoclonal antibodies to cell surface antigens. The first identification of the protein now termed MCP was in 1981, with the characterization of an antigen present on human trophoblast recognised by a monoclonal anti-body, H316 [709]. The protein displayed a characteristic two-band pattern of 60 kDa and 69 kDa on SDS-PAGE and the same antibody also detected protein with a similar banding pattern in leukocytes, epithelial cells and spermatozoa [710, 711]. Others identified a widely distributed antigen display-ing a similar electrophoretic mobility, using the monoclonal antibody

E4.3, which was termed Hu Ly-m5 antigen[712]. In 1985, an antigen was identified using yet another monoclonal antibody, TRA-2-10, which gave a similar two-band pattern, was present on most human cells (except erythrocytes) and was encoded by a gene on chromosome 1[713]. Finally, a monoclonal antibody, GB24, was raised against human trophoblast microvilli. It also cross-reacted with the acrosomal region of sperm and with leukocytes[714–716]. The ligand recognised by GB24 was thus termed trophoblast-leukocyte common antigen (TLX ag). Purcell and co-workers compared several of these antibodies and demonstrated that all recognised the same antigen, showing that MCP, TLX ag and Hu Ly-m5 antigen were one and the same[717–719].

*Function*

MCP is a cofactor for the first two fI-mediated cleavages of C3b or C4b[148, 707]. It does not accelerate decay of the C3 convertase and as such it complements the activity of DAF (Figure 3.36). MCP protects cells by an intrinsic mechanism, only inactivating C3b bound to the same cell, indicating a strict requirement for correct orientation of MCP to C3b on the membrane[148, 720, 721]. It has been suggested that MCP preferentially inactivates convertases of the alternative pathway[722], and preferentially inactivates C3b molecules that are covalently bound to certain other membrane proteins, including other C3b (C3b dimer)[720]. It has been reported that in the absence of fI, MCP can actually stabilize the C3 convertase of both

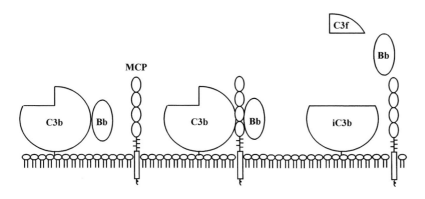

**Figure 3.36. Inhibition of the C activation pathways by membrane cofactor protein (MCP).**

MCP is a cofactor for fI-mediated inactivation of C3b and C4b, resulting in formation of iC3b (but not C3c and C3dg), C4c and C4d. In contrast to DAF, MCP exhibits a high-affinity interaction with C3b which is abrogated following cleavage of C3b and formation of iC3b, enabling MCP to 'recycle' and interact with other cell-bound C3b.

activation pathways leading to enhanced C3 deposition [148]. This may indicate that MCP binds the convertase with high affinity which is abrogated when C3b or C4b is cleaved and inactivated. High-affinity binding of MCP to C3b, enabling its purification by affinity chromatography, distinguishes it from DAF which at best is only retarded on flow through a C3b affinity column. This *in vitro* observation reflects an aspect of 'design' essential for the function of DAF and MCP as either decay-accelerating factor or as a cofactor for fI. DAF must have a low affinity for binding C3b or C4b, in contrast to the high-affinity binding to the convertase, as it must release intact C3b/C4b following dissociation of the catalytic domain, Bb or C2a; these catalytic subunits cannot rebind to C3b/C4b and are inactive. On the other hand, MCP can bind strongly to C3b/C4b as its activity results in cleavage of C3b/C4b with the formation of the inactive molecules iC3b/iC4b for which MCP has only low affinity and is hence released.

*Distribution*

MCP is expressed on all circulating cells, including platelets, granulocytes, T cells, B cells, NK cells and monocytes, but is absent from human E [704, 708, 723]. Leukocytes express between 4,000 and 10,000 copies of MCP per cell and platelets about 600 copies per cell [719, 724]. Some neoplastic haematopoietic cell lines express much higher levels of MCP than their non-neoplastic counterparts. MCP is expressed on virtually every cell and tissue so far studied [725] including human fibroblasts, epithelial and endothelial cells [726], kidney [659, 727], skin [728], respiratory tract [729], eye [730] and brain [731, 732]. MCP is abundantly expressed on sperm [279, 715], trophoblast [710, 733, 734] and in the genital tract where it has an important role in fertility (discussed in detail in Chapter 6). The relative amounts of the two protein bands on western blots of MCP are a stable characteristic which varies between individuals. In the normal human population, 65% express predominantly the larger isoform on their peripheral blood cells, 29% demonstrate approximately equal amounts of the two major isoforms, and 6% express predominantly the lower form. Family studies indicate that the relative amounts of the two forms of MCP expressed are controlled in an autosomal codominant fashion [735]. MCP is also present in biological fluids: a prostasome-associated form of MCP has been described in seminal plasma [736, 737] and soluble forms of MCP have been demonstrated in plasma and tears [738–740]. The soluble MCP isolated from human serum (present at about 50 ng/ml) contains three bands when analysed by SDS-PAGE, at 29 kDa, 47 kDa and 56 kDa [739]. These forms of MCP, which are present in greater quantity in the serum of some patients with cancer (with either solid tumours or haematological malignancies), have been separately purified and shown to exist as monomers all possessing cofactor activity, although all are less active than their membrane counterparts.

*Molecular cloning and structural analysis*

The MCP cDNA was cloned from a human monocytic (U937) library in 1988, using a probe based on the amino-terminal protein sequence of MCP purified from the T-cell line, HSB2 (Figure 3.37)[147]. The derived protein sequence confirmed that MCP contained the structural SCR unit found in many other C regulatory molecules, and *in situ* hybridization studies showed that the *mcp* gene was localized to human chromosome 1 in the RCA cluster[147, 317]. The predicted MCP structure is very similar to DAF in that there are, from the amino terminus of the molecule, four SCR domains followed by a Ser/Thr/Pro-rich region (STP region) (Figure 3.38). Unlike DAF, however, MCP has transmembrane and cytoplasmic domains. The original cDNA clone isolated by Lublin and colleagues in 1988 encoded a protein of 384 amino acids. This included a 34-amino acid signal peptide, followed by four SCR units (250 amino acids), a 29-amino acid STP region, a short domain of unknown function (13 amino acids), the transmembrane domain (34 amino acids) and finally a cytoplasmic tail of 23 amino acids. There were three potential sites for N-glycosylation within the SCR domains of the mature protein at Asn-49 (SCR1), Asn-80 (SCR2) and Asn-239 (SCR4). The STP region, like that in DAF, was predicted to be heavily O-glycosylated. The predicted molecular weight was 42 kDa, suggesting that glycosylation at some or all of these sites contributed substantially to the molecular weight of the mature protein.

**Figure 3.37. Derived amino acid sequence of MCP (STP-BC/CYT-2 isoform).**

```
MEPPGRRECPFPSWRFPGLLLAAMVLLLYSFSDACEEPPTFEAMELIGKP    16
KPYYEIGERVDYKCKKGYFYIPPLATHTICDRNHTWLPVSDDACYRETCP    66
YIRDPLNGQAVPANGTYEFGYQMHFICNEGYYLIGEEILYCELKGSVAIW   116
SGKPPICEKVLCTPPPKIKNGKHTFSEVEVFEYLDAVTYSCDPAPGPDPF   166
SLIGESTIYCGDNSVWSRAAPECKVVKCRFPVVENGKQISGFGKKFYYKA   216
TVMFECDKGFYLDGSDTIVCDSNSTWDPPVPKCLKVSTSSTTKSPASSAS   266
GPRPTYKPPVSNYPGYPKPEEGILDSLDVWVIAVIVIAIVVGVAVICVVP   316
YRYLQRRKKKGKADGGAEYATYQTKSTTPAEQRG                   350
```

The amino-terminal signal peptide is underlined. The first residue of the mature protein is **C** in bold. Numbering is for the mature protein.

**GenBank Accession No:**   Y00651

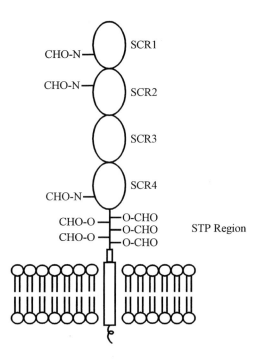

**Figure 3.38. Structure of MCP.**

MCP is comprised from four SCR domains arranged contiguously from the amino terminus of the protein. These are followed by a heavily glycosylated region which is rich in Ser/Thr/Pro residues (STP region). In contrast to DAF, MCP is a transmembrane protein with a cytoplasmic domain. Between the STP-rich region and the transmembrane domain is a short region of unknown function.

One of the many interesting features of MCP is the presence of different isoforms on the cell surface (Figure 3.39). The original cDNA clone did not explain the size heterogeneity of MCP evident on SDS-PAGE, and searches for further cDNAs revealed a large number of additional clones generated through alternative splicing [358, 706, 741, 742]. Three different STP regions have been identified in MCP (A, B and C) which can be spliced in or out of the molecule. There are also two different cytoplasmic tails, termed CYT-1 (16 amino acids) and CYT-2 (23 amino acids), which can exist with any of the STP variants. The original cDNA clone isolated by Lublin and colleagues in 1988 encoded the STP-BC isoform with CYT-2 [147]. Whilst a multitude of different MCP isoforms might be produced by alternative splicing, there are only four common isoforms. These are STP-C (smaller isoform evident on SDS-PAGE; 51–58 kDa) and STP-BC (larger isoform; 59–68 kDa), each of which can occur with either

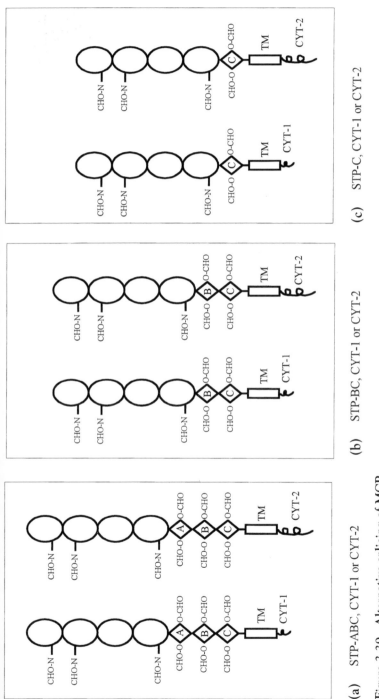

**(a)**  STP-ABC, CYT-1 or CYT-2

**(b)**  STP-BC, CYT-1 or CYT-2

**(c)**  STP-C, CYT-1 or CYT-2

**Figure 3.39.  Alternative splicing of MCP.**

Alternative splicing of the *mcp* gene results in isoforms which differ both in the amount of glycosylated region and in the cytoplasmic tail. There are three STP regions (STP-A, B and C) each encoded on a separate exon, these may be spliced in or out of the gene. Alternative splicing of exon 13 (see text) also generates two different cytoplasmic tails, termed CYT-1 and CYT-2. Whilst a multitude of different isoforms have been identified in various tissues, there are only four common isoforms. These contain STP-BC or STP-C with either of the two cytoplasmic tails. **(a)** Rare isoforms; contain all STP regions. **(b)** Common isoforms; mobility on SDS-PAGE 51–58 kDa. **(c)** Common isoforms; mobility on SDS-PAGE 59–68 kDa. TM, transmembrane domain.

of the cytoplasmic tails (CYT-1 or CYT-2) attached at the carboxy terminus. All of these isoforms may be expressed in the same cell. Whilst STP-B contains only 15 amino acids, the abundant O-glycosylation in this region of the protein accounts for the large molecular weight difference between the STP-BC and STP-C isoforms evident by SDS-PAGE analysis. Indeed, pulse chase experiments in the monocyte cell line U937 indicate that the two main precursor forms prior to glycosylation have molecular weights of 41 and 43 kDa [743]. Heterogeneity in glycosylation probably accounts for the broad bands demonstrated by individual isoforms on SDS-PAGE. The cytoplasmic tail attached to MCP contributes little to the variation in molecular weight of the isoforms as CYT-1 and CYT-2 differ by only seven amino acids and contain no carbohydrate.

Inclusion of the STP-A region (STP-ABC isoform) generates MCP with a molecular weight of 74 kDa. This isoform is rarely expressed in normal cells but is more commonly found in association with neoplasm [742, 744]. Monoclonal antibodies have been raised against peptides derived from STP-A and STP-B and used to assess expression and distribution of the different protein isoforms by flow cytometry, immunohistochemistry and Western blot [745]. The STP-B-containing isoform was widely expressed, whereas the STP-A-containing isoform was restricted mainly to the intestine and to some tumour cell lines. There are several atypical forms of MCP on specific cell types. For example, MCP on neutrophils exhibits an unusual structure in that it migrates as a very broad band on SDS-PAGE, regardless of the MCP phenotype on other cells of the individual. This has been demonstrated to be due to differences in post-translational O-linked glycosylation [723, 746]. Spermatozoal MCP is also unusual in that it migrates as a protein some 10–20 kDa smaller than leukocyte MCP, primarily due to a lack of glycosylation [742, 747, 748]. An unusual cDNA clone which might encode sperm MCP has been obtained both from a testis library and by RT-PCR from RNA isolated from semen, although the possibility that these clones were derived from cells other than spermatozoa is difficult to eliminate [742, 749]. In this clone, deletion of the exon encoding the second half of the transmembrane domain results in a truncated transmembrane domain and a unique 'tail', recently termed CYT-4. A shift in the reading frame means that the amino acid sequence of CYT-4 bears no resemblance to CYT-1 or CYT-2. An isoform with CYT-4 (STP-C; CYT-4) has been expressed on CHO cells; the expressed molecule had no O-linked carbohydrate and was mostly retained in the endoplasmic reticulum [749]. Interestingly, cells expressing this isoform formed rosettes with hamster eggs, indicating a potential role in the fertilization process. A second clone with a tail similar to CYT-4 has also been described in placenta [741, 742]. In this case, differential use of splice sites in the exon encoding the second half of the transmembrane domain results in a tail (CYT-3) which is identical to that of CYT-4 except that CYT-3 has nine additional amino acids. Whether the additional sequence in CYT-3 forms part of the transmembrane domain or represents a tail distinct from CYT-4 remains to be shown.

Whilst the cytoplasmic domains are not essential for C inhibitory function (see below), it has recently become evident that they influence MCP in other ways. Both cytoplasmic tails have been shown to associate with intracellular kinases, this interaction being in part dependent on the juxtamembrane Tyr-X-X-Leu motif present in both tails [750]. Whilst the potential role of MCP as a signalling molecule is still under investigation, the role of the two different tails in processing of the protein has been well characterized. U937 cells contain two MCP precursor proteins which are processed into the mature form of the protein with different rates, one with a t½ of 20 minutes and the other with a t½ of 90 minutes [743]. Further studies using transfectants in pulse chase experiments have been used to analyse the different processing rates of MCP precursors, and have revealed that isoforms expressing the CYT-1 tail are processed into the mature protein in 10–13 minutes, whereas those with CYT-2 are processed more slowly (35–40 minutes). MCP mutants in which the tail has been entirely deleted are processed very slowly (over 160 minutes). By deleting stretches of amino acids within CYT-1, it has been demonstrated that the terminal four amino acids (Phe-Thr-Ser-Leu; FTSL) are responsible for rapid exit from the endoplasmic reticulum [751]. Over 40 proteins have been discovered to date that have this rapid-transit motif at the carboxy terminus of the protein.

Preferential expression of particular isoforms of MCP has been demonstrated in various tissues regardless of the phenotype on other cells. For example, fetal heart expresses predominantly the smaller isoform (STP-C) [752], whereas salivary gland and kidney express predominantly the larger isoform [725, 727, 753]. There are reports that brain expresses exclusively the STP-C isoform [725, 753] although others have reported the presence of both the STP-C and STP-BC isoforms in brain [731, 732]. There are also several reports of a predominance of CYT-2 in the brain, salivary gland and kidney [732, 753]. In addition to processing times, these variably spliced domains are known to influence cofactor activity and infectivity with measles virus (see below).

The *mcp* gene is approximately 43 kb in length and contains 14 exons and 13 introns (Figure 3.40) [358]. Exon 1 encodes the 5′ untranslated region and the signal peptide, exon 2 encodes SCR1, exons 3 and 4 encode SCR2, and exons 5 and 6 encode SCRs 3 and 4 respectively. The three

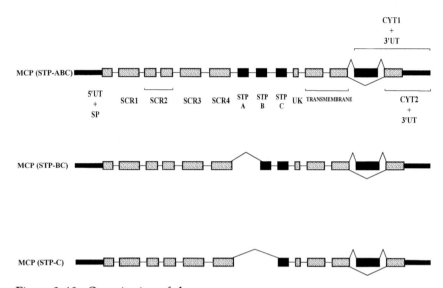

**Figure 3.40. Organization of the *mcp* gene.**

The *mcp* gene is located in the RCA cluster on chromosome 1. It spans approximately 43 kb and comprises 14 exons. The contribution of each exon to the structure of the mature protein is indicated in the figure and discussed in the text. Alternative splicing which results in the common isoforms of MCP (STP-BC and STP-C) is indicated in the bottom two diagrams. Alternative splicing of exon 13 also results in different cytoplasmic tails: CYT-1 or CYT-2. The shorter cytoplasmic tail (CYT-1; 16 amino acids) is encoded by exon 13; this contains a stop codon which converts exon 14 into part of the 3′ untranslated region. If exon 13 is spliced out, exon 14 encodes the second cytoplasmic tail (CYT-2; 23 amino acids) and the 3′ untranslated region. The 5′ and 3′ untranslated regions (5′UT and 3′UT respectively) are indicated by narrow black boxes, and the alternatively spliced exons are indicated as thick black boxes. SP, signal peptide; STP, Ser/Thr/Pro-rich region; CYT, cytoplasmic tail; UK, short region of unknown function.

STP-rich regions are encoded separately on exons 7–9 in the order A, B, C. Interestingly, comparison of the nucleotide sequences of STP-A and B indicate that they probably arose through intragenic duplication as they show 70% homology and the surrounding intronic sequences also demonstrate high levels of homology. Exon 10 encodes the short region of unknown function, and exons 11 and 12 the transmembrane domain. The shorter cytoplasmic tail (16 amino acids) is encoded by exon 13; this contains a stop codon which converts exon 14 into part of the 3′ untranslated region. If exon 13 is spliced out, exon 14 encodes the second cytoplasmic tail (23 amino acids) and the 3′ untranslated region.

*Binding/active sites*

As with DAF, the SCR domains in MCP are the site of functional activity (Figure 3.41). Deletion mutants have been used to determine which SCRs contribute to ligand binding and cofactor activity [754]. Deletion of SCR1 had little effect on C3i (C3b) binding and cofactor activity whilst C4b binding was reduced. The mutant protein lacking SCR2 still bound C3i but lacked cofactor activity whereas the binding of C4b was abolished, suggesting that C3b and C4b bind to different sites on MCP. SCR3 and 4 were demonstrated to be essential for both ligand binding and cofactor activity. The recent molecular cloning of MCP from New World monkeys has revealed an interesting finding – they express a form of MCP that lacks SCR1 [755, 756]. These forms retain cofactor activity for fI-mediated cleavage of C3b, underlining the minor

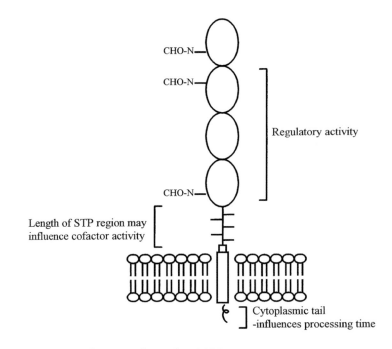

**Figure 3.41. Binding sites located in MCP.**
C regulatory activity has been located in SCR2-4 of MCP. Isoforms which contain variable amounts of STP region may also differ in their ability to regulate C. The different cytoplasmic tails do not influence C-regulatory ability but rather they influence the speed with which MCP is processed. Isoforms with CYT-1 (which contains a Phe/Thr/Ser/Leu motif) are processed rapidly. The amino-terminal SCR is not implicated in functional activity, but has been implicated in binding to measles virus.

role played by SCR1 in C regulation. Functional assays on cells trans-fected with MCP mutants lacking none, one, two or all three of the STP regions suggest that these glycosylated regions may also have a functional role. In one study, the smaller molecules (STP-C or those with no STP regions) provided better protection from C attack via CP activation, whilst those with long STP regions (particularly STP-ABC) provided better protection against AP activation[757]. This is in apparent conflict with a recent report which suggests that isoforms containing both the B and C STP regions bind C4b more efficiently than those containing only STP-C, and that the larger isoform provides enhanced protection against CP-mediated cytolysis[758]. The precise roles of the individual STP regions remains to be clarified. A GPI-anchored MCP has also been engineered and tested for cofactor activity and appears to function as efficiently as the transmembrane counterpart, illustrating that the transmembrane region and cytoplasmic tails are not required for cofactor activity[696].

### Role in reproduction

MCP and other C regulators may play a role in successful fertilization and reproduction (discussed in detail in Chapter 6). Spermatozoa are protected from C activation in the female genital tract by the presence of MCP, CD59 and DAF both on the sperm surface and in seminal plasma[737, 759, 760]. MCP is also located at the fetomaternal interface on the trophoblast epithelium where it helps protect the fetus from maternal comple-ment[733, 761]. On sperm, MCP is located exclusively on the inner acrosomal membrane and is exposed only following the acrosomal reaction. Once exposed, it might have a key role in the fertilization process[279, 280]. Indeed, abnormalities of spermatozoal MCP, or low levels of expression, have been associated with infertility[762].

### Interaction with micro-organisms

MCP has been identified as a cellular receptor for measles virus[289–291]. The measles virus-binding domain has been located at the amino-terminal end of the MCP molecule, with binding sites identified in both SCR1 and SCR2, although other domains within MCP, including the STP regions and transmembrane domain, may influence the susceptibility of cells to infection (discussed in detail in Chapter 7)[763–765]. The N-linked carbo-hydrate moieties, particularly that located on SCR2, appear necessary for measles virus binding[766–768]. In addition to its role in measles virus infec-tion, recent data indicate that MCP may also be a keratinocyte receptor for the pathogenic Gram-positive bacterium *Streptococcus pyogenes* (dis-cussed in Chapter 7)[302].

# OTHER REGULATORS OF THE ACTIVATION PATHWAYS

## Factor I

*History and Function*
As described in the preceding sections of this chapter, a wide range of CRPs are present either in plasma or on the cell surface, whose roles are to 'police' the cell membrane and prevent amplification of the C cascade on self-cells. They have two modes of action: they accelerate decay of the C3 and C5 convertases and/or they act as essential cofactors for enzymatic degradation of C3b and C4b, the central components of the convertase enzymes. Enzymatic inactivation of C3b or C4b proceeds by cleavage of the molecule in the $\alpha'$ chain. The enzyme that fulfils this function is factor I (fI), a serine protease synthesized primarily in the liver and by monocytes, and present in an active form in plasma at about 35 μg/ml. Prior association of C3b or C4b with any one of the CRP which express cofactor activity, fH (in the case of C3b), C4bp (in the case of C4b), CR1 or MCP, is a prerequisite for inactivation by fI. Factor I was one of the first C inactivators recognised in serum and its crucial role for keeping 'tick over' of the AP in check, evident in situations of fI deficiency or depletion, helped elucidate the mechanisms of AP activation and the feedback cycle (discussed in Chapter 1). A C3b-inactivating factor, originally termed C3b-inactivator (C3b-INA) was first described in guinea pig serum in 1966 by Tamura & Nelson and partially characterized the following year [769, 770]. In 1968, Lachmann & Muller-Eberhard partially purified a factor in human serum that converted C3b within EAC14b2a3b to an antigenically distinct form that bound conglutinin, a collagen-containing bovine lectin. It was proposed that the alteration in structure or conformation was due to enzymatic cleavage. They termed this converting factor 'conglutinogen-activating factor' or KAF [771], and the following year its identity with C3b-INA was recognised [772, 773]. It was later demonstrated that conglutinin bound to a carbohydrate epitope on the $\alpha'$ chain of iC3b, masked in uncleaved C3b [774].

The central role of fI in homeostasis of the AP was an intrinsic feature of the 'tick over hypothesis' proposed by Nicol & Lachmann in 1973 (Figure 3.42) [370]. They demonstrated that removal of fI from plasma by immunochemical depletion (by absorption with Fab$_2$ anti-fI) resulted in uncontrolled amplification of the AP, consumption of both fB and C3, and a consequent deficiency of these components in serum. This work extended previous observations in a patient who was deficient in fI and who suffered from recurrent pyogenic bacterial infections due to profound

FEEDBACK CYCLE

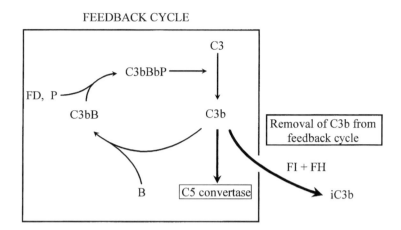

**Figure 3.42. Function of fI in maintaining homeostasis of the AP.**
In the presence of cofactors (fH, C4bp, CR1 or MCP) fI cleaves and inactivates both C3b and C4b. This is illustrated in the preceding sections of this chapter. Both fI and its fluid phase cofactor, fH, also play a major role in maintaining homeostasis of the AP. In the absence of these regulatory proteins, low-level activation of the AP would result in uncontrolled amplification leading to profound deficiency of both C3 and fB. By binding to fluid-phase C3b and inactivating it, fH and fI remove C3b from the 'feedback cycle' thus preventing uncontrolled fluid-phase C activation.

deficiency of plasma C3 and fB[775, 776]. The consequences of fI deficiency are described in Chapter 5. It was soon demonstrated that fI cleaved C3b in the α' chain and that the resulting fragments closely resembled those previously shown to be produced by the action of trypsin on C3, C3c and C3d[777, 778]. The pattern of cleavage and inactivation of C3b by fI was precisely defined in 1980 by Harrison & Lachmann who demonstrated that inactivation and formation of iC3b actually proceeded via two closely spaced cleavages at sites 17 amino acids apart which results in the release of a small (3-kDa) fragment termed C3f[365]. iC3b can be further degraded by fI in the presence of CR1, but not the other cofactors, producing the fragments C3c and C3dg[366, 526]. Cleavage and inactivation of C4b proceeds slightly differently in that two cleavages convert C4b to C4c and C4d. The two cleavage sites are situated on either side of the thioester bond and cleavage proceeds so rapidly that the intermediate, iC4b, is virtually undetectable[367, 458, 779].

Factor I cannot function in the absence of a cofactor. This important characteristic, essential for the functioning of the C system, was demonstrated by several groups who showed that fH (originally termed C3b inactivator accelerator; A·C3b-INA) was required for the cleavage of C3b by fI and that both factors had to be present simultaneously for cleavage to occur [155, 361]. Around the same time, Shiraishi & Stroud demonstrated that a high molecular weight cofactor, later shown to be C4bp, was required for the fI-mediated cleavage of fluid-phase C4b into C4c and C4d [458, 459]. Although there are reports that C4bp can act as cofactor for C3b cleavage and that fH can support cleavage of C4b, the weight of evidence indicates that neither of these alternative reactions occurs to a significant degree under physiological conditions (see preceding sections) [361, 362, 364, 366–369, 459, 460, 462]. As the other CRPs were identified and characterized it became evident that multiple cofactors existed that could potentiate the activity of fI. MCP and CR1 act as cofactor for cleavage of both C3b and C4b, but only CR1 supports the further cleavage of iC3b with formation of C3c and C3dg [366, 518, 519, 705]. The precise nature of the C3b–cofactor–fI interactions has yet to be defined. Factor I has a very weak affinity for binding to C3b which is greatly enhanced in the presence of cofactors and inhibited in the presence of P [409, 780–782]. Whilst fH augments binding of fI to C3b, the converse is not true; fI has no influence on fH binding. This contrasts with the binding of fB and P to C3b in which all three proteins interact with each other and binding is cooperative (section 3.3.2). Whilst a low-affinity interaction between fH and fI has been suggested [782, 783], it seems likely that binding of fH to C3b invokes a conformational change within C3b resulting in exposure of a high-affinity binding site for fI [409, 782].

### Structural analysis

Factor I was purified to homogeneity in 1977 by a combination of ammonium sulphate precipitation, anion exchange chromatography and gel filtration [154]. The purified protein had a molecular weight of approximately 90 kDa and consisted of two chains, termed heavy (50 kDa) and light (38 kDa). Partial amino acid sequencing of the fI light chain indicated that it contained the serine protease domain and this was confirmed when the full amino acid sequence was obtained by analysis of a cDNA clone from a liver library [339, 784–786]. The single-chain precursor form of fI consists of 583 amino acids (Figure 3.43). The first 18 amino acids constitute the signal peptide, the heavy chain consists of 322 amino acids and the light chain of 243 amino acids. The four carboxy-terminal amino acids of the heavy chain (Arg-Arg-Lys-Arg) are removed during proteolytic processing of the precursor molecule [787]. The predicted molecular weights of the two chains are 35.4 kDa (heavy) and 27.6 kDa (light) indicating that carbohydrate constitutes about 27% by weight of the protein; indeed, three potential N-glycosylation sites are present in each chain. The predicted size of the polypeptide backbone was in good agreement with the

**Figure 3.43.  Derived amino acid sequence of fI.**

```
MKLLHVFLLFLCFHLRFCKVTYTSQEDLVEKKCLAKKYTHLSCDKVFCQP    32
WQRCIEGTCVCKLPYQCPKNGTAVCATNRRSFPTYCQQKSLECLHPGTKF    82
LNNGTCTAEGKFSVSLKHGNTDSEGIVEVKLVDQDKTMFICKSSWSMREA   132
NVACLDLGFQQGADTQRRFKLSDLSINSTECLHVHCRGLETSLAECTFTK   182
RRTMGYQDFADVVCYTQKADSPMDDFFQCVNGKYISQMKACDGINDCGDQ   232
SDELCCKACQGKGFHCKSGVCIPSQYQCNGEVDCITGEDEVGCAGFASVA   282
QEETEILTADMDAERRRIKSLLPKLSCGVKNRMHIRRKRIVGGKRAQLGD   332
LPWQVAIKDASGITCGGIYIGGCWILTAAHCLRASKTHRYQIWTTVVDWI   382
HPDLKRIVIEYVDRIIFHENYNAGTYQNDIALIEMKKDGNKKDCELPRSI   432
PACVPWSPYLFQPNDTCIVSGWGREKDNERVFSLQWGEVKLISNCSKFYG   482
NRFYEKEMECAGTYDGSIDACKGDSGGPLVCMDANNVTYVWGVVSWGENC   532
GKPEFPGFYTKVANYFDWISYHVGRPFISQYNV                    565
```

The amino-terminal signal peptide is underlined as are the four residues (RRKR) removed during processing. The first residue of the mature protein is **K** in bold. Numbering is for the mature protein.

**GenBank Accession No:**   J02770, Y00318, M25615

size of the precursor protein synthesized by cells in the presence of tuni-camycin to inhibit glycosylation or produced in a rabbit reticulocyte lysate system (65 kDa)[787]. The light chain bears general sequence homology with other serine proteases around the catalytic site and in the conserva-tion of cysteine residues. The heavy chain has a mosaic structure consist-ing of four domains (Figure 3.44). The first is a 65–70-amino acid domain homologous to domains found in C6 and C7 and termed the fI module (fIM; also termed the fI/C6/C7 module). The second is a Cys-rich region consisting of about 100 amino acids which is homologous with the so-called scavenger receptor (Sc) domain, present in the scavenger receptor and also in CD5, CD6 and several other proteins[788, 789]. The final two domains are class A low density lipoprotein receptor (LDLr) modules con-sisting of about 40 amino acids each. Although fI consists of a number of discrete domains, the protein does not adopt an elongated conformation as seen in the RCA proteins; rather, the protein has a compact structure with an overall length of between 12 and 15 nm[409, 790]. Visualization of fI by electron microscopy revealed an asymmetrical bi-lobed protein, each lobe having a diameter close to 5 nm; the larger domain (heavy chain) was responsible for binding C3b while the smaller contained the serine pro-tease activity[409].

The gene for fI is located on chromosome 4q25 and spans 63 kb, although over half of this span is accounted for by the first intron (36 kb) [339, 791, 792]. The gene comprises 13 exons, the first encodes the 5′ untrans-lated region, the signal peptide and the first amino acid of the heavy chain.

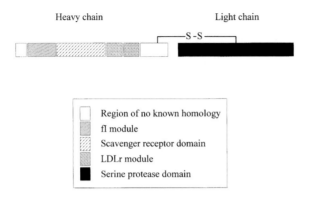

**Figure 3.44. Structure of fI.**

FI consists of two chains: heavy (50 kDa) and light (38 kDa) linked together by a single disulphide bond. The serine protease activity resides in the light chain whereas the heavy chain has a 'mosaic' structure, being organized into domain structures also found in a variety of other proteins with a range of different functions.

Sequence coding for the light chain starts in the middle of exon 9. The domain structure of the heavy chain is reflected in its intron/exon organization. The second exon encodes the fIM repeat, the third and fourth exons encode the Sc repeat and the LDLr repeats are encoded separately on exons 5 and 6. Surprisingly the intron/exon structure of the gene encoding the serine protease domain is unlike that in other C serine proteases (such as fB, C1r and C1s) but more closely resembles the organization seen in trypsin and chymotrypsin.

# Properdin

*History*

In contrast to the other regulators of the C cascade, properdin (P) has a unique role as a positive regulator of the AP, stabilizing the otherwise labile C3 convertase. Its importance to efficient AP function is exemplified in individuals lacking P, who frequently present with life-threatening pyogenic bacterial infections, often involving attacks of meningococcal disease. P has a long and stormy history, being first described in 1954 by Louis Pillemer, who was studying the mechanism by which C3 was inactivated by incubation with zymosan (a yeast cell wall polysaccharide) [793]. He proposed a non-specific serological mechanism for activation of C which contrasted with the antibody-dependent initiation of the 'classical' pathway. He termed this system the 'properdin system' and demonstrated that P acted in an $Mg^{2+}$-dependent manner with other components of the C cascade. Unfortunately, Pillemer's initial observations and proposals of an alternative activation pathway were discredited by other investigators of that time, who suggested that anti-zymosan antibodies were actually responsible for C activation [44]. Sadly, Pillemer died in 1957, many years prior to the vindication of his work. His ideas and findings were neverthe-less pursued by his colleagues who demonstrated in the subsequent few years that activation of the properdin system required a hydrazine-sensitive factor (then termed properdin factor A; now known to be C3) and a heat-labile factor (properdin factor B; fB) [794, 795]. In 1968, following the advent of classical chromatography techniques, P was purified and demonstrated to be antigenically distinct from antibody [796]. This went some way to resolving the conflict, although it was several years before the new system became accepted. In the early 1970s it was observed that certain activators could trigger C leading to C3 depletion, whilst com-ponents of the CP, C4 and C2, were unaffected [797–800]. This implied that an alternative pathway (AP) existed for C3 activation and, as the intrica-cies of this new pathway unfolded over the subsequent few years, the bio-chemical and functional similarities of its components to those described by Pillemer and his colleagues more than a decade earlier soon became clear [373, 801–804].

*Function*

In 1971, Götze & Müller-Eberhard described the 'C3 activator system' – a pathway for C activation which functioned as an alternative to the CP [802]. As discussed above, it became apparent during the early 1970s that the C3 activator system and the properdin system were one and the same [801, 804, 805]. However, the role of P in the AP was initially unclear. Götze & Müller-Eberhard proposed that P was essential for formation of the 'initial' C3 convertase required to trigger amplification of the pathway [806]. They sug-gested that 'activation' of P, by interaction with C-activating surfaces, resulted in modulation of native C3 such that it could activate factor D (fD), thereby promoting activation of fB. The functional C3 convertase thus formed cleaved further C3 and the AP was initiated. Fearon and col-leagues came to similar conclusions, suggesting that fD existed in a precur-sor state and that P was responsible for its conversion to an active form capable of cleaving fB [807]. It was soon established that the function of P in the AP convertase was not to mediate 'activation' of fD, but rather to bind to and stabilize the complex of C3b and Bb, increasing the half-life of this otherwise very labile enzyme from just a few minutes to about half-an-hour (Figure 3.45) [49, 780, 808–810]. In fact, the AP can function in the absence of P, but with a much lower efficacy [811]. In addition to its role in convertase

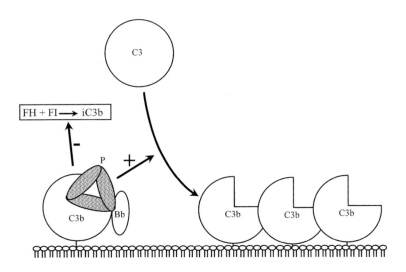

**Figure 3.45. Function of properdin (P) in stabilization of the AP C3 convertase.**

P binds to the C3 convertase, C3bBb, extending the half-life of the otherwise very labile enzyme from around 2 minutes up to 30 minutes. In addition to this 'stabilizing' role, P inhibits fI-mediated inactivation of C3b, possibly by competing for the same binding site on C3b.

stabilization, P also inhibits binding of fI to C3b, possibly by competing for the same binding site [780, 781].

*Distribution/biosynthesis*

P is present in plasma at about 5 µg/ml [812, 813]. Its expression in tissues is very limited; indeed this hampered initial cloning and sequencing as no cDNA could be isolated from liver or spleen libraries. P was eventually cloned from a library generated from PMA-stimulated U937 monocytic cells. The primary source of serum P may be peripheral blood monocytes [395, 814], although it has recently been reported that T cells [815] and the HepG2 hepatoma cell line can synthesize P [816]. Neutrophils stimulated with various substances such as FMLP, C5a, IL-8 and TNF-α are also capable of secretion of P, apparently by release from stores in intracellular granules [817].

In accordance with its role in the AP, individuals deficient in P suffer from bacterial infections. It is common for affected individuals to present with severe, fulminant meningococcal meningitis. P deficiency is inherited in an X-linked manner; the gene has been mapped to the short arm of the X chromosome [818]. Three variants of P deficiency have been described, distinguished by the plasma level of properdin detected immunochemically. Type I deficient individuals have no detectable protein. Type II and Type III individuals have detectable, but non-functional, P in plasma at either very low levels (Type II) or normal levels (Type III). The non-functional forms of P do not form the polymers which are essential to properdin action and do not stabilize fluid-phase alternative pathway convertases [819, 820]. P deficiency is discussed in depth elsewhere in this book (Chapter 5).

*Molecular cloning and structural analysis*

Initial sedimentation studies on purified P suggested that the protein had a molecular weight of about 220 kDa [796]. Further analysis under dissociating conditions revealed that P was formed from multiple identical subunits of about 56 kDa and contained about 10% carbohydrate [821–826]. Hydrodynamic analysis and electron microscopy indicated that P exists primarily as cyclic polymers constructed from two, three or four monomers linked in a head-to-tail fashion, although higher cyclic oligomers were also visualized by electron microscopy (Figure 3.46) [827]. Each monomer within the cyclic oligomer appeared as a flexible rod-like structure (26 nm × 2.5 nm) linked through its ends to the others, presumably through interaction of

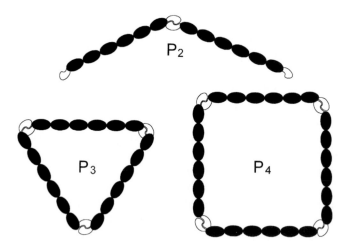

**Figure 3.46. Oligomeric forms of P formed in plasma.**

P exists in plasma as polymers consisting of two, three or four P monomers. These associate to form 'cyclic' forms of the protein in which the carboxy-terminal domain of one monomer interacts with the amino-terminal domain of another. The most active form of P is $P_4$, followed by $P_3$ and then $P_2$.

the carboxy- and amino-terminal domains of adjoining monomers. This structural organization was further corroborated by neutron and X-ray scattering studies which demonstrated that the P-trimer could indeed exist in a 'triangular' form [828]. In serum, the different oligomeric forms $P_2$:$P_3$:$P_4$, existed in an approximate 1:2:1 ratio and the activities decreased in the order $P_4 > P_3 > P_2$ [812].

It was initially believed that P existed in two forms in plasma, native (or precursor) and activated, and that the latter arose from the former by binding to the AP convertase or to an activating surface. Early studies reported that native and activated P differed antigenically and in molecular weight, and it was even suggested that P was activated by proteolytic cleavage [824, 829]. However, it was subsequently shown that the two forms of P had identical subunit compositions and that activation was the result of conformational change [822, 830]. The 'active' form which was usually obtained following purification procedures and led to consumption of C3 when added to serum, could also be generated from native inactive P by freeze/thawing [822]. Clarification of the relationship between native and activated P came in 1987 when Farries and co-workers, demonstrated that formation of 'active' P was not a physiological process but an artefact of purification. Using gel filtration to separate different forms of P it was shown that plasma contained only native P, and that activated (high molecular weight) P could be produced by freeze/thawing [831]. Electron microscopy of the 'activated' form of the protein indicated amorphous aggregates rather than the discrete cyclic structures previously identified. Similarly, in 1989 Pangburn demonstrated that chromatography on a TSK-400 gel filtration column or cation exchange could be used as a final 'clean-up' method (following PEG precipitation and QAE Sephadex chromatography) to isolate and identify different P oligomers existing in plasma [812]. Purification on gel filtration columns was dependent on the increasing size of the oligomers, whilst separation by cation exchange chromatography depended on the unusually high pI of P which resulted in greater retention of the higher oligomers. Using this methodology, Pangburn demonstrated that plasma P consisted of only three forms, dimer, trimer and tetramer, and that higher oligomers of P (containing greater than four monomers) were artefacts generated upon freeze/thawing or long-term storage of the smaller oligomers (dimers, trimers and tetramers). Aggregated P thus formed could activate C and consume C3 when added to serum. Despite this apparent resolution of the situation, new confusion has been created by the recent finding that during activation of the AP, C3b and 'active' (aggregated?) P can form a covalent complex [832, 833]. The precise nature of the interaction of P with the convertase and the physiological relevance of 'active' P thus requires clarification.

A partial amino acid sequence of P was obtained by sequencing fragments obtained from digestion with proteolytic enzymes and cyanogen bromide [822, 834]. The complete sequence was obtained following the isolation of a cDNA clone from a library generated from PMA-treated U937 cells [835]. Cloning of the P cDNA confirmed an intriguing finding that had emerged from analysis of amino acid composition: virtually half of P is composed of just four amino acids, Glu, Pro, Gly and Cys (Figure 3.47) [823, 834]. These residues are distributed fairly evenly throughout the whole protein and their predominance can be accounted for by the substantial amount of β sheet and β-turn structures evident from Fourier transform infrared (FTIR) spectroscopy studies [836]. This peculiar composition may account for the rather unusual electrophoretic pattern of P when analysed by non-reducing SDS-PAGE; it runs as a triplet of one fast-migrating band followed by two slightly slower, closely spaced bands. These diverse mobilities are not due to glycosylation differences, but are probably due to alternative disulphide-bonding between the multiple cysteines present in the subunits [821].

The mature protein consists of 442 amino acids. Initial analysis indicated that residues 50–410 were arranged into six domains each containing some 60 amino acids, known as thrombospondin repeats (TSRs) [835]. This protein domain was first described in the cell matrix adhesion molecule, thrombospondin, hence its name [837]. It contains conserved Trp, Cys, Ser and Arg residues and has also been identified in components of the terminal C pathway, C6, C7, C8 and C9 and in the circumsporozoite protein common to a number of malaria parasites [23, 838]. A modification of the P domain structure described above has recently been proposed based on comprehensive analysis of inter-domain sequence homologies [839]. Alignment of the amino acid sequences from the TSRs in P indicates that

**Figure 3.47. Derived amino acid sequence of P.**

```
MITEGAQAPRLLLPPLLLLLLTLPATGSDPVLCFTQYEESSGKCKGLLGGG    23
VSVEDCCLNTAFAYQKRSGGLCQPCRSPRWSLWSTWAPCSVTCSEGSQLR     73
YRRCVGWNGQCSGKVAPGTLEWQLQACEDQQCCPEMGGWSGWGPWEPCSV    123
TCSKGTRTRRRACNHPAPKCGGHCPGQAQESEACDTQQVCPTHGAWATWG    173
PWTPCSASCHGGPHEPKETRSRKCSAPEPSQKPPGKPCPGLAYEQRRCTG    223
LPPCPVAGGWGPWGPVSPCPVTCGLGQTMEQRTCNHPVPQHGGPFCAGDA    273
TRTHICNTAVPCPVDGEWDSWGEWSPCIRRNMKSISCQEIPGQQSRGRTC    323
RGRKFDGHRCAGQQQDIRHCYSIQHCPLKGSWSEWSTWGLCMPPCGPNPT    373
RARQRLCTPLLPKYPPTVSMVEGQGEKNVTFWGRPLPRCEELQGQKLVVE    423
EKRPCLHVPACKDPEEEEL                                   442
```

The amino-terminal signal peptide is underlined. The first residue of the mature protein is **D** in bold. Numbering is for the mature protein.

**GenBank Accession No:** M83652, S49355

the sixth TSR differs from the other five in that it contains a unique region with no homology to the other TSRs. The carboxy- and amino-terminal portions of TSR6 conform to the TSR consensus sequence, whereas the middle portion (about 25 amino acids) is unique and contains a potential N-linked glycosylation site. Assuming that the disulphide bonds hold TSR6 in a similar conformation to the other TSRs, it could be imagined that the anomalous region 'loops out' from TSR6 forming a sub-domain containing carbohydrate [839]. The position of the inserted region corresponds to the position of an intron which splits the two exons encoding TSR6 (see below). TSR6 is the only repeat in P that is encoded on two exons, implying that the insertion of the intron was responsible for the extra coding sequence [840]. Schematic representations of the two models of P structure are illustrated in Figure 3.48. In both models of P domain structure, the amino-terminal domain is distinct and highly charged. In the original model the carboxy-terminal region also forms a discrete domain; however, in the modified model it consists purely of a short stretch (eight amino acids) of charged residues.

The gene for P has been localized to the X chromosome, region Xp11.23-Xp21.1 [818]. The gene spans approximately 6 kb and is organised into ten exons (Figure 3.49). The first two exons encode the 5′UT region and most of the signal peptide, and the remaining three amino acids of the signal peptide are encoded on exon 3 along with the amino-terminal non-TSR region. TSRs 1–5 are encoded on a single exon each (exons 4–8), exon 9 encodes the first half of TSR6 and exon 10 encodes the second half of this TSR, the carboxy-terminal region and the 3′UT region [840].

*Binding/active sites*
P exhibits a weak affinity for surface-bound C3b which increases following formation of the pro-enzyme C3bB, and is markedly enhanced following formation of the functional convertase (C3bBb) [781]. As might be expected if P and fB express binding sites for each other, they show positive cooperativity in their binding to C3b. Physiological concentrations of fB increase binding of P to zymosan-C3b some four-fold; similarly physiological concentrations of P increase binding of fB to the same surface seven-fold [841]. P and fB are closely associated in the AP convertase as clearly demonstrated in cross-linking studies using a linking reagent with a short bridging length (0.6 nm) [842]. Recent mutational analyses of P have shed new light on the functional activities of the different TSRs [839]. Recombinant forms of the protein were prepared lacking either TSR 3, 4, 5 or 6. The TSR3-deletion mutant retained activity whereas TSR4 and TSR5 deletion mutants showed decreased or no binding to the C3 convertase and were non-functional. Mutant P lacking any one of TSRs 3–5 was still capable of

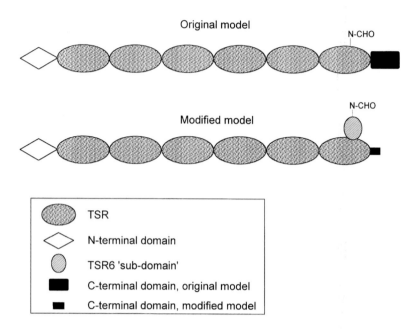

Figure 3.48. Structural organization of P.

The P protein consists in the main of a repeating domain structure called the thrombospondin repeat (TSR). In the original model of P structure there are six TSRs arranged contiguously and located between residues 50 and 410. The carboxy- and amino-terminal regions form distinct domains at either end of the molecule. In the modified model of P structure, a short stretch of amino acids in the middle portion of TSR6 (approximately 25 residues in length) forms a distinct domain which 'protrudes' from TSR6. The amino-terminal domains are identical in the two models, but the carboxy-terminal domain in the 'modified' model is very short, consisting of a short stretch of charged amino acids.

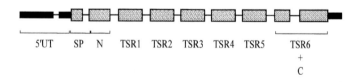

Figure 3.49. Organization of the *p* gene.

The *p* gene is located on the X chromosome. It spans approximately 6 kb and comprises 10 exons. The contribution of each exon to the structure of the mature protein is indicated in the figure and discussed in the text. The 5′ and 3′ untranslated regions (5′UT and 3′UT respectively) are indicated by narrow black boxes. SP, signal peptide; TSR, thrombospondin repeat; N, N-terminal domain; C, C-terminal domain.

forming oligomers, although in some cases only dimers were formed. P lacking TSR6 and the charged C-terminal domain was not able to form oligomers and was non-functional; it is unclear whether this is due to lack of oligomerization or deletion of functional domains. The roles of TSRs 1 and 2 in functional activity have not been tested. It is interesting to note that TSR4 contains the motif VTCG (Val-Thr-Cys-Gly) implicated in cell binding in other proteins[843, 844]. It is unclear whether this motif has any functional role in P.

## Anaphylatoxin inactivator (carboxypeptidase N)

*Introduction*
At a site of C activation, cleavage of C3, C4 and C5 during formation of the C3 and C5 convertases results in formation and release to the fluid phase of small (74–77-amino acid) polypeptides originating from the amino-terminal ends of the α chains of each of the molecules. These polypeptides, termed C3a, C4a and C5a respectively, are powerful ana-phylatoxins, binding to receptors on the surface of mast cells and basophils resulting in degranulation and release of vasoactive mediators[845]. C5a is also the major chemotactic factor for neutrophils, stimulating their adher-ence to vascular endothelial cells and margination through vessel walls; the concentration gradient of C5a guides the neutrophils to the site of C activation. Furthermore, C5a stimulates neutrophil activation and is capable of triggering the oxidative burst and production of leukotrienes. C5a is by far the most powerful anaphylactic molecule, with C3a being of rather lower potency[846]. No biological function of C4a has yet been demonstrated in man. All three anaphylatoxins are structurally similar, having between 31% and 39% amino acid identity[847–849]. Each is gener-ated through cleavage of an Arg-X bond in the parent molecule which leaves a carboxy-terminal Arg residue in the released polypeptide. Removal of this Arg residue by a plasma enzyme, termed anaphylatoxin inactivator or carboxypeptidase N (CPN), completely abrogates the func-tion of C3a and dramatically reduces that of C5a. These forms of the ana-phylatoxins lacking the carboxy-terminal Arg residue are termed C5a des-Arg, C4a des-Arg and C3a des-Arg. C5a des-Arg, while much less potent than C5a, retains some of its chemotactic ability and can still activ-ate neutrophils and stimulate granule release from basophils[850–852].

The inactivation of the anaphylatoxins in plasma is extremely quick and efficient. Whilst early work demonstrated that anaphylatoxins could be generated from C3 and C5 *in vitro*, all efforts to demonstrate anaphyl-actic activities of these peptides in plasma failed, leading some workers to question whether the anaphylatoxins could mediate inflammatory effects *in vivo*. The identification of CPN and demonstration of its efficient inactivation of the anaphylatoxins provided the explanation for lack of activity of C3a and C5a in plasma. The crucial importance of CPN is evidenced *in vivo* by the lethal effects of C activation in animals in which CPN has been inhibited. In one study, guinea pigs were injected with DL-2 mercaptomethyl-3-guanidinoethylthiopropionic acid, an analogue of Arg and a specific inhibitor of CPN, resulting in a lack of functional CPN. Subsequent activation of C in these animals using intravenously delivered cobra venom factor (CVF) resulted in death within minutes, whereas control animals treated with CVF alone survived[845, 853]. Histological examination of lungs from the animals revealed that the animals died from asphyxia with evidence of vasoconstriction, bronchoconstriction, severe pulmonary congestion and oedema and interstitial infiltration of mononuclear cells.

*History*
CPN has many activities in plasma other than the inactivation of anaphylatoxins; hence it has been discovered in many different guises and attributed many different names. These include kininase I, arginine carboxypeptidase, protaminase, creatine kinase conversion factor and several others. It was first identified in 1962 as an enzyme which inactivated bradykinin by removal of the carboxy-terminal Arg residue, earning it its original name, kininase I[854]. Nearly a decade later it was realised that this same enzyme was also capable of inactivating anaphylatoxins. In 1969, a factor was identified in human plasma which had the ability to inactivate anaphylatoxins (hence termed 'anaphylatoxin inactivator')[855]. This was partially purified and characterized in 1970[856]. The anaphylatoxin inactivator was present in plasma at 30–40 µg/ml and its molecular weight was estimated as 310 kDa from sedimentation studies, and 290 kDa by SDS-PAGE analysis. The purified protein had the ability to inactivate the anaphylatoxins C3a and C5a, and also bradykinin. These workers proposed that their anaphylatoxin inactivator was identical to the carboxypeptidase originally identified in 1962 by Erdös & Sloane[854]. In 1973 Vallota & Müller-Eberhard finally managed to purify active anaphylatoxins from human serum by first removing CPN from the serum using immune adsorption[857]. The purified anaphylatoxins were active as judged by their ability to produce wheals when injected into human skin. Over the subsequent few years CPN was purified to homogeneity by several groups by anion exchange chromatography and arginine-Sepharose affinity chromatography[858–860]. It was shown that the enzyme had an unusually high Leu content (16%) and was comprised of subunits of molecular weights 83 kDa and 50 kDa. These subunits formed heterodimers which themselves dimer-

ized to give a tetrameric protein of total molecular weight 280 kDa. The subunits could be separated by treating the enzyme with 3 M guanidine prior to gel filtration[861]. The two subunits were not antigenically related, and the enzymatic activity resided only in the smaller subunit. However, this activity was rapidly lost by the purified 50 kDa subunit when incubated at 37°C or at acid pH, indicating that the presence of the larger subunit preserved the activity. The current consensus is that the larger subunit acts as a 'carrier' molecule, stabilizing the enzyme in plasma and preventing its loss by glomerular filtration.

*Structural analysis*

Amino acid sequencing of tryptic peptides derived from human CPN gave a partial sequence which showed high homology to CPN from other species (43% identity to bovine CPN), but less homology to other carboxypeptidases such as carboxypeptidase B (18% identity)[862]. The cDNA encoding the small catalytic subunit was cloned in 1989. The unglycosylated catalytic subunit consists of 438 amino acids. Comparison of the sequence with other carboxypeptidases demonstrated that mechanistically important residues were conserved[863]. The cDNA sequence encoding the larger 'stabilizing' subunit was isolated from a liver cDNA library the following year. This subunit comprises 536 amino acids, from which a polypeptide backbone of approximately 59 kDa could be predicted[864]. This indicated that carbohydrate might contribute as much as 30% to the molecular mass of the protein and analysis of the sequence indicated that there were seven potential sites for N-glycosylation and a Ser/Thr rich region which might be abundantly O-glycosylated. The sequence contains an unusual feature which accounts for the high proportion of Leu residues evident from amino acid composition studies. There are 12 Leu-rich repeats in the molecule, arranged contiguously from amino acid 68 through to 355. Each repeat consists of 24 residues with a consensus of conserved Leu residues (seven in total), a Pro residue and an Arg. This repeat was first identified in Leu-rich $\alpha_2$ glycoprotein[865] and has subsequently been identified in a variety of proteins with a wide range of functions including human and bovine proteoglycans[866, 867] and the ribonuclease inhibitor of human placenta[868]. The function of the Leu-rich region is still not clear but probably relates to its protein-protein binding interactions, a feature common to several of the proteins containing this domain.

# Factor J

Factor J (fJ) was originally described in 1989 but remains one of the least well characterized inhibitors of C. The original reports described the purification from urine of a factor capable of inhibiting C-mediated lysis of EAC14; the same factor was subsequently demonstrated in plasma and termed fJ [869, 870]. The activity of fJ was shown to be due to an inhibitory effect on C1, inhibiting by binding C1 and preventing the association of the C1 subcomponents [869]. Analysis by SDS-PAGE indicated that the protein had an apparent molecular weight of 20 kDa. Further investigation indicated that it was a cationic molecule with a pI of 9.6, and was heavily glycosylated (40% by weight). The carbohydrate was essential for its function as enzymatic deglycosylation abrogated C-inhibitory action [871]. In addition to its effect on the CP, it appears that fJ may also inhibit the AP, either by restricting formation of the AP C3 convertase or by accelerating its decay [872]. It has no effect on fD activity [873]. Interestingly, an antibody raised against fJ purified from urine recognises an antigen present on the surface of certain cell lines, including the K562 erythroleukaemia line and U937 monocytic line [874]. A small sub-population of peripheral blood lymphocytes (mainly B cells) also showed low level expression of this antigen. The antibody immunoprecipitated a 65-kDa protein from U937 cells (analysed by SDS-PAGE under reducing conditions) which, when purified, had the ability to interfere with C-mediated lysis of EAC14. Whilst fJ awaits further characterization, it is possible that it represents one of the group of highly charged molecules, such as heparin, which are known to interfere with the C cascade in a rather non-specific manner.

# Chapter 4

# REGULATION IN THE TERMINAL PATHWAY

## INTRODUCTION

The terminal or membrane attack pathway of C represents the final common pathway for both the classical and alternative activation pathways. Unlike the activation pathways, the terminal pathway is not a proteolytic cascade. Membrane attack involves an almost unique sequence of events during which five soluble plasma proteins assemble sequentially into a membrane-associated complex, which becomes progressively more deeply inserted into the membrane as the later components are recruited until the complex traverses the membrane. The complete transmembrane complex (membrane attack complex; MAC) contains a single molecule each of C5b, C6, C7 and C8 but multiple copies (as many as 12) of C9 (Figure 1.5). Within the MAC is a 'pore' through which ions and water can enter causing the cell to swell and eventually to burst. The prevailing view is that this pore is physically constituted by the multiple copies of C9 which are suggested to line a rigid, water-filled channel [69]. However, some still contend that the pore is a less rigidly constrained area of lipid disorganization (a 'leaky patch') in the immediate vicinity of the MAC [70, 876]. Metabolically inert targets such as liposomes or aged erythrocytes are rapidly and efficiently lysed by the MAC, the kinetics implying that only a single functional lesion is necessary to lyse each target [3]. In metabolically active targets, such as fresh erythrocytes, platelets and nucleated cells, ion pumps enable the target to survive the presence of a limited number of MACs on the membrane, lysis occurring when the MAC burden exceeds the capacity of the target to maintain ionic equilibrium. Even in the absence of cell lysis, the ionic fluxes cause cell activation events which may profoundly influence the fate of the target.

The first line of defence against C attack on host cells are the regulators of the activation pathways described in Chapter 3. Experiments involving the neutralization of activation pathway regulators on endothelia *in vivo* in rats clearly demonstrate that the 'tick over' C activation to which endothelium is continually exposed is effectively controlled at this first line [877, 878]. However, there are situations *in vivo* where the intensity of C attack is sufficient to breach this defence. The terminal pathway can also be initiated directly on cells at the site of C activation in a 'bystander' manner by binding, to adjacent innocent host cells, of C5b67 complexes generated on a legitimate target. For these eventualities it is essential that cells possess a second line of defence in the terminal pathway in order to minimize the potentially lethal consequences of MAC deposition.

As with regulation in the activation pathways, regulators of the terminal pathway are present both in plasma and on membranes. These regulators either prevent the binding of C5b67 to the membrane or interfere with the incorporation of the final component, C9, into the membrane-associated complex. Fluid-phase inhibitors act almost exclusively on the C5b67 complex, whereas membrane inhibitors bind the forming MACs which have evaded the fluid-phase regulators and prevent the formation of the fully functional MAC pore.

Another form of defence against MAC damage, used by many different cell types, involves the physical removal of the MAC from the membrane with restoration of membrane integrity, a process termed 'recovery' which acts in a manner complementary to the 'resistance' confered by the regulatory proteins to provide the maximum degree of protection for the host cell [879].

## FLUID-PHASE REGULATORS OF THE TERMINAL PATHWAY

The C5b67 complex assembled on the C5 convertase is released into the fluid phase and must find and bind the lipid membrane (Figure 1.5). The process is intrinsically inefficient because the hydrophobic membrane binding site in the C5b67 complex is highly unstable with a half-life estimated to be approximately 0.1 s. This instability will itself ensure that most C5b67 complexes never bind membrane and that the deposition of MAC is tightly restricted to the site of C activation. The capacity of the C5b67 complex to bind membrane is further restricted by inhibitors in the fluid-phase. Perhaps the most important of the fluid-phase inhibitors is the next component in the pathway, C8 [166]. If C8 binds the C5b-7 complex in the fluid phase, the membrane binding site is masked and the resulting C5b-8 complex cannot bind membranes. For an MAC to form, the C5b-7 complex must therefore bind membrane before binding C8. In addition to C8, several other serum factors interfere with the binding of the C5b-7 complex to membrane and these are discussed below and illustrated in Figure 4.1.

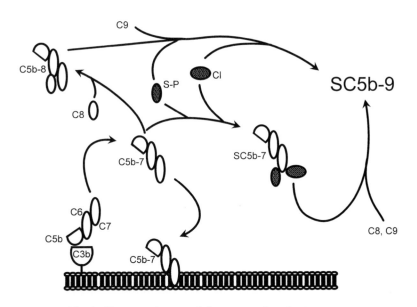

**Figure 4.1. Fluid-phase regulators of the terminal pathway.**
C5b, while bound to the C5 convertase binds C6 and C7 and is then released from the convertase. The C5b-7 complex is released to the fluid phase where it transiently expresses a hydrophobic membrane-binding site. Should the complex encounter a membrane during the lifespan of this site it binds and forms a nidus for the assembly of the membrane attack complex (MAC). However, should either S-protein (S-P) or clusterin (Cl) bind the C5b-7 complex the membrane-binding site is blocked. C8 and C9 will still bind to form the fluid-phase terminal complex SC5b-9. Binding of C8 to the C5b-7 complex in the fluid phase will also mask the membrane-binding site and lead to formation of the SC5b-9 complex.

## S-protein

*History*
In 1967, Holmes [880] purified, by adsorption onto glass beads, a plasma protein which had powerful cell adhesion properties. This protein was further characterized in the early 1980s and shown to have cell adhesion properties similar to those of fibronectin and was thus termed *cell-spreading factor* or *vitronectin* [881, 882]. Antibodies were developed against vitronectin and revealed the widespread distribution of the protein in biological fluids and tissues [881, 883]. Vitronectin was cloned in 1985 and analysis of the predicted structure revealed that the protein utilized cell adhesion sites identical to those in fibronectin (Figure 4.2) [884, 885]. Independent work by the Müller-Eberhard group had identified a serum protein, present in the fluid-phase terminal C complex, which they termed

**Figure 4.2.  Derived amino acid sequence of S-protein.**

```
MAPLRPLLILALLAWVALADQESCKGRCTEGFNVDKKCQCDELCSYYQSC    31
CTDYTAECKPQVTRGDVFTMPEDEYTVYDDGEEKNNATVHEQVGGPSLTS    81
DLQAQSKGNPEQTPVLKPEEEAPAPEVGASKPEGIDSRPETLHPGRPQPP   131
AEEELCSGKPFDAFTDLKNGSLFAFRGQYCYELDEKAVRPGYPKLIRDVW   181
GIEGPIDAAFTRINCQGKTYLFKGNQYWRFEDGVLDPDYPRNISDGFDGI   231
PDNVDAALALPAHSYSGRERVYFFKGKQYWEYQFQHQPSQEECEGSSLSA   281
VFEHFAMMQRDSWEDIFELLFWGRTSAGTRQPQFISRDWHGVPGQVDAAM   331
AGRIYISGMAPRPSLTKKQRFRHRNRKGYRSQRGHSRGRNQNSRRPSRAM   381
WLSLFSSEESNLGANNYDDYRMDWLVPATCEPIQSVFFFSGDKYYRVNLR   431
TRRVDTVDPPYPRSIAHYWLGCPAPGHL                         459
```

The amino-terminal signal peptide is underlined. The first residue of the mature protein sequence is **D** in bold. Numbering is for the mature protein.

**GenBank Accession No:**  X03168

*S-protein* (S for site-specific)[886, 887]. S-protein purified from serum was an 80-kDa polypeptide composed of two disulphide-linked chains. The purified protein inhibited formation of the MAC by binding the C5b-7 complex in the fluid phase[887, 888]. S-protein also bound avidly to immobilized heparin[889, 890]. The identity of S-protein with vitronectin was realised in 1986 by two groups who independently pointed out many common features, obvious in retrospect, and demonstrated antigenic cross-reactivity and shared functions of the proteins[891–893]. These reports were confirmed by comparisons of the newly available cDNA sequences of S-protein[894] and vitronectin[884].

*Distribution*

S-protein is present in plasma and in a variety of other biological fluids, including urine, seminal plasma and amniotic fluid[895]. S-protein is abundant in plasma with a concentration of between 240–540 µg/ml (6–10 µM)[896]. The primary site of biosynthesis of S-protein is the liver, although many other tissues and cell types, including platelets, monocytes and glial cells can also synthesise S-protein[897–901]. S-protein behaves as a positive acute-phase reactant, plasma levels increasing some two-fold following surgical stress or other inflammatory insults[902]. Studies in rabbits suggest that S-protein is relatively short-lived in the circulation with a calculated plasma half-life of about 8 hours[903]. Deposition of S-protein is seen in many tissues, notably the skin, the kidney and in blood vessel walls[904–908]. S protein may be deposited in association with C5b-9, particularly in pathology, but is also seen deposited independently of C5b-9 and associated with connective tissue elements or other fibrillar structures. Whether S-protein deposited at these sites plays any physiological role is uncertain, although it has been suggested that the protein might be responsible for providing a link between the various connective tissue elements or in the local protection of tissues from immune damage [909].

*Structure*
S-protein is an abundant (240–540 μg/ml) plasma glycoprotein (10–15% w/w carbohydrate) with a molecular weight on SDS-PAGE of 78–83 kDa. S-protein is highly acidic, running at a pI of 3.9 on isoelectric focusing. When run on SDS-PAGE under non-reducing conditions, various higher molecular weight species are present, suggesting the presence of dimers and higher aggregates linked by disulphide bonds. When run on SDS-PAGE gels under reducing conditions, the high molecular weight aggregates disappear and a variable proportion of the protein is broken down into two chains of molecular weights 65 kDa and 12 kDa, suggesting that S-protein exists in part as a disulphide-bonded dimer in plasma (Figure 4.3). The relative proportions of cleaved and uncleaved S-protein in plasma varies between individuals, some having almost 100% cleaved, others having 100% uncleaved and a third group having approximately equal amounts of the two forms. These proportions are stable within individuals, suggesting that cleavage represents a stable, inherited characteristic. The molecular basis of this variable susceptibility to cleavage is discussed below. Whether cleavage influences function is unclear [887, 893].

Early purification procedures were hampered by the adhesive properties of the molecule and by its tendency to aggregate during purification. In later purification protocols these same adhesive properties were exploited by using columns of glass beads to which the protein non-specifically bound; hence the name *vitronectin* [910, 911]. S-protein contains one free thiol group per molecule protein, providing an explanation for its propensity to aggregate [890].

*Function*
The primary biological role of S-protein is as an adhesion molecule, mediating cell attachment in tissues in a manner similar to that of fibronectin [883, 910]. In cultured cells S-protein is found associated with membrane at areas of tight interaction (focal adhesions) between the cells; indeed S-protein, due to its high affinity for binding plastic and glass surfaces, is a major cell attachment factor in cell culture media supplemented with bovine or human serum [912]. The cell-attachment activity of S-protein is provided by an RGD (Arg-Gly-Asp) domain, found in many adhesive glycoproteins, which interacts with the vitronectin receptor

(αVβ3 integrin) and other cell membrane receptors of the integrin family [910, 911, 913]. The heparin-binding properties of S-protein enable the molecule to bind tightly to proteoglycans in connective tissue matrix and appropriate carbohydrate ligands on cells [914]. Evidence suggests that the heparin-binding site may be hidden in the native molecule and exposed only when S-protein is complexed with the thrombin–antithrombin III complex generated during blood coagulation [915]. The resultant complex can bind surfaces through the heparin-binding and RGD sites in S-protein. Binding to endothelium may exert a local procoagulant effect and play a role in haemostasis and wound healing [894, 916]. S-protein also interferes with the anticoagulant activities of heparin, providing a second important interaction with the coagulation system. Heparin mediates its anticoagulant effect by forming a complex with antithrombin III which enhances the capacity of antithrombin III to inactivate thrombin. S-protein binding of antithrombin III interferes with the formation of this complex, thereby neutralizing heparin-mediated anticoagulation [892, 914]. S-protein also binds and forms complexes with several other serine protease inhibitors (serpins) of the coagulation pathway, including plasminogen activator inhibitors I and II and heparin cofactor II, and hence may influence coagulation at multiple stages [917, 918]. These alternative complexes also express heparin-binding activity. S-protein expresses binding sites for the naturally occurring opioid peptide β-endorphin and binding of heparin markedly increases the capacity for binding this peptide [919, 920]. The physiological relevance of this interaction remains uncertain.

S-protein also has several roles in immune defence. Various bacteria bind S-protein [921]. Indeed, specific receptors for S-protein have been identified on bacteria of the genus *Streptococcus* [922]. Binding of S-protein to the bacterial surface has been suggested to provide an additional anchorage for attachment of the bacterium to cells which express integrin receptors. In most cases this will aid bacterial clearance but in some situations it may assist bacteria in gaining entry into cells.

In the C system S-protein functions to regulate MAC activity by binding the fluid-phase C5b-7 complex, forming the non-lytic SC5b-7 complex. S-protein binds tightly at or near the metastable membrane binding site in C5b-7, thereby preventing the association of the complex with the membrane [163, 887, 888]. Binding of S-protein to C5b-7 was first suggested to involve the heparin-binding domain, but regions remote from this domain have recently been implicated [923]. The relative contributions of the various binding sites in S-protein to the association with C5b-7 remains unclear. The SC5b-7 complex once formed can still bind the late terminal components C8 and C9 and the complex so formed, SC5b-9, is an excellent fluid-phase index of ongoing C activation. The complex contains one or two copies of S-protein, single copies of C5b, C6, C7 and C8 but multiple copies (two to four) of C9 [893, 924, 925] (see also *clusterin*, below). In contrast to the MAC, C9 molecules in SC5b-9 do not form cylindrical structures; however, conformational changes (unfolding events) do occur which expose neo-epitopes in C9 apparently identical to those exposed in the MAC [926, 927].

*In vitro*, S-protein inhibits C9 binding into the forming MAC and C9 polymerization [163, 928]. It has been suggested that this represents a second mechanism by which S-protein inhibits C lysis, acting not on the fluid phase C5b-7 but on the membrane-associated C5b-8 complex [928, 929]. Inhibition of C9 binding and polymerization appears to involve sites distinct from those binding C5b-7, although here too the literature is confusing. It has been suggested that the heparin binding domain in S-protein binds a negatively charged region in C9, thereby preventing C9 polymerization [929]. The observation, described above, that this site is probably not exposed in 'native' S-protein, and the fact that other sites in S-protein have also been implicated in binding C9, cast doubt on the relevance of this finding [923]. To further compound the confusion, in a recent study of interactions of S-protein with the individual components of the C5b-9 complex it was found that S-protein bound C5b and C8 but not C9 or the other components [930]. The physiological relevance of this second mechanism of MAC inhibition, acting on membrane-associated complexes, is unproven. However, it has been shown that small amounts of S-protein are found in association with the MAC in tissues [907, 931].

S-protein has also been implicated as an inhibitor of the T cell lytic protein perforin. Perforin contains significant homology with the terminal C components and addition of S-protein to purified perforin *in vitro* inhibited target cell lysis [932]. Like the C9 interaction described above, interaction of S-protein with perforin has been suggested to involve the heparin-binding domain which is proposed to bind and block negatively charged regions on the perforin molecule homologous to those in C9. Given the well-documented mechanism of target killing by cytolytic T cells, involving the application of perforin directly onto the target membrane at the site of cell–cell contact, S-protein is unlikely to mediate any significant regulatory influence on T cell killing *in vivo*, although the possibility remains that S-protein has a role in protecting bystander cells from perforin 'leaking' from the targeted site.

Genetic deficiency of S-protein has not been reported. However, S-protein-deficient mice have recently been produced by gene targeting [933]. The homozygous S-protein-deficient mice display no detectable phenotype, develop normally and are fully fertile, indicating that, at least in mice, S-protein is not essential for survival and implying that there is considerable redundancy in both C regulation and cell/tissue adhesion systems.

*Molecular cloning and identification of functional domains*

The cloning of S-protein/vitronectin confirmed that the protein consisted of a single polypeptide chain of 478 residues [884, 894]. The sequence encoding the mature protein is preceded by a 19-amino acid signal peptide. The coding sequence contains several interesting structural motifs or domains (Figures 4.2 and 4.3). At the amino terminus is the entire sequence of the serum-derived peptide, somatomedin B (residues 1 to 44). Somatomedin B has been ascribed mitogenic properties but this finding has subsequently been refuted and the function of the peptide remains obscure [934]. It is also uncertain whether, as originally proposed, serum somatomedin B is derived from S-protein; the demonstration that this structural unit is also found in other proteins makes this less likely [935]. The somatomedin B domain is highly structured, containing four

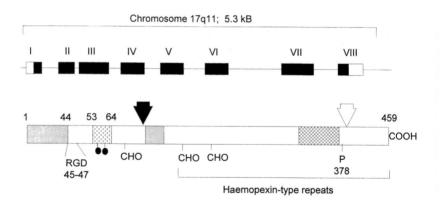

**Figure 4.3. Structural organization of S-protein.**

Top: The S-protein gene extends over 50 kb on chromosome 17q11 and is composed of eight exons. The contributions of each exon to the structure of the mature protein are detailed in the text.

Bottom: The mature protein consists of several domains. Between residues 1 and 44 is a domain encoding somatomedin B; the RGD adhesion domain is encoded at residues 45–47. Between residues 48 and 130 is a Cys-free stretch unique to S-protein and containing two sulphated Tyr residues (Tyr-56, Tyr-59; shown as filled circles). The remainder of the molecule consists primarily of haemopexin domains. The filled arrow indicates a Glu-rich region involved in intermolecular bonding. The open arrow indicates the point at which a variable proportion of S-protein is cleaved post-translationally. Glycosylation sites are indicated by CHO.

intra-domain disulphide bonds. The Arg-Gly-Asp (RGD) cell attachment sequence immediately follows the somatomedin B domain (residues 45–47). This sequence is crucial to the cell adhesion properties of the protein[936]. Between residues 48 and 130 of the molecule is a stretch unique to S-protein which is free of Cys residues, giving this portion of the molecule a very mobile conformation. Within this unique stretch, residues 53–64 comprise a highly acidic region containing two sulphated Tyr residues (Tyr-56 and Tyr-59)[937]. Glu-93 and several adjacent residues are involved in the formation of cross-links between individual S-protein monomers and this same region of the molecule has also been implicated in collagen binding[938].

Between residue 132 and the carboxy terminus at residue 459, S-protein is composed predominantly of domains which are homologous to the repeating unit first described in the haem binding protein, haemopexin. The haemopexin domain is also found in all members of the matrix metalloproteinase family of proteins known to play important roles in tissue remodelling[939]. Two separate haemopexin domains are present in S-protein, each containing multiple haemopexin repeats[940]. A Pro residue at position 268 forms a hinge between the two domains. Within the second haemopexin domain near the carboxy terminus is a 40-residue stretch rich in basic (positively charged) amino acids which forms the heparin binding domain implicated in the binding of S-protein to C5b-7 and the thrombin–antithrombin III complex[917, 941]. However, others have suggested that sites elsewhere in the molecule are responsible for binding C5b-7[923, 929]. A very recent report describes the identification of two additional heparin-binding sites, one between Asp-82 and Cys-137 and the other in the second repeat within the first haemopexin domain between Lys-175 and Asp-219[942]. Native S-protein expresses negligible heparin binding activity, suggesting that some or all of these heparin-binding sites are hidden, only exposed following either denaturation, binding to surfaces or by association of S-protein in complexes with thrombin and antithrombin III or other serine protease inhibitors. At the carboxy terminus of the original heparin-binding domain, there is a phosphorylation consensus sequence which permits the post-translational phosphorylation of Ser-378, just proximal to the cleavage site at Arg-379/Arg380[943]. Phosphorylation is catalysed by a cAMP-dependent protein kinase released into plasma from activated platelets. The influence of phosphorylation on function of S-protein remains unclear, although it has recently been demonstrated that phosphorylated S-protein exhibits a decreased binding affinity for plasminogen activator inhibitor 1 and is more readily incorporated into the SC5b-9 complex, implying that functional differences do exist[903, 944].

Immediately distal to the phosphorylation site is the site at which S-protein is post-translationally cleaved. Although S-protein is synthesized as a single-chain precursor in hepatocytes [945], it was recognised early on that a variable proportion of the purified protein was cleaved to yield disulphide-bonded subunits of 65 kDa and 12 kDa [946]. Cleavage occurs between Arg-379 and Ala-380 and recently, it has been shown that the variability in amount of cleaved product is an inherited characteristic, a single amino acid difference at position 381 dictating the susceptibility to proteolytic cleavage (Figure 4.3). In individuals homozygous for Thr at position 381, most of the circulating S-protein is cleaved whereas in those homozygous for Met at this position, very little is cleaved [947, 948]. The allelic distribution in Caucasians is approximately 50%, making this a useful marker in linkage studies; in some 25% of the population almost all the S-protein is uncleaved (Type 1-1), in 25% virtually all is cleaved (Type 2-2) and in 50% half is cleaved (Type 1-2) [949]. The functional differences between the cleaved and uncleaved forms are incompletely resolved.

The gene for S-protein has been mapped by fluorescence *in situ* hybridization (FISH) to the centromeric region of chromosome 17q, a location remote from other C regulators and from evolutionarily related genes such as the collagenases and stromelysin [950]. The gene extends over 5.3 kb and comprises eight exons and seven introns (Figure 4.3) [951]. The first exon of the gene encodes the 19-residue signal peptide and the first two residues of the mature protein. The second exon encodes most of the somatomedin B domain (residues 3–42). Exon III encodes the remaining two residues of the somatomedin B sequence, and the RGD sequence and extends through to the end of the cysteine-free region (in all, covering residues 43–131). Exons IV–VIII encode the haemopexin domains and include the originally defined heparin-binding site (encoded in exon VII). The conserved gene structure found in S-protein, haemopexin and several other proteins, including the matrix metalloproteinases and interstitial collagenase, has provoked the suggestion that they belong to a novel gene family termed 'pexin'. The murine S-protein gene has also been character-ized [952]. Although smaller (3 kb), the gene organization is identical to that of the human gene.

## Clusterin

*History*
Clusterin has been identified in many different guises over the last 15 years, leading to a bewildering number of names and functions (Table 4.1). Clusterin was first isolated from ram testicular fluid [953, 954]. The

Table 4.1. *The multiple personalities of clusterin.*

| Name | Species | Proposed role |
| --- | --- | --- |
| Clusterin | Ram | Aggregation of germ cells |
| Sulphated glycoprotein 2 (SGP-2) | Rat | Aggregation of germ cells |
| SP-40,40 | Human | C inhibition |
| Apolipoprotein J (Apo-J) | Human | Cholesterol transport |
| Cytolysis inhibitor (CLI) | Human | C inhibition |
| T64 | Quail | mRNA induced in virus-infected retinal cells |
| TRMP-2 | Rat | mRNA induced in prostate following androgen withdrawal |
| Secretogranin IV | Human | Marker for neuroendocrine cells |
| Glycoprotein III (gp-III) | Human | Constituent of chromaffin granules |
| Glycoprotein-80 (gp-80) | Dog | Secretory product of glomerular epithelial cells |

purified protein was shown to induce clustering of Sertoli cells, hence the name. Over the next few years species analogues were purified from bovine, rodent and avian sources and from a variety of tissues (Table 4.1). In most instances a role in reproduction or cell turnover was suggested. Human clusterin emerged from a very different experimental approach. As part of a search for novel components of immune deposits in the nephritic kidney, Murphy and co-workers generated a panel of mono-clonal antibodies using extracts of nephritic kidney as immunogen and screened for specific recognition of the diseased tissue [955]. Among these antibodies were several which recognised a heterodimeric molecule con-taining two chains each of approximately 40 kDa molecular weight which they termed serum protein 40 kDa,40 kDa (SP-40,40). Amino-terminal sequencing of the two chains of clusterin immunoaffinity purified from serum demonstrated that the chains were indeed distinct and that the molecule represented a novel heterodimeric serum protein. Antibodies against SP-40,40 bound in tissues in the same distribution as those detect-ing terminal C complexes and the protein was also present in SC5b-9 complexes generated *in vitro*, one molecule being incorporated into each SC5b-9 complex [908, 955, 956]. Indeed, clusterin could be efficiently purified from the *in vitro* generated SC5b-9 complex in a form identical to that present in serum [956, 957].

*Distribution*

Early reports suggested that clusterin was present in serum at a concentration of around 35–105 µg/ml [955]. However, more recent reports suggest that the true serum level of clusterin is much higher (250–420 µg/ml), the artefactually low values originally obtained being a result of the adhesive nature of the protein [896]. The primary source of serum clusterin is the liver [958]. Clusterin concentration in human seminal plasma is some tenfold that in serum (2–15 mg/ml). The primary source of seminal plasma clusterin appears to be the Sertoli cells in the testis which abundantly express the protein, although the epididymis also strongly expresses clusterin. Although the majority of tissues can express clusterin, constitutive expression is, with the above exceptions, low. However, a variety of stimuli, including inflammatory cytokines, markedly enhance expression in these other sites. Clusterin expression is also enhanced following cell death or injury in diverse tissues [959–961].

*Structure*

Human clusterin is a heterodimeric serum glycoprotein composed of two non-identical subunits termed α and β which are of approximately the same size (35–40 kDa) on SDS-PAGE. The two chains are derived from a single chain precursor and are disulphide-bonded [962]. The molecule is heavily glycosylated, containing some 25–30% carbohydrate by weight in six N-linked groups, three in each chain. Purified clusterin aggregates spontaneously to form dimers and higher molecular weight complexes. Such complexes can also be seen in serum clusterin and are held together by inter-molecular non-covalent bonds. Like S-protein, clusterin is a very 'sticky' molecule which binds readily to many different types of surface. Clusterin also binds proteoglycans through heparin-binding sites.

*Function*

The multi-functional nature of clusterin is made very apparent by review of the literature which reveals that the same protein has emerged in many different guises and with numerous proposed functions (Table 4.1) [164]. Clusterin is very broadly expressed in tissues and clusterin expression is induced or upregulated following cell injury in diverse tissues [959–961, 963]. Indeed, increased expression of clusterin has been used as a marker for cell damage, apoptosis or cell division at various tissue sites. In most situations, clusterin appears to act as an aid to cell survival and is expressed by the surviving cells rather than by the dying cells [964, 965]. In numerous tissues, expression of clusterin correlates closely with the onset of apoptotic cell death, either as a consequence of tissue damage or as part of tissue remodelling [966]. The reasons for the large increase in clusterin expression at sites of cell death remain unclear but it has been proposed that clusterin is cytoprotective [959, 961]. The capacity of clusterin to act as a 'molecular dustbin', binding numerous potentially toxic products of cell breakdown and mediating their efficient clearance, is likely to be its most important role *in vivo*. The enormous literature describing the role of clusterin as a marker for cell death or division in many tissues and species is outside the scope of this book, but is well reviewed in the works referenced above. By way of example of these other roles, the relevance of clusterin expression to cell survival in the brain has attracted particular attention. Clusterin expression in brain is markedly increased in response to diverse injuries and a neuroprotective role for the expressed protein has been proposed [967–969]. One fascinating suggestion is that upregulation of clusterin has a role in protecting against neuronal damage in Alzheimer's disease [970].

Clusterin was first discovered in the male reproductive system and its potential role in reproduction has continued to attract interest. Clusterin is present in human seminal plasma at high concentration (2–15 mg/ml) and low seminal plasma clusterin concentrations are associated with infertility, particularly in association with azoospermia, provoking the suggestion that clusterin is important in protection of the spermatozoa [971]. However, it is equally possible that the low sperm numbers are a cause rather than consequence of the low clusterin concentration. Normal spermatozoa do not have clusterin on their membranes but the majority of morphologically abnormal and/or immotile spermatozoa in ejaculates stain strongly for clusterin and the majority of these are aggregated [972]. The process by which clusterin becomes bound to the sperm membrane and whether this is secondary to or a cause of the sperm immotility and aggregation remain to be determined. A truncated form of clusterin, recognised only by antibodies against the α-chain is present on the acrosome membrane (exposed following capacitation) in normal spermatozoa. Clusterin is clearly important in the reproductive system but its key roles remain to be elucidated.

Clusterin in plasma is associated with high density lipoproteins (HDL). Indeed, one of the guises in which clusterin emerged was as a novel apoliprotein termed apoliprotein J which was found in association with apoA-1 in HDL particles [973–975]. A role for clusterin in the clusterin–apoA-1 complex in lipid transport has been suggested [976].

Clusterin is a potent stimulator of cell adhesion and aggregation (clustering, as the name implies), as first demonstrated for Sertoli cells derived from boar testis [953, 954]. The aggregation activity is not limited to Sertoli cells; clusterin has similar aggregation activities to fibronectin, causing clustering of many different cell types, including human endothelial and epithelial cells[977]. Clusterin also binds certain bacteria and induces bacterial aggregation. Binding to *Staphylococcus aureus* appears to occur through specific receptors on the bacteria and binding of clusterin has been proposed to be a virulence determinant in this organism[978]. Coating with clusterin increases the 'stickiness' of the bacterium, enhancing its capacity to invade tissues.

Soon after the discovery of SP-40,40/clusterin in nephritic kidneys in association with C components, it was realised that clusterin purified from serum or from SC5b-9 complexes was an efficient inhibitor of MAC formation in reactive lysis assays[956, 979]. The data strongly suggested that clusterin, like S-protein, bound the nascent C5b-7 complex, preventing its association with membranes. Murphy and co-workers directly compared S-protein and clusterin and showed that clusterin was the better inhibitor on a molar basis and that the effects of the two inhibitors were additive[979]. Of the isolated MAC components, clusterin binds C7, the $\beta$ chain of C8 and the b domain of C9 (the carboxy-terminal portion of C9 generated on thrombin cleavage)[980]. However, the artificial conditions utilized in these *in vitro* assays disallow extrapolation to the forming MAC. In these studies, the $\alpha$ and $\beta$ subunits of clusterin were separated and both were found to bind C9, indicating that binding sites were present on each subunit. There is remarkably little evidence that inhibition of C by clusterin is of physiological relevance. Perhaps the best evidence comes from a study of glomerular injury in the isolated perfused rat kidney[981]. In this model, perfusion with human plasma depleted of clusterin caused more glomerular injury than did undepleted plasma.

*Molecular cloning and identification of functional domains*
Cloning of clusterin cDNA from a human liver library confirmed that the two chains were encoded as a single-chain precursor, the mature two-chain protein being generated by a post-translational cleavage event at an internal bond between Arg-205 and Ser-206, dividing the molecule into a 205-amino acid $\alpha$ chain and a 222-amino acid $\beta$ chain (Figures 4.4 and 4.5)[982]. The two chains are held together by a unique five-disulphide bond motif[962]. The molecule contains a high abundance of $\alpha$-helical regions. All of the functional domains defined to date are in the $\alpha$ chain. Near the amino terminus of the $\alpha$ chain is a long $\alpha$-helical stretch (residues 1–76) which is implicated in the aggregation of clusterin monomers. This is followed by a Cys-rich domain (residues 77–98), homologous with the thrombospondin type I domains found also in the terminal C components C7, C8 and C9, which has been implicated in the C inhibitory functions of clusterin. Cloning of clusterin analogues from other species revealed high inter-species conservation: 72% between human and bovine and

**Figure 4.4. Derived amino acid sequence of clusterin.**

```
MMKTLLLFVGLLLTWESGQVLGDQTVSDNELQEMSNQGSKYVNKEIQNAV      28
NGVKQIKTLIEKTNEERKTLLSNLEEAKKKKEDALNETRESETKLKELPG      78
VCNETMMALWEECKPCLKQTCMKFYARVCRSGSGLVGRQLEEFLNQSSPF     128
YFWMNGDRIDSLLENDRQQTHMLDVMQDHFSRASSIIDELFQDRFFTREP     178
QDTYHYLPFSLPHRRPHFFFPKSRIVRSLMPFSPYEPLNFHAMFQPFLEM     228
IHEAQQAMDIHFHSPAFQHPPTEFIREGDDDRTVCREIRHNSTGCLRMKD     278
QCDKCREILSVDCSTNNPSQAKLRRELDESLQVAERLTRKYNELLKSYQW     328
KMLNTSSLLEQLNEQFNWVSRLANLTQGEDQYYLRVTTVASHTSDSDVPS     378
GVTEVVVKLFDSDPITVTVPVEVSRKNPKFMETVAEKALQEYRKKHREE      427
```

The amino-terminal signal peptide is underlined. The first residue of the mature protein is **D** in bold. Numbering is for the mature protein.

**GenBank Accession No.** X14723

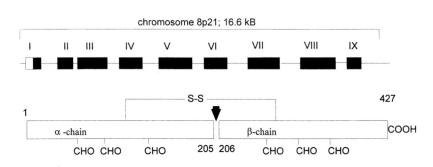

**Figure 4.5. Structural organization of clusterin.**

Top: The S-protein gene extends over 17 kb on chromosome 8p21 and is composed of nine exons. The contributions of each exon to the structure of the mature protein are detailed in the text.

Bottom: The mature protein consists of two chains, generated by post-translational cleavage between residues 205 and 206 (arrowed). The two chains remain disulphide-bonded. Both chains are heavily glycosylated (indicated by CHO).

49% between human and avian. The region between residues 87 and 150 is extremely highly conserved between species (95–99%) and may be critical to clusterin structure and/or function. Within this very highly conserved region is a putative dinucleotide binding site (residues 105–135), the role of which is undefined.

Comparison of the cDNA sequence of clusterin with that of a novel component of high density lipoproteins (HDL) termed *apolipoprotein J* (apoJ), isolated and cloned in 1990 [973, 974] demonstrated that these molecules were identical. Similar comparisons with the sequence of the major secreted product of rat Sertoli cells (and hence an abundant protein in rat seminal plasma), *sulphated glycoprotein 2* (SGP-2), revealed very high homology, indicating that clusterin and SGP-2 were species analogues of the same protein [962, 983].

The clusterin gene was mapped by FISH to chromosome 8p21 in close proximity to the gene for lipoprotein lipase [984]. The gene was single-copy and composed of nine exons distributed over 16.6 kb (Figure 4.5) [985]. A surprising finding was that the gene sequence differed markedly from the cDNA sequence reported for clusterin in the 5′ untranslated region (encoded by exon I). No clear explanation for this finding has yet emerged, but it has been suggested that alternative exon I sequences may exist unrecognised in the clusterin gene. Polymorphisms in the clusterin gene have been demonstrated which, while rare in Caucasians, occurred with increased frequency in African-Americans [986]. Whether these polymorphisms, which in some cases caused major differences in glycosylation of the molecule, are of functional significance remains to be tested. No deficiencies of clusterin have yet been reported.

## 'Non-specific' fluid-phase inhibitors of the MAC

Both S-protein and clusterin inhibit MAC formation by binding the nascent C5b-7 complex in the fluid phase. A variety of other factors present in serum also act at this vulnerable stage of MAC assembly by binding and blocking the membrane-binding site. The important role of C8 as a 'double-agent' has already been mentioned. C8 binding to the membrane-associated C5b-7 complex is essential for MAC formation but binding of C8 to the fluid-phase C5b-7 complex prevents association of the complex with membrane [166]. Serum lipoproteins represent another important inhibitor of C5b-7 in the fluid phase [165, 888]. Both high density (HDL) and low density lipoproteins (LDL) have been shown to mediate this activity, probably through direct binding to the hydrophobic interaction site in C5b-7. It is not clear whether the inhibitory effect of HDL is mediated in part through clusterin contained within the lipoprotein complex. An effect of HDL, or lipoproteins derived from HDL, on MAC assembly subsequent to the membrane binding of C5b-7 has also been proposed [987]. The role of clusterin, which is a major component of a subclass of HDL, in this inhibition and the biological relevance of the phenomenon remain uncertain.

Heparin is a polyanionic anticoagulant with significant C-inhibiting activity [988]. Heparin binds the C5b6 complex *in vitro* and prevents formation of C5b-7, a property shared with other highly polyanionic molecules [989]. Polyanions inhibit C at multiple points in the reaction sequence and have been proposed as potential therapeutics to inhibit C activation *in vivo*[990].

# MEMBRANE REGULATORS OF THE TERMINAL PATHWAY

Once bound, the C5b-7 complex is protected from the fluid-phase regulators which target the membrane-binding site. Rescue of cells is then dependent on membrane proteins which specifically interfere with the further assembly of the MAC. Two such proteins have been described to date, of which the first described, homologous restriction factor (HRF), remains very poorly characterized. The second membrane inhibitor of the terminal pathway to be described, CD59, has been the focus of intense interest over the past few years.

## Homologous restriction factor (HRF)

*History*

The phenomenon of *homologous restriction*, whereby cells are less susceptible to lysis by C of the same species than to lysis by C from other species has been known for at least 80 years[137]. In 1970, Lachmann & Thompson used a reactive lysis system (requiring only the purified components of the terminal pathway) to demonstrate that at least part of the restriction occurred in the teminal pathway[991, 992]. This data was built upon by Yamamoto[993] who showed that the species source of C9 was critical, suggesting that homologous restriction occurred late during assembly of the MAC. These results were later confirmed and extended by Hansch and co-workers[140], who showed that the species source of both C8 and C9 influenced homologous restriction, further implicating the last steps in MAC assembly as critical in homologous restriction.

The membrane factor(s) responsible for homologous restriction of MAC remained elusive until 1986 when two groups independently reported the isolation from human erythrocyte membranes of a protein which, when reincorporated in lipid bilayers, conferred resistance to C reactive lysis[149, 150]. Both groups identified a protein in detergent or phenol extracts of human erythrocyte membranes which bound C8 and/or C9. Zalman used this property to isolate the protein on columns of immobilized C9 and obtained a protein which, when isolated from time-expired erythrocytes had a molecular weight of 38 kDa, but had a molecular weight of 65 kDa when isolated from fresh erythrocytes; the protein was termed *homologous restriction factor* (HRF). Hansch and co-workers isolated a protein of 65 kDa by chromatofocusing, which they termed *C8-binding protein* (C8bp). The purified proteins incorporated into targets (liposomes or chicken erythrocytes) and protected against C reactive lysis. Antiserum raised against HRF rendered human erythrocytes some 20-fold more sensitive to reactive lysis.

*Distribution*
HRF has been reported to be present on erythrocytes, leukocytes, platelets, glomerular cells, thyroid epithelial cells and on a large variety of nucleated cell lines [994–996]. The protein expressed on nucleated cells was slightly smaller than that on erythrocytes (50 kDa vs. 65 kDa). Erythrocytes and leukocytes in the haemolytic disorder, paroxysmal nocturnal haemoglobinuria (PNH), were deficient in HRF [997, 998]. A soluble form of HRF was found in urine which was purified and shown to retain C-inhibiting capacity [999]. HRF has also been found at low concentration in serum [1000].

*Structure*
HRF purified from fresh erythrocytes had a molecular weight on SDS-PAGE of 65 kDa whereas that from nucleated cells and platelets was about 50 kDa [150, 994]. The observed differences were suggested to be due to differential glycosylation but this has not been formally demonstrated. Protein from both sources incorporated spontaneously into the membranes of other cells or liposomes, indicating that HRF was linked through a glycosyl phosphatidylinositol (GPI) anchor [997, 998]. This was confirmed by the demonstration that the molecule was removed from cells by treatment with phosphatidylinositol-specific phospholipase C (PIPLC) which specifically cleaves GPI anchors [1001]. Evidence based upon antibody cross-reactivity has suggested that HRF bears some antigenic similarity to the terminal components to which it binds [1002]. This cross-reactivity extends to perforin and is suggested to involve cysteine-rich domains unmasked upon reduction, as only antisera generated to the reduced proteins showed cross-reactivity [1003].

*Function*
HRF inhibited C only during the assembly of the MAC. HRF had no effect on the numbers of C5b-7 sites formed but reduced C9 binding to targets, suggesting that it acted late in MAC assembly. Of the terminal components, HRF bound purified C7 and C9 with low affinity and C8 or the isolated C8αγ subunit with high affinity [1002]. It was suggested that HRF bound the γ chain in C8; however, the observation that C8 lacking the γ

chain was lytically active and homologously restricted would contradict this possibility [1004, 1005]. The inhibitory activity of the purified protein was reported to be exclusive to human C, causing no inhibition of C from a wide variety of other species, including primates [1006].

Erythrocytes and other blood cells in PNH are exquisitely susceptible to C damage, in part due to a deficiency of DAF [670, 673]. However, PNH erythrocytes are also more susceptible to reactive lysis [1007, 1008], suggesting a deficit in regulation of MAC assembly and provoking an examination of HRF expression in PNH. Indeed, PNH cells are deficient in HRF, and incorporation of HRF into PNH erythrocytes protected against reactive lysis [998]. It is now known that the underlying deficit in PNH is in the machinery for synthesis of GPI anchors, affected cells lacking all proteins anchored in this manner, including DAF, HRF and CD59 (see Chapter 5) [1009, 1010].

The presence of HRF in the lytic granules of cytotoxic T cells provoked the suggestion that it played a role in regulating the lytic protein perforin [1011]. However, this suggestion has been fiercely contested by others who have shown that perforin lysis exhibits no species restriction and that GPI anchor-negative (and hence HRF-negative) cells are not more susceptible to perforin lysis [1012–1014].

A major deficiency in the HRF story is the almost complete lack of structural information. More than 12 years after its discovery, there is still no protein or DNA sequence available for this apparently broadly expressed molecule. Many in the C field are now beginning to question whether HRF exists at all or whether all the data reviewed above is artefactual, perhaps generated because of trace contamination of protein preparations with the more recently described inhibitor CD59, and of antisera with reactivities against CD59. Those individuals who have worked with HRF tell of an evanescent activity, difficult to retain during purification and storage. Nevertheless, it is now clear that only the cloning of the molecule will convince the many sceptics.

## CD59

*History*
The history of CD59 is very different from that of HRF. First described in 1988, it is now thoroughly characterized in terms of distribution, function and structure. Sugita *et al.* [1015] described the isolation from human erythrocyte (E) membranes of a protein of molecular weight 18 kDa (hence called P-18) which strongly inhibited the reactive lysis of guinea pig E. The purification procedure utilized classical column chromatography

followed by preparative SDS-PAGE and western blotting. The blot was cut into strips and proteins were eluted in detergent-containing buffer. Reactive lysis-inhibiting activity was found only in the strip corresponding to a molecular weight of 18 kDa. The remarkable stability in the face of denaturation, heating and storage implied in this purification scheme is an intrinsic feature of the molecule which has greatly aided progress on its characterization. The protein inhibited lysis even if target cells were washed after incubation with P-18 and before exposure to C attack, suggesting that P-18 incorporated in the target membrane. Within a year this same inhibitor was independently rediscovered in at least three other laboratories, and was given a variety of names. Holguin and co-workers utilized classical chromatography techniques to isolate the protein and called it *membrane inhibitor of reactive lysis* (MIRL) [1016]; Okada and associates used a monoclonal antibody (IF-5), which had previously been shown to enhance the lysis of neuraminidase-treated human erythrocytes by human C, to isolate the protein and called it *homologous restriction factor-20* (HRF-20) [1017, 1018]; Davies and colleagues used a monoclonal antibody raised against human leukocytes (YTH-53.1), which had been shown to enhance C lysis of human cells, to isolate the protein and called it *CD59 antigen* (CD59) [1019]. This last name originated from the placing of the antibody YTH53.1 and another antibody, MEM-43 [1020], in a cluster termed CD59 at the 4th Leukocyte Antigens Workshop. The most widely accepted name for the molecule is CD59 but the names noted above and several others (protectin, H19, MAC inhibitory factor) still emerge from time to time to confuse the literature.

*Distribution*
CD59 is widely and abundantly expressed, present on all circulating cells, endothelia, epithelia and on cells in most organs examined [1019–1022]. Erythrocytes express approximately 25,000 copies per cell [1023], and many nucleated cell types appear to express significantly more. The molecule is GPI-anchored on all cell types so far examined, although the degree to which the anchor is cleaved by PIPLC varies between cell types, erythrocyte CD59 being particularly resistant to PIPLC release [1024]. Expression of CD59 is upregulated almost five-fold on the K562 erythroleukaemia cell line and three-fold on a human endothelial cell line upon incubation with phorbol esters [1025–1027]. On renal mesangial cells, non-lethal C attack caused increased expression of CD59, apparently mediated through C5a [1028]. Increased expression of CD59 has been reported on some tumour cell lines *in vitro* and on solid tumours *in vivo* [1029–1031]. It is tempting to speculate that the overexpression of CD59 might play a role in protecting the tumour from immune attack. In marked contrast, many leukaemic cell

lines and human leukaemias express little or no CD59, yet continue to proliferate despite this deficit [1032].

CD59 is also found in biological fluids. Urine contains CD59 at a concentration of between 1 and 4 μg/ml [1019, 1033]; urine CD59 has lost the GPI anchor and is incapable of incorporating into target membranes. Fluid-phase forms of CD59 are also present in amniotic fluid, seminal plasma, breast milk and cerebrospinal fluid [662, 759, 1034, 1035]. In the first three of these fluids the bulk of the CD59 is associated with lipid in a form which retains its GPI anchor and can incorporate into lipid membranes.

*Structure*
Erythrocyte CD59 runs on SDS-PAGE as a broad band of apparent mass 18–23 kDa. CD59 purified from nucleated cells and platelets runs in the same molecular weight range but shows a clear banding pattern with multiple bands separated by about 1 kDa. After enzymatic deglycosylation, CD59 from all sources runs as a sharp band of molecular weight 12 kDa, indicating that the molecule is heavily glycosylated and that the broad molecular weight range is due to variable glycosylation [1036]. The protein is GPI-anchored as demonstrated by the capacity of the purified protein to incorporate into membranes and the efficient removal of CD59 from cells by treatment with PIPLC [1019].

*Function*
The capacity of CD59 to incorporate into membranes of target cells and protect against C lysis greatly aided studies of its mode of action. In all of the original reports of the isolation of CD59 it was stated that inhibition of C occurred late in the assembly of the MAC, preventing C9 incorporation and polymerization. Parker and co-workers suggested that CD59 also inhibited C5b-7 site formation although others found no inhibition at this step [1024]. The details of the mechanism by which CD59 inhibited MAC assembly emerged from experiments utilizing cellular and fluid-phase MAC intermediates into which CD59 was incorporated at various stages in the assembly process [1023, 1037]. CD59 did not bind to C5b-7 sites on erythrocytes but bound tightly to the C5b-8 complex. Although a single C9 still bound the CD59–C5b-8 complex, incorporation of C9 and the subsequent recruitment of additional C9 molecules to form the lytic lesion was blocked (Figure 4.6). Association of CD59 in the complex was sufficiently strong to withstand detergent extraction of the membrane. Further analysis of the interaction of CD59 with the forming MAC demonstrated specific and strong binding to the α chain of C8 and to the carboxy-terminal b domain of C9 [1038].

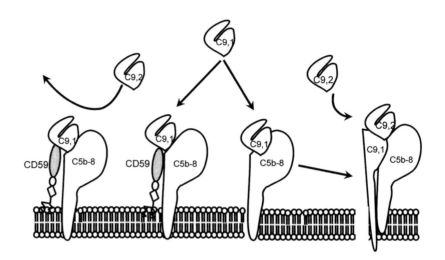

**Figure 4.6. Inhibition of MAC assembly by CD59.**

The C5b-8 complex, formed on the membrane as described elsewhere, expresses a binding site to which the first C9 (C9,1) attaches. In the absence of CD59 (to right) this first C9 will unfold and insert into the membrane, causing disruption of the bilayer and exposing a binding site for recruitment of a second C9 molecule (C9,2). In the presence of CD59 (to left), CD59 binds tightly to the C5b-8 complex. Although the first C9 (C9,1) binds it cannot unfold and insert. The membrane is not disrupted and no binding sites for additional C9 molecules are exposed.

*Molecular cloning and structural analysis*

The cDNA for CD59, identified by several groups, encoded a 128-amino acid polypeptide which bore no resemblance to any of the other C regulators or components; the only significant homology was with the murine Ly-6 antigens which shared 26% amino acid identity (Figures 4.7 and 4.8) [1018, 1019, 1039, 1040]. In fact, multiple sizes of CD59 cDNA were identified, but

**Figure 4.7. Derived amino acid sequence of CD59.**

```
MGIQGGSVLFGLLLVLAVFCHSGHSLQCYNCPNPTADCKTAVNCSSDFDA     25
CLITKAGLQVYNKCWKFEHCNFNDVTTRLRENELTYYCCKKDLCNFNEQL     75
ENGGTSLSEKTVLLLVTPFLAAAWSLHP                          103
```

The amino-terminal signal peptide and carboxy-terminal signal for GPI anchor addition are underlined. The first amino acid of the mature protein is **L**. The single N-linked glycosylation site is indicated by **N**. The site of GPI anchor addition is at residue 77, indicated by **N**. Numbering is for the mature protein.

**GenBank Accession No.**   X15861; X16447

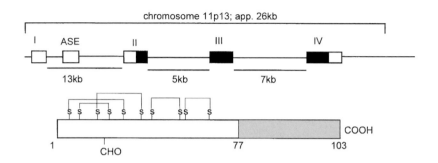

**Figure 4.8. Structural organization of CD59.**

Top: The CD59 gene extends over 26 kb on chromosome 11p13 and is composed of five exons. The contributions of each exon to the structure of the mature protein are detailed in the text.

Bottom: The mature protein consists of a single chain of 77 amino acids; the C-terminal signal peptide (hatched) is removed during addition of the GPI anchor. The five intra-chain disulphide bonds essential to function are shown.

all were shown to be derived from the same coding sequence, differing only in the degree of polyadenylation. The first 25 residues of the cDNA were characteristic of a hydrophobic signal peptide, and protein sequencing confirmed that the first residue of the mature protein was Leu[1041]. The predicted site of GPI anchor addition was at Asn 77, the final 26 amino acids forming the signal for anchor addition. The mature 77-residue protein contained ten Cys residues, arranged in five intra-chain disulphide bonds linking Cys-3 to Cys-26, Cys-6 to Cys-13, Cys-19 to Cys-39, Cys-45 to Cys-63 or Cys-64 and Cys-63 or Cys-64 to Cys-69[1041]. These disulphide bonds thus create several locked loop structures and effectively divide the molecule into an amino-terminal 'domain' (up to Trp-40) containing three intra-domain disulphide bonds and a carboxy-terminal 'domain' (from Trp-40) containing two disulphide bonds. The mature protein had a predicted molecular weight of about 11 kDa, much smaller than the apparent mass on SDS-PAGE, suggesting heavy glycosylation. A single potential N-glycosylation site (N-X-S/T) at Asn-18 was shown by protein sequencing to be occupied. A second potential site at Asn-8 contained Pro at the X position, rendering it unusable as a site for glycosylation[1042, 1043].

Further analyses revealed numerous other proteins of diverse source and function which shared limited homology with CD59, notably in the conservation of the disulphide bonding pattern, now termed the Ly-6 multigene family[1044, 1045] (Table 4.2). The 're-invention' of the same small, cysteine-rich structural motif through evolution suggests peculiar

**Table 4.2.  *The Ly-6 multi-gene family.***

The members of the family share a common domain structure with strong conservation of five intra-domain disulphide bonds. Most but not all are GPI-anchored to membranes.

| Name | Species | Properties |
| --- | --- | --- |
| Ly-6 antigens | Mouse | T-cell activation; ? other roles? |
| Sgp-2 | Squid | Major glycoprotein in squid brain |
| HVS-15 | Herpesvirus | No protein data |
| Urokinase-type plasminogen activator receptor (uPAR) | Human | Receptor for uPAR |
| Snake venom α neurotoxins | Snake | Family of secreted venom toxins, bind acetylcholine receptor |
| Phospholipase A2 inhibitor | Snake | Isolated from Thai cobra plasma |
| E48 antigen | Human | Analogue of Ly-6, involved in keratinocyte adhesion |
| RIG-E | Human | Analogue of Ly-6, induced by retinoic acid in leukaemic cells |
| SP-10 | Human | Sperm acrosomal protein |
| CD59 | Human | C inhibition |

advantages, probably in relation to stability of the motif rather than to any shared function of the proteins.

The gene for CD59 was localized to the short arm of chromosome 11, p14-p13 by a variety of techniques [1046, 1047]. Bickmore and colleagues analysed a panel of somatic cell hybrids carrying various fragments of chromosome 11 to map the *cd59* gene to 11p13 in close proximity to the gene encoding a cell surface antigen termed MIC 11, identified by a single monoclonal antibody. It has recently been shown that the MIC 11 antigen is in fact an epitope on CD59 [1048]. The *cd59* gene occupied approximately 26 kB and was composed of four exons, the N-terminal signal peptide being encoded in exon II and the coding sequence for the mature protein being contained within exons III and IV (Figure 4.8) [1049, 1050]. The gene structure was strikingly similar to that of other genes in the Ly-6 family. Very recently, a short fifth exon has been identified lying in the long intronic sequence between the original exons I and II [1051]. This exon is only included in a small proportion (10–20%) of the mRNA product of the gene and is termed the *alternatively spliced exon* (ASE). Inclusion of the ASE has no effect on the structure of the protein product and there are no obvious regulatory elements in the sequence; the role of this sequence thus remains unknown.

A single case of complete deficiency of CD59 has been described [1052]. This individual presented with a PNH-like syndrome and multiple thrombotic episodes. The consequences of CD59 deficiency in this individual will be further described in a later chapter. The defect was caused by a base deletion in the coding region of the CD59 gene which resulted in premature termination at residue 16 of the mature protein [1053].

*Role of carbohydrate*
The role of the carbohydrate group in CD59 has recently attracted much attention. CD59 is a small molecule containing a very large N-linked carbohydrate group (accounting for 8–9 kDa of the total apparent molecular mass of 18–23 kDa). Early studies utilizing enzymatic deglycosylation of CD59 reported that removal of carbohydrate markedly reduced functional activity, suggesting an important role in C-inhibitory function [1054]. However, several groups have subsequently engineered and expressed CD59 in which the N-glycosylation site is absent; this molecule has a molecular mass of only 11 kDa on SDS-PAGE yet retains full or even enhanced C-inhibitory function, demonstrating that the large carbohydrate group is not required for C inhibition [1055–1058]. Indeed, one recent report describes the expression of a soluble form of CD59 in which the glycosylation site has been mutated; the resultant molecule was some seven-fold more active as an inhibitor than the glycosylated form [1059]. On erythrocytes, the N-linked carbohydrate group is enormously complex, containing over 140 different structures and occupying as much space as the protein core (Figure 4.9) [1060]. It is difficult to believe that this large, complex structure has no role in CD59 function or stability. What then is the function of the large carbohydrate group? It has been suggested that the carbohydrate is essential for some of the adhesive properties of CD59 (described below) [1061], but this remains controversial. Our unpublished data indicate that the carbohydrate may also play a role in protecting CD59 from proteolysis on the cell membrane.

*Active site of CD59*
The quest for the site(s) in CD59 which enable it to bind strongly in the forming MAC and inhibit lysis, the *active site(s)*, has occupied several research groups (including our own) over the last few years. Early studies using peptides derived from the CD59 sequence implicated the region between residues 27 and 38 as the active site [1062]. However, these results have been difficult to replicate and do not correlate well with more recent data. A knowledge of the three-dimensional structure of a molecule is an enormous aid to the identification of surface-exposed features which might function as active sites and provides a logical basis for targeting specific

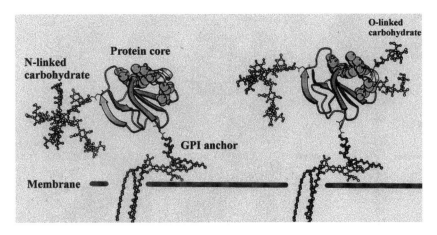

**Figure 4.9. Schematic of structure and mutational analysis of CD59.**

The three-dimensional structure of CD59 (based on NMR analysis) together with the GPI anchor structure, the N-linked carbohydrate group (to left) and, in the second view, the putative O-linked groups (to right). The putative active site residues are shown as spheres on the upper face of the molecule. Modified from Rudd *et al.* (1997), *Journal of Biological Chemistry* **272**; 7229–7244.

areas of a molecule. Two groups have independently solved the three-dimensional structure of CD59 by NMR spectroscopy, working either with soluble CD59 purified from urine [1063, 1064] or with recombinant soluble CD59 generated in CHO cells [1065]. The two structures were very similar and revealed CD59 to be a rather flat, disc-shaped molecule containing two anti-parallel β-sheet regions, a short helical region and a carboxy-terminal region lacking in secondary structure (Figure 4.9). The structures obtained were remarkably similar to previously solved structures of other Ly-6 family members.

Guided by the three-dimensional structures, molecular engineering has been used to search for the active site either by mutating individual residues predicted to be exposed in the CD59 molecule or by swapping regions between CD59 and related molecules. Our mutagenesis studies have implicated two regions in CD59 as putative active sites: one a groove between the two anti-parallel β-sheets on the upper face (away from the membrane) of the molecule and the other a loop between Cys residues 19 and 26 which projects laterally at the base of the surface groove [1057]. Substitution of residues within these regions reduced or ablated the capacity of the expressed molecule to inhibit the MAC (Figure 4.10). Of particular note, mutations in and around Trp-40 not only ablated inhibitory function but also eliminated the epitope(s) recognized by all available monoclonal

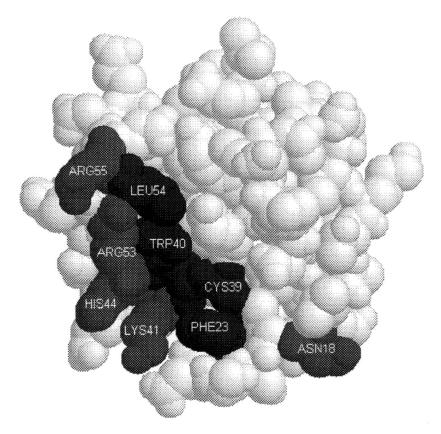

**Figure 4.10. The putative active site of CD59.**
View from above of CD59, based on the structures obtained from NMR analysis. The cluster of amino acid residues implicated in C-inhibitory function by antibody-binding data and/or the effects of point mutation are dark shaded and identified. Mutations of other residues remote from this cluster had little or no effect on function. Modified from Bodian *et al.* (1997), *Journal of Experimental Medicine* **185**; 507–516.

antibodies which inhibited CD59 function, whereas antibodies which did not influence CD59 function still bound. Petranka and co-workers also introduced a series of point mutations in CD59 and observed effects on function and binding of a series of monoclonal antibodies[1066]. Mutation of Arg53→Ser abrogated binding of all available blocking monoclonal antibodies. This residue is immediately adjacent to Trp40, identified as key for binding of blocking antibodies in our study. Taken together these data strongly suggest that this region is involved in MAC inhibition. Petranka and co-workers also mutated individual Cys residues to remove single disulphide bonds and disrupt loops in the molecule. With the exception of substitutions involving the 6–13 and 64–69 bonds, each of these mutations prevented expression of the protein. Substitution of Tyr61→Gly yielded an inactive molecule, although replacement of this residue with several other amino acids had no effect on function. This result is difficult to reconcile with the other data as Tyr61 is on the opposite face of the molecule to the groove around Trp40.

Tomlinson and co-workers have adopted a different approach. They have chosen to substitute regions of CD59 with the same region derived either from Ly-6 or from the rat analogue of CD59 (rat CD59) to create chimaeric molecules [1067, 1068]. The human CD59 to rat CD59 swaps are based upon the observation that human CD59 inhibits rat C very poorly whereas rat CD59 inhibits human and rat C to similar degrees. Swap mutants in human CD59 containing the regions of rat CD59 responsible for activity and/or species selectivity will then behave like rat CD59 and inhibit rat (and human) C efficiently. These experiments have implicated the region around the groove between the two anti-parallel β-sheets on the upper face of the molecule in conferring the observed differences in cross-species activities in rat and human CD59. Of note, this region is almost identical to that identified in the point mutation studies described above as the likely active site of CD59. A similar approach has been adopted by Sims and co-workers who have made a series of swap mutants utilizing human and rabbit CD59 [1069]. This species combination was chosen because each of these CD59s is very inefficient at inhibiting C of the other species in the combination. Swap mutants identify a region between residues 42 and 58 of the human CD59 sequence as the main determinant of species selectivity, supporting the results obtained by Tomlinson and colleagues using rat/human chimaeras.

Identification of the complementary site(s) on C8 and C9 to which CD59 binds has also been the subject of intense research. Tomlinson and co-workers identified a peptide derived from the hinge region of C9 (residues 247–261) which bound CD59 on human erythrocytes and blocked the C-inhibitory activity, rendering the cells susceptible to C lysis [1070]. The peptide was just as effective as a CD59-blocking monoclonal antibody, suggesting that the peptide displayed a high affinity for CD59. No further data on this intriguing peptide have been published. As noted above, Sims and co-workers have reported the inability of rabbit C8 or C9 to interact with human CD59. They have searched for regions of poor sequence conservation between human and rabbit C8/C9 and created chimaeric C8 or C9 molecules in which these poorly conserved regions are transplanted from one species' C8/C9 to the other [1071–1073]. These elegant studies have effectively narrowed down the putative binding sites to short homologous regions in the α-chain of C8 (residues 334–385) and the b domain of C9 (residues 334–415).

*Other roles of CD59*

One of the earliest reports of the molecule now known as CD59 described a 20-kDa surface protein detected on erythrocytes by the monoclonal antibody H19 [1074]. The H19 antibody inhibited the binding of erythrocytes to T cells, suggesting that the H19 antigen (subsequently shown to be identical to CD59) was involved in cell adhesion. Evidence emerged indicating that CD59 was a ligand for CD2 on T cells, binding to a site on CD2 distinct from that utilized by the adhesion molecule CD58 [1075–1077]. An essential role for the N-linked carbohydrate in the binding of CD59 to CD2 has been suggested [1061]. However, others, using highly sensitive methods for the detection of binding events, have been unable to demonstrate any binding affinity of purified or recombinant CD2 for purified or recombinant CD59, making this an area of controversy [1078].

The homology of CD59 with the murine Ly-6 antigens which are known to be involved in T-cell activation, led workers to examine whether CD59 was involved in activation of human T cells [1079]. Indeed, cross-linking of CD59 was shown to provide a co-stimulus for T-cell activation. The mechanism by which this GPI-anchored molecule, tethered only to the outer lamella of the membrane and making no direct contact with the cell interior, mediates cell activation provided a puzzle which is still only partially solved. Horejsi and co-workers showed that CD59 on T-cell lines (and other nucleated cells) was co-associated with other GPI-anchored molecules in membrane clusters which also contained tyrosine kinases [277, 1080]. The evidence that these clusters are the signalling complex through which GPI-anchored molecules signal remains incomplete but compelling. CD59 on other nucleated cell types is also in clusters and indeed on many cell types can mediate events associated with activation. Evidence has recently been presented indicating that CD59 on nucleated cells is in the form of a homodimer and that this homodimerization is important in signalling through CD59 [1081].

On neutrophils, cross-linking of CD59 triggers calcium store release and generation of reactive oxygen species [699]. If a 'foreign' CD59 is incorporated into a human cell, it is at first randomly distributed on the membrane and incapable of signalling cell activation; however, within a few hours at 37°C the protein becomes clustered and acquires the capacity to signal, further implicating the clusters in signalling and showing that exogenously added molecules can enter the clusters [1082]. The acquisition of signalling capacity by a 'foreign' CD59 is difficult to reconcile with the concept of specific transmembrane signal-transducing molecules as an essential component of signalling through CD59.

Another possible 'alternative function' ascribed to CD59 was the inhibition of lysis by the T-cell cytolytic protein perforin. The logic for this suggestion was similar to that described above for HRF; perforin shares homology with the terminal complement components and some cell types are very resistant to perforin lysis. However, experiments comparing the protective effect of incorporated CD59 on lysis of cells by MAC and by purified perforin detected no inhibition of perforin lysis, effectively eliminating this possibility [1083].

## Other membrane factors regulating the MAC

In early reports on the isolation of CD59, some noted a capacity of the isolated protein to inhibit C5b-7 site formation [1016]. Others found no such activity, suggesting that this additional activity was due to a contaminant in the preparation. These same workers later identified the component involved as the major erythrocyte sialoglycoprotein glycophorin A [1084]. Purified glycophorin A inhibited in a reactive lysis system by restricting the binding of the C5b-7 complex. Abnormalities in glycosylation of glycophorin A have been implicated in the pathogenesis of a rare haemolytic disorder termed *hereditary erythroblastic multinuclearity with a positive acidified serum lysis test* (HEMPAS) [1085].

The observation that glycophorin A inhibited lysis by the MAC had actually been made at least a decade prior to these studies, making glycophorin A the first membrane regulator of the lytic pathway to be described [1086–1088]. Despite this long history, the physiological relevance of C inhibition by glycophorin A to erythrocyte survival in the circulation remains unproven. An interesting twist in the glycophorin story is the recent report that C5b-6 complexes bind glycophorin through sialic acid groups on the molecule and that desialylation of glycophorin reduces C5b6 binding and erythrocyte lysis [1089]. Purified glycophorin supplied in the fluid phase will thus inhibit C lysis by competing with membrane glycophorin for a limited supply of C5b6, but the membrane protein itself appears to enhance rather than inhibit MAC deposition and consequent lysis.

## RECOVERY FROM C MEMBRANE ATTACK

Since the earliest studies on C more than a century ago, it has been recognized that nucleated cells are in general much more difficult to kill with C than are erythocytes. Many possible reasons were suggested for this resistance, including low antigen density, low binding of specific antibody, lower capacity for binding early C components and influences of the cell cycle (reviewed in [1090]). In the late 1970s and early 1980s attention began to focus on the composition of the membrane and particularly the lipid content, continuing the theme whereby resistance was considered a 'passive' phenomenon rather than an active response of the target cell [1091–1093]. Nevertheless, circumstantial evidence was accumulating that the target played a more active role in resisting lysis. In the 1950s, Goldberg & Green had noted that nucleated cells attacked by C underwent major changes in architecture long before the membrane was ruptured [1094, 1095]. Others went on to show that it was possible to rescue cells attacked with an otherwise lethal dose of C by incubating with cyclic nucleotides, implying a signalled killing process [1096–1098].

We had chosen the neutrophil to examine the effects of C membrane attack on intracellular $Ca^{2+}$ levels. Neutrophils were extremely resistant to lysis by MAC but even at levels of attack which caused no lysis, large and transient increases in intracellular $Ca^{2+}$ were seen [1099, 1100]. The largest component of the $Ca^{2+}$ rise was due to influx, presumably through the MAC pore, beginning within seconds and lasting for around a minute. The transient nature of the $Ca^{2+}$ flux implied that the MAC pore was itself transient, provoking us to examine the fate of the MAC. Perhaps because of the serendipitous choice of cell type, it was relatively easy to demonstrate that MACs were rapidly cleared from the cell membrane in a $Ca^{2+}$-dependent manner, primarily by shedding [1099, 1100]. We termed this process 'recovery' and suggested that it was an important survival mechanism in nucleated cells. A comprehensive analysis of recovery in neutrophils demonstrated that it was possible to assemble as many as 25,000 MACs on the neutrophil surface without lysis and that once formed, MACs were rapidly shed from the cell membrane in a temperature- and $Ca^{2+}$-dependent manner, the clearance half-time being about 2 minutes at 37°C [171]. The process could be visualized by scanning and transmission electron microscopy and membrane vesicles densely decorated with MACs were present in supernatants from recovering neutrophils (Figure 4.11).

That this phenomenon was not unique to neutrophils became apparent from work in several other groups. Shin and co-workers showed the rapid disappearance of functional MACs from the membranes of Ehrlich tumour cells and provided evidence that the route of loss in this cell line was at least in part through endocytosis of complexes [1101–1103]. Sims & Wiedmer reported that platelets, which though enucleate are metabolically active, also resisted lysis by removing MAC on membrane vesicles [1104, 1105]. As with the neutrophil, recovery in both of these cell types was dependent on $Ca^{2+}$. Many other nucleated cell types and even fresh erythrocytes have

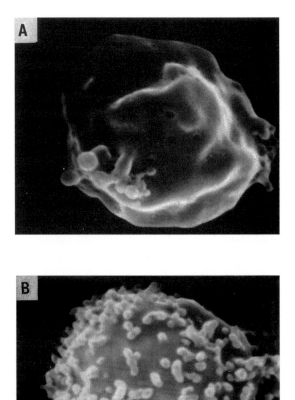

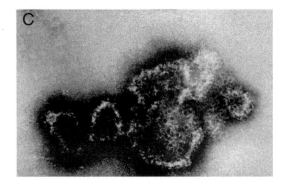

subsequently been shown to physically remove MACs from the membrane and recover from C attack (reviewed in [879, 1106]).

The mechanism underlying recovery by shedding or endocytosis of MAC remains poorly understood. In neutrophils and platelets there is evidence of selectivity in the shedding process – the membrane lipid and protein composition of the shed vesicles differs markedly from that of the parent cell [1105, 1107]. The data imply that MACs are either formed in, or migrate into, specialized microdomains in the cell membrane where they accumulate in high density prior to shedding. However, the nature of these microdomains is unclear. Lipid and protein analyses indicate that the shed vesicles are enriched in cholesterol and GPI-anchored molecules [1107], provoking the suggestion that the GPI-anchored inhibitors of the MAC (CD59, HRF) might be involved in the accumulation of MACs in GPI-rich patches and in creating an interface between the processes of resistance to and recovery from MAC attack [700, 1082].

There is limited evidence that recovery processes occur *in vivo*. In the cerebrospinal fluid of individuals with demyelinating disease we found membrane fragments derived from the myelin-producing cell, the oligodendrocyte, which were densely coated with MAC [1108]. Others have suggested that the vesicles or 'microparticles' shed by platelets during recovery from MAC attack *in vitro* are also generated *in vivo* and play an important role in coagulation [1109, 1110]. It seems likely that these rather poorly defined 'recovery' processes contribute a great deal to nucleated cell survival following C attack *in vivo*, perhaps conferring protection of importance equal to or even greater than that provided by the membrane regulators.

## CONCLUDING REMARKS

Although C activation is well-policed during the activation stages there is, nevertheless, a need for fail-safe mechanisms to prevent lysis of cells by

---

**Figure 4.11. Recovery from C membrane attack.**

Neutrophils exposed to non-lethal C attack respond by shedding membrane vesicles richly decorated with MACs.
**a.** Scanning electron micrograph of a quiescent neutrophil.
**b.** Scanning electron micrograph of a neutrophil 2 minutes after initiating C attack; note the appearance of abundant cell-surface projections.
**c.** Transmission electron micrograph of vesicles shed from C attacked neutrophil; numerous vesicles with diameters in the range 0.1–1 μm are present in the supernatant, which are densely covered with ring-like lesions with an appearance typical of the MAC.

MAC, formed either because of 'leakage' in control of the activation steps or because of direct bystander deposition on self-cells of C5b-7 complexes formed by activation of C on appropriate targets. Diffusion of C5b-7 away from the active site is regulated by the fluid-phase regulators S-protein and clusterin and any MAC which do form despite this fluid-phase control are, on most host cells, efficiently handled by CD59 and by the physical removal of MAC from the membrane. Lysis of host cells *in vivo* will thus occur only in exceptional circumstances, in the face of overwhelming C attack or where one or more of the regulators is in short supply on a particular cell type. Nevertheless, there may be substantial non-lethal consequences to the cell of C membrane attack which, while stopping short of immediate killing, will alter cell function and contribute to pathology.

*Chapter 5*

# DEFICIENCIES OF COMPLEMENT REGULATORS

## INTRODUCTION

Isolated deficiencies of each of the components of C occur with varying frequencies and degrees of severity. Isolated deficiencies of many of the fluid-phase and membrane C regulators have also been described and are the subject of this chapter (Table 5.1). Deficiencies of fluid-phase regulators of the activation pathways all cause loss of control of C activation to differing degrees. With the exception of deficiency of C1 inhibitor (C1inh), all are rare.

Table 5.1. *Deficiencies of C-regulatory proteins.*
Abbreviations: PNH, paroxysmal nocturnal haemoglobinuria; DAF, decay-accelerating factor; fH, factor H; fI, factor I; C4bp, C4b binding protein.

| Protein | Prevalence | Symptoms |
|---|---|---|
| *Fluid-phase* | | |
| C1inh | Hundreds | Angiedema |
| fH | 16 cases | Pyogenic infections |
| fI | 23 cases | Pyogenic infections |
| C4bp | 2 cases | Uncertain |
| | | |
| *Membrane* | | |
| PNH | Hundreds | Haemolysis, thrombosis |
| DAF | 4 families | Gastrointestinal tract symptoms |
| CD59 | 1 individual | Severe PNH-like illness |

## DEFICIENCY OF C1 INHIBITOR (C1inh)

Hereditary angiedema (HAE), caused by deficiency of C1inh, is by far the most frequent inherited C deficiency to present to the clinician. HAE is relatively common because clinical disease occurs in those heterozygous for the deficiency. This unusual feature was noted by William Osler in 1888 in a family study which provided the first clear description of the syndrome[1111]. The clinical syndrome is characterized by repeated episodes of painless, non-pruritic swelling of the skin and/or mucous membranes (Figure 5.1). Involvement of the gut mucosa causes attacks of severe abdominal pain which often accompany the cutaneous signs and may mimic a surgical emergency; involvement of the mucosa of the airways can cause laryngeal or tracheal obstruction and is potentially life-threatening[1112, 1113]. Attacks usually begin in infancy but some individuals present much later in life. Attacks tend to be milder and less frequent prior to puberty but increase in severity and frequency thereafter. The incidence is equal in males and females and the autosomal dominant mode of inheritance is usually obvious in families. A substantial proportion of cases will be the result of new mutations in the gene and lack a family history. The factors precipitating an attack remain uncertain but stress, infection and trauma (often trivial) have all been implicated. Attacks develop over about 24 hours and if left untreated will usually resolve over the next 3 or 4 days. Atypical attacks may last much longer and involve multiple regions of the body, either simultaneously or sequentially.

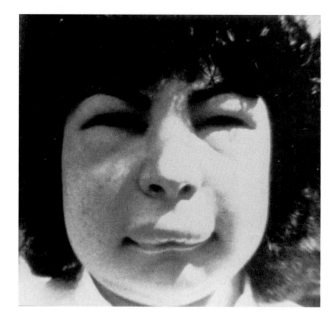

**Figure 5.1. Hereditary angiedema (HAE).**

Note marked swelling of upper lip and lower eye lids. Reproduced with permission from: *Clinical Immunology* (eds. J. Brostoff, G.K. Scadding, D. Male and I.M. Roitt); Gower Medical Publishing, 1991.

The structure and function of C1inh is fully discussed in Chapter 3 and will not be reiterated here except to emphasise that C1inh is a serine protease inhibitor (serpin) and is the only plasma regulator of activated C1. In the great majority (at least 85%) of HAE patients, plasma levels of C1inh measured by immunochemical methods are low, the defective gene producing no protein (Type I HAE; CRIM-negative). Plasma levels are typically less than 25% of normal rather than the 50% which might be predicted for a heterozygous deficiency; this is because the reduced synthesis rate cannot compensate for the normal or increased catabolic rate [219, 1114, 1115]. A minority (15% or less) of patients have normal or elevated levels of immunochemical C1inh but the bulk of the protein is functionally inactive, the product of the abnormal gene (Type II HAE; CRIM-positive). Mutations in these individuals are found at the catalytic site of C1inh, usually at the key P1 residue [1116]. In both type I and type II HAE the plasma level of C1 is normal because the actual amount of C1 activation occurring is not increased. However, the C1 convertases which are formed are inefficiently regulated, resulting in decreased plasma levels of C2 and C4 due to chronic, low-level activation of the classical pathway (CP). Levels of C1inh, C4 and C2 are all further reduced during acute attacks.

The *c1inh* gene has been mapped to chromosome 11 and the gene structure and normal gene sequence determined [157, 228]. Many different mutations in C1inh have been found in different kindreds. Most kindreds with Type II HAE have a point mutation at or near to the reactive site, the most common mutation being at the essential P1 residue Arg-444 [1116, 1117]. The resultant protein lacks antiprotease activity. In Type I HAE a variety of major alterations in the gene have been described, causing deletions or mutations which prevent transcription of the gene or give rise to truncated products which are often not secreted [1118, 1119]. A 'hot-spot' for mutations is found in a region close to the fourth exon which is extremely rich in tandem *Alu* repeats, a known cause of genetic instability [244, 245]. The molecular genetics of HAE are well reviewed elsewhere [1120].

The underlying cause of attacks is a profound deficiency of functional C1inh in the affected tissues, a consequence of the reduced plasma level and local consumption due to activation in the tissues of C and other proteolytic cascades which consume C1inh [219, 220]. Which of the many products of these proteolytic cascades is the main trigger to tissue oedema is still uncertain. Small peptides with powerful kinin activity derived from the carboxy terminus of C2b have been implicated. However, others have provided evidence implicating activation of the contact system with the local generation of bradykinin from high molecular weight kininogens as a major contributing factor (Figure 5.2) [1121]. There is no evidence that mast

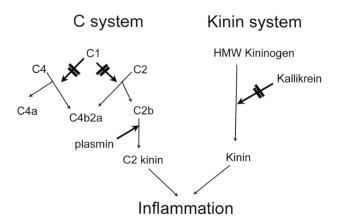

**Figure 5.2. Generation of kinins in HAE.**
C1inh regulates not only the C system but also several other plasma mediator systems such as the clotting and fibrinolytic systems and the kinin system. In deficiencies of C1inh, control in all these systems is disturbed. Bradykinins and other kinins are generated from cleavage of high molecular weight kininogen. A C2 kinin is generated by the action of plasmin on the C2b fragment of C2. The relative importance of these sources of kinins to pathology in HAE is the subject of debate. The points at which C1inh acts are indicated by double bars.

cell degranulation and histamine release play any role in the causation of oedema in HAE.

Therapy of acute attacks of HAE involves the rapid restoration of plasma C1inh levels by infusion either of fresh frozen plasma or purified C1inh preparations [1122, 1123]. The safety of C1inh concentrate has recently been improved by the introduction of a heat-treatment step which effectively eliminates the risk of transmission of hepatitis C [1124]. Occasionally, individuals treated with plasma as replacement therapy will experience an exacerbation of disease (due to increased supply of C and other substrates) but in most cases the acute attack resolves rapidly with replacement therapy. Anti-histamines, corticosteroids and adrenaline are not effective in acute HAE. Prophylactic therapy in HAE involve strategies to increase C1inh synthesis or to decrease consumption. In adult males, anabolic steroids are an effective therapy, causing an increased plasma level of C1inh through increased synthesis [1113, 1125, 1126]. Danazol is particularly effective and is widely used in HAE. The androgenic effects of the anabolic steroids complicate their use in women, although in most cases few problems are encountered. Use of anabolic steroids in prepubertal children is contraindicated because of their virilizing effects and because they may limit growth by causing premature closing of the bone epiphyses. The protease inhibitors ε-aminocaproic acid (EACA) and tranexamic acid reduce consumption of C1inh by inhibiting some of the proteases which might otherwise consume the molecule [1127–1129]. EACA can be used in males and females of any age, although efficacy is often poor and side-effects, notably nausea, frequently occur and limit the dose which can be given. Prophylactic therapy with C1inh given by regular infusions has recently been advocated for the more severe cases of HAE [1123].

Patients with disseminated malignancies, particularly lymphomas and leukaemias, occasionally develop a syndrome of angiedema which closely resembles HAE and has been termed Type I acquired angiedema (Type I AAE) [1130–1132]. Successful therapy of the underlying disease is almost always accompanied by resolution of the symptoms of angiedema, indicating that the tumour load is itself causal. Current evidence indicates that chronic, low-level activation of the CP on tumour cells causes consumption of C1inh and the CP components C1, C4 and C2. The reduced plasma level of C1 helps to distinguish acquired from hereditary angiedema. However, it has recently been shown that the majority of patients with angiedema secondary to leukaemia or lymphoma have autoantibodies against C1inh which catalyse the consumption of the protein (see below) [1133]. Acquired angiedema may also occur in the absence of overt malignancy. In some of these cases the underlying malignancy may become apparent months or even years after presentation. However, acquired angiedema may occur independent of malignancy, usually in association with non-organ-specific autoimmune diseases such as systemic lupus erythematosus or rheumatoid arthritis, but occasionally in the absence of detectable symptoms of autoimmunity (Type II AAE) [1134, 1135]. In autoimmune disease, the greatly increased immune complex load, by increasing consumption of C1inh, may contribute to disease but most patients with Type II AAE have autoantibodies against C1inh that render the protein inactive, often by holding the molecule in a configuration which renders it susceptible to cleavage by plasma proteases [1136, 1137]. Anti-C1inh antibodies (which may be paraproteins) and a cleaved form of C1inh with an apparent molecular mass of about 96 kDa are found in the plasma.

Acute attacks in both types of AAE often respond well to replacement therapy with C1inh. Plasmapheresis, by removing the causative autoantibody, may also be effective in the short term. The strategy for long-term therapy differs from that in HAE in that approaches aimed at increasing synthesis are unlikely to be helpful. Instead, therapy is targeted at the underlying disease, anti-tumour therapy in Type I AAE and immunosuppressive therapy (steroids or other appropriate agents) to reduce autoantibody production in Type II AAE.

## DEFICIENCIES OF FLUID-PHASE C3 CONVERTASE REGULATORS

The C3/C5 convertases in the CP and alternative pathway (AP) are controlled from the fluid phase by factor I (fI) in concert with C4bp and factor H (fH) respectively. Deficiencies of fH and fI have been reported and give rise to very similar clinical pictures; only two kindreds with C4bp deficiency have been reported to date.

### Deficiencies of fI and fH

The history of the discovery of fI and its deficiency go hand in hand. A 25-year-old male with Klinefelter's syndrome and a history of multiple and severe pyogenic infections was described by Rosen and co-workers in 1970 [776, 1138]. The clinical picture was suggestive of immunoglobulin deficiency. However, his immunoglobulin levels were normal, prompting a search for other immunological deficits. His serum haemolytic C activity was essentially zero and this was shown to be due to a profound deficiency of C3 and factor B. Further investigation revealed the presence of high amounts of C3b and other C3 breakdown products in the plasma, indicating that this was not a primary deficiency of C3 and/or fB but was probably caused by a missing regulator of C3 activation which was termed C3 inactivator (Figure 5.3). Investigation of the index patient allowed the characterization of the C3 inactivator as a plasma protein inhibitor of both activation pathways [1138, 1139] and it was later shown that infusion of the

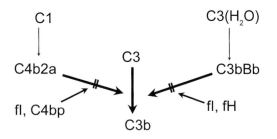

**Figure 5.3. C activation in deficiencies of fluid-phase C3 regulators.**
The activation pathways are held in check in the fluid phase by fI working with fH in the alternative pathway (AP) and C4bp in the classical pathway (CP). Deficiency of fI causes loss of control in both activation pathways with C activation and profound depletion of C3. Deficiency of fH causes loss of control only in the AP but the end result, C3 depletion, is the same.

purified C3 inactivator protein (hereafter termed fI) corrected the deficit in the patient[134].

Deficiency of fI is rare, only some 23 individuals from nine kindreds having been reported in the world literature[1140]. In all cases, profound secondary deficiency of C3 and fB occurred leaving the individual susceptible to pyogenic infections. The majority have presented with meningitis, either aseptic or due to *N. meningitidis* or *S. pneumoniae*[74, 1141–1143]. Less common presentations include infections of the respiratory tract or urinary tract, immune complex disease and, rarely, renal disease due to the presence of nephritic factors in the plasma (see below)[1140, 1144, 1145]. The laboratory findings include profoundly depressed complement hemolytic activity (CH50) in both activation pathways, with low levels of C3 and fB secondary to consumption. C2 levels are usually normal. The diagnosis can be confirmed by measurement of the plasma concentration of fI using an immunodiffusion or ELISA technique.

FI deficiency is inherited as an autosomal recessive; the gene for fI has been mapped to chromosome 4 and a restriction fragment length polymorphism (RFLP) in fI has been demonstrated in several kindreds. The molecular basis of the deficiency in two kindreds has recently been solved[1146]. One kindred (two siblings) was homozygous for a point mutation in the gene which caused the replacement His400>Leu while the other kindred was compound heterozygote for the above mutation and for a splice site mutation resulting in the deletion of the fifth exon of fI.

Whereas deficiency of fI results in uncontrolled activation of both activation pathways, deficiency of fH causes a selective loss of control in the AP alone (Figure 5.3). Nevertheless, the consequences are almost identical. Plasma levels of C3 and fB are profoundly reduced due to consumption, leaving the individual susceptible to pyogenic infections. Deficiency of fH is even less common that fI deficiency. The first case to be described was that of an 8-month-old Asian boy presenting with the haemolytic uraemic syndrome (described below)[1147]. The index case and his healthy 3-year-old brother had 5–10% of the normal plasma level of fH which was functionally active and thus must be considered to be incomplete deficiencies. The parents, who were first cousins, had approximately 50% of the normal plasma level of fH. The nature of the deficiency was comprehensively studied in an Algerian family resident in Germany. Two brothers in this family presented with severe glomerulonephritis with unusual pathological features in early life (4 months and 14 months). CH50, C3 and fB levels were profoundly depressed and a deficiency of fH was confirmed by immunochemical analysis. Like the first kindred, the deficiency was incomplete and the parents, again cousins, were heterozygous for the deficiency[1148].

To date a total of 16 homozygous fH-deficient individuals have been described from eight kindreds [377]. Most have presented with pyogenic infections or with meningococcal infections [375, 1149]. However, fH deficiency differs from deficiency of fI in that severe renal disease and/or lupus is often the major clinical feature [1142, 1150]. The hallmark renal lesion accompanying fH deficiency is membranoproliferative glomerulonephritis (MPGN) [1151]. Some individuals with fH deficiency have a syndrome characterized by intermittent attacks of severe intravascular haemolysis, thrombocytopenia and renal dysfunction, the haemolytic uraemic syndrome (HUS) [376]. The mechanism by which fH deficiency precipitates this complex syndrome is unclear. HUS is much more commonly associated with acute diarrhoeal illnesses caused by coliform bacteria and is a frequent cause of renal failure in children. The presence of this unusual syndrome in the absence of gastrointestinal infections, in association with a low C3 level and/or CH50, or in several members of a family, is very suggestive of an underlying fH deficiency and warrants thorough investigation.

The recent description of a strain of pigs with high incidence of fH deficiency has shed further light on the pathology in humans [1152]. Pigs of the Norwegian Yorkshire breed were known to have a very high incidence of piglet loss due to severe renal disease. It was shown that these animals developed hypocomplementaemia and fulminant MPGN Type II secondary to a subtotal deficiency of fH. In these pigs fH deficiency was inherited as a simple autosomal trait with complete penetrance. All affected animals developed hypocomplementaemia and lethal MPGN type II and died of renal failure in infancy. Renal damage was already present at birth in the majority of animals. Transfusion of normal pig plasma or purified pig fH was protective.

The gene for fH is close to the RCA cluster on chromosome 1 and deficiency of fH is inherited as an autosomal recessive. The molecular defects causing fH deficiency in the various kindreds have not yet been elucidated; however, a recent preliminary report has described two different mutations in the gene in patients with HUS and partial deficiencies in fH [1527].

Deficiencies of fI and fH cause profound secondary deficiencies of C3 and fB, eliminating C function distal to this level. Therapy of these conditions is thus extremely unsatisfactory. All that can be done is to guard against infection by the judicious use of prophylactic antibiotics and aggressively treat any infections that do occur. Despite these attentions, the morbidity and mortality associated with deficiencies of fI and fH remains high. The index fI deficiency case was successfully treated with partially purified fI [134]. Subsequent cases have been treated with some success using fresh frozen plasma as a source of fI; however, the half-life of fI *in vivo* is short (24–48 h), limiting the use of replacement therapy to the emergency situation. A potential problem of replacement therapy in the individual with a complete deficiency of any plasma protein is that the infused protein is seen as foreign and an immune response mounted[1153].

## Deficiency of C4bp

C4bp is the fluid-phase cofactor in regulation of the CP convertases. Deficiency of C4bp is extremely rare. A single kindred was reported where the index case (a 28-year-old caucasian female) presented with angiedema and symptoms suggestive of Behçet's disease[475]. She had a diminished C3 level and C haemolytic activity and further investigation revealed that her C4bp level was about 15% of normal. Family studies identified two relatives (father and sibling) who had similarly depressed C4bp levels but were healthy. A recent case report described a pregnant female presenting with purpura fulminans secondary to meningococcal infection who had a very low C4bp level[1154]. Other healthy family members were also reported to be deficient but details were not presented. In addition to its role in the C system, C4bp contributes to the clotting cascade by binding protein S through the C4bp β-chain and providing a reservoir of protein S. Deficiency of C4bp might therefore adversely affect not only the C system but also haemostasis. Indeed, investigation of protein S levels in the first C4bp-deficient kindred revealed that the amount of free protein S in the plasma was increased, although no abnormalities of coagulation were found[476].

## Deficiency of properdin

Properdin (P) is unique in that it is a positive regulator, acting to stabilize the alternative pathway convertases C3bBb and C3bBbC3b, thereby substantially increasing the half-life of these intrinsically unstable complexes and protecting against negative regulators. In the absence of P, the efficiency of AP activation is severely compromised. P deficiency was first described by Sjoholm and co-workers in a kindred presenting with severe, fulminant menigococcal infections in multiple family members[1155]. All affected family members were male and it has since become apparent that P deficiency is inherited in an X-linked manner, the only deficiency in the C system known to be inherited in this manner. Family studies have permitted the mapping of the gene for P near the centromere on the short arm of the X chromosome[840, 1156–1158]. The major clinical problem experienced by P-deficient males is infection with *Neisseria*, an indication of the important role of the AP in dealing with these organisms. Typically, individuals present with severe, fulminant meningococcal meningitis which may be rapidly fatal. Most present in their teens and infection is often with uncommon serogroups of *Neisseria*. In those who survive the initial infection recurrences are rare, probably because antibodies generated during the first exposure permit CP activation on meningococci in subsequent encounters. Although the risk to the individual of further life-threatening attacks is low, making the diagnosis of P deficiency in those presenting with severe meningococcal disease is of great importance. It will allow the identification of siblings and other family members who are deficient; these can then be immunized prophylactically to eliminate the high risk of fatal infection in the future. The recent development of microsatellite markers for properdin deficiency should aid carrier detection in the future [1159, 1160]. Some individuals with P deficiency present with infections other than those caused by *Neisseria* or with symptoms of immune complex disease [1161, 1162], but these are of minor significance in comparison with the risk of fatal meningitis.

P deficiency is not rare; although precise data on prevalence in various racial groups are not available, P deficiency is a frequent finding in retrospective studies of meningococcal infection and is certainly under-ascertained in the general population. Almost 100 cases in about 30 kindreds have been reported in the literature. The vast majority of cases have been in Caucasians; a few have been reported in Sephardic Jews but none to date in other racial groups [1142]. This may reflect a degree of ascertainment bias, but the absence of blacks and orientals from the list is very suggestive of true racial differences in incidence.

Three variants of P deficiency have been described, distinguished by the plasma level of P detected immunochemically. Type I-deficient individuals have no detectable protein, whereas Type II individuals have low but detectable (<10% of normal) plasma levels and Type III individuals have normal plasma levels (25 mg/l) of P. In Types II and III the protein is functionally abnormal, does not form the polymers which are essential to P action and does not stabilize fluid-phase AP convertases [819, 1163]. The presence of a dysfunctional protein present at low or normal plasma levels in many P-deficient individuals suggests that point mutations have occurred in the gene involving residues crucial to function. Indeed, recent gene sequencing studies have revealed the basis of the deficiency in kindreds with each type of deficiency. A single point mutation in exon 9 of the gene which causes an aminoacid substitution (Tyr→Asp) has been found in a kindred with Type III deficiency [1164]; a point mutation in exon 4, causing an Arg→Trp substitution has been detected in a kindred with Type II deficiency, and a kindred with Type I deficiency had a point mutation in exon 5 which introduced a stop codon early in the coding sequence [1165].

## Factors causing pathological stabilization of the C3 convertase

Natural and accelerated decay of the C3 convertases are important facets of control. Any factor causing abnormal stabilization of the convertase is likely to cause pathology. Between 10% and 30% of individuals presenting with membranoproliferative glomerulonephritis (MPGN) have in their plasma an autoantibody against the convertase termed nephritic factor (NeF). NeF may bind and stabilize the C3bBb complex (termed C3NeF) or the C4b2a complex (C4NeF) and both may coexist in the same patient [1166]. The complex so formed is resistant to the actions of the natural regulators and thus persists, consumes C3 and generates a state of secondary C3 deficiency [1167].

Many other conditions apart from MPGN can be associated with the presence of NeF; poststreptococcal glomerulonephritis [1168]; SLE [570]; urticarial vasculitis [1169]; partial lipodystrophy [1170]. In all cases, hypocomplementaemia is a feature. The syndrome of partial lipodystrophy is worthy of further mention. Individuals present with profound loss of subcutaneous fat from the face and extremities, usually in association with renal disease (Figure 5.4). The reasons for the association between C, NeF and fat loss have remained obscure but recent evidence that fat cells generate an endogenous C system has provoked the intriguing suggestion that C3 convertases stabilized on the fat cell suface by NeF cause fat cell lysis [1171].

## DEFICIENCIES OF FLUID-PHASE MEMBRANE ATTACK PATHWAY REGULATORS

Two plasma proteins, S-protein and clusterin, are purported to regulate the membrane attack pathway from the fluid phase. These proteins, described at length in Chapter 4, are multifunctional proteins which, together with many other plasma constituents, have a limited role in restricting the binding of the C5b-7 complex to the target membrane. Deficiencies of S-protein or of clusterin have not been described. Two mutually exclusive conclusions can be drawn from this fact: either the proteins are indispensable and deficiency is always lethal *in utero* or the proteins are of minor importance and deficiency, because it is symptomless, goes unnoticed. Our bias is towards the former view – that the proteins (particularly clusterin) perform vital roles but that these vital roles are not those performed in

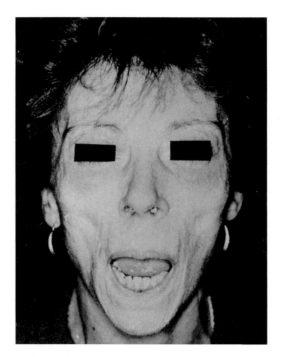

**Figure 5.4. Partial lipodystrophy.**
The wasted appearance typical of this syndrome is due to the profound loss of sub-cutaneous fat from the face (shown here) and upper torso, often with relative sparing of fat deposits elsewhere. Reproduced with permission from Font *et al.* (1990) *Journal of the American Academy of Dermatology*, **22**; 338.

relation to the C system. S-protein-deficient mice have recently been produced by gene targeting [933]. The homozygous S-protein-deficient mice display no detectable phenotype, develop normally and are fully fertile, indicating that, at least in mice, S-protein is not essential for survival.

## DEFICIENCIES OF MEMBRANE REGULATORS OF C

As described comprehensively elsewhere in this volume, proteins are pre-sent on self-cells to restrict activation of C by interfering with the system at multiple stages. Assembly and stability of the convertases of the activa-tion pathways is controlled by decay-accelerating factor (DAF; CD55), membrane cofactor protein (MCP; CD46) and complement receptor 1 (CR1; CD35) and formation of the membrane attack complex is regulated by homologous restriction factor (HRF) and CD59. Isolated complete deficiencies of individual membrane regulators are rare.

## Paroxysmal nocturnal haemoglobinuria

The commonest syndrome to be associated with deficiencies of membrane regulators is *paroxysmal nocturnal haemoglobinuria* (PNH) in which the synthesis of glycosyl phosphatidylinositol (GPI) membrane anchors is defective in a clone of circulating cells, resulting in a deficiency on these cells of all proteins linked to the membrane in this manner (Table 5.2). The most striking feature of the clinical picture in PNH is haemolysis causing haemoglobinuria, which is particularly noticeable first thing in

**Table 5.2. *GPI-anchored proteins on human cells.***
The list is not exhaustive but indicates the major groups of proteins linked by GPI anchors to the surface of human cells.

| Group | Protein | Role |
|---|---|---|
| Hydrolytic enzymes | Alkaline phosphatase | Ectophosphatase |
| | 5′ nucleotidase (CD73) | Nucleotide salvage |
| | Acetylcholinesterase | Breakdown of acetylcholine |
| | Aminopeptidase P | Peptidase in kidney |
| | Lipoprotein lipase | Lipid handling in adipose tissue |
| C regulators | Decay-accelerating factor (DAF; CD55) | Inhibits C3 convertase |
| | Homologous restriction factor (HRF) | Inhibits membrane attack complex (MAC) |
| | CD59 | Inhibits MAC |
| Cell adhesion molecules | NCAM (CD56) | Neuronal adhesion molecule |
| | Heparan sulphate | Cell matrix adhesion |
| Other immune molecules | CD14 | Receptor for LPS/LBP |
| | FcγRIII (CD16) | Receptor for IgG (only neutrophil form GPI) |
| | HSA (CD24) | B cell regulation? |
| | Blast-1 (CD48) | Ligand for CD2? |
| | LFA-3 (CD58) | Ligand for CD2 |
| | Campath-1 (CD52) | Lymphocyte (target in therapy) |
| Other receptors | uPAR (CD87) | Receptor for uPA |
| | Folate receptor | Folate uptake |
| Others | Carcinoembryonic Ag | Tumour marker (cell adhesion) |
| | Scrapie prion protein | Physiological role uncertain; role in spongiform encephalopathies |
| | Thy-1 (CD90) | Unknown; thymocytes, brain |

the morning, hence the name. Haemoglobinuria is most obvious in the morning because the urine is most concentrated at this time and also because haemolysis may be increased overnight due to the mild respiratory acidosis which commonly occurs during the hours of sleep. Blood cells other than erythrocytes are also involved; abnormal activation of neutrophils and platelets can cause thrombotic episodes which may be life-threatening.

PNH is a clonal disease in which a single haemopoietic stem cell mutates in a manner which causes the loss of the capacity to synthesis GPI anchors; the mutation arises *de novo* in the majority of cases. For reasons which remain unclear, this deficit confers upon the mutant cell a significant growth advantage, causing a clone of GPI-deficient stem cells and their progeny to arise and expand to the point where they may become the dominant cells in the circulation [1010, 1172]. Because of the pluripotent nature of the stem cell, GPI-deficient erythrocytes, platelets and leukocytes are all produced. The percentage of circulating cells derived from the PNH clone varies enormously – from a few percent to all cells. Individuals with aplastic anaemia in which blood cell numbers decline precipitously are occasionally 'rescued' by the emergence of a PNH clone which provides circulating cells, all of which are abnormal but nevertheless perform vital functions. The uncontrolled expansion of the GPI-deficient clone is typical of a premalignant state and a significant proportion of PNH patients will eventually develop haemopoietic malignancies, notably acute myeloblastic leukaemia.

Kinoshita and others have recently unravelled the molecular basis of PNH [1009, 1010, 1173]. In all cases so far studied the mutation is in the gene for an enzyme essential to the synthesis of the GPI anchor, which is termed PIG-A (Figure 5.5). Many different mutations in this gene have been described but all result in defective or absent PIG-A and a failure of anchor synthesis at this stage. The frequency with which mutations in this gene occur and give rise to PNH is in part explained by its location on the

**Figure 5.5. Pathway of GPI anchor biosynthesis.**

Panel A shows the essential core structure of the GPI anchor which consists of a molecule of phosphatidylinositol (PI), a glycan core consisting of glucosamine (Glu) and three mannose (Man) moieties and an ethanolamine (Etn) moiety which attaches to the carboxy terminus of the protein.

Panel B summarizes the key steps in assembly of the GPI anchor. The precise point at which the enzyme encoded by the *pig-A* gene acts in the pathway is not defined, but the block in anchor biosynthesis in PNH patients occurs early at or close to the point of addition of GlcNAc to PI. Additional abbreviations: PE, phosphatidylinositol; Ac, acetyl; GlcN, glucosamine; Dol, dolichol.

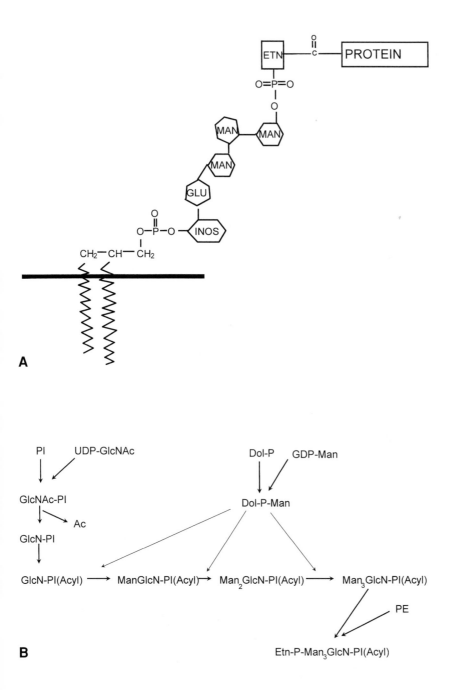

**A**

**B**

X chromosome. This means that *pig-A* is effectively a single copy gene in both sexes – only one X chromosome present in the stem cell in males and only one functional copy due to X-inactivation in females. A point mutation at an important site in this 'single copy' gene will thus cause complete deficiency of the PIG-A enzyme.

The list of proteins anchored through GPI and hence absent on PNH cells is long (Table 5.2) [680]. Among them are three membrane regulators of C: DAF, HRF and CD59. Absence of these proteins renders all the circulating cells highly susceptible to C damage. Erythrocytes are particularly susceptible because human erythrocytes, unlike those in other species, do not express MCP; in PNH they are therefore completely without protection (CR1 on erythrocytes is present at low copy number and contributes little to protection of the cell from C). The physiological 'tick over' activation of C deposits small amounts of C3 convertases on erythrocytes which on normal cells would be efficiently regulated but in PNH initiates further activation of C which can lead to haemolysis. The classical test for PNH, the Ham test, exploits the exquisite sensitivity of PNH erythrocytes to C lysis; test E are exposed to serum which has been acidified to enhance C activation and particularly the generation of C5b6 [1174, 1175]. The resultant degree of haemolysis correlates with the percentage of PNH cells in the test sample [1176]. This venerable test is now rapidly being replaced by the direct assessment of the levels of GPI-anchored molecules on blood cells using specific antibodies (Figure 5.6) [1177].

Although the C sensitivity of PNH erythrocytes is the cardinal clinical sign, susceptibility of other circulating cells to C damage is of more concern. Platelets and neutrophils lacking DAF and CD59 are less easily lysed than erythrocytes but are damaged by C, leading to activation of the cells and increased adhesiveness. Sticky platelets and neutrophils will form microthrombi and block small blood vessels in organs such as the brain, lungs and kidney. Stroke is the major cause of morbidity in PNH.

---

**Figure 5.6. CD59 deficiency on paroxysmal nocturnal haemoglobinuria (PNH) erythrocytes assessed by flow cytometry.**

Erythrocytes were stained for CD59 using the monoclonal anti-CD59 antibody BRIC229 followed by FITC-conjugated second antibody and analysed by flow cytometry. Cell number is on the Y axis and relative fluorescence intensity on the X axis. Results from two patients with PNH (A,B) and from a normal individual are shown. Patient A has more than 90% of cells CD59-ve whereas erythrocytes in patient B are approximately 21% CD59-ve. Using the same gates, normal erythrocytes are 97% CD59+ve (C).

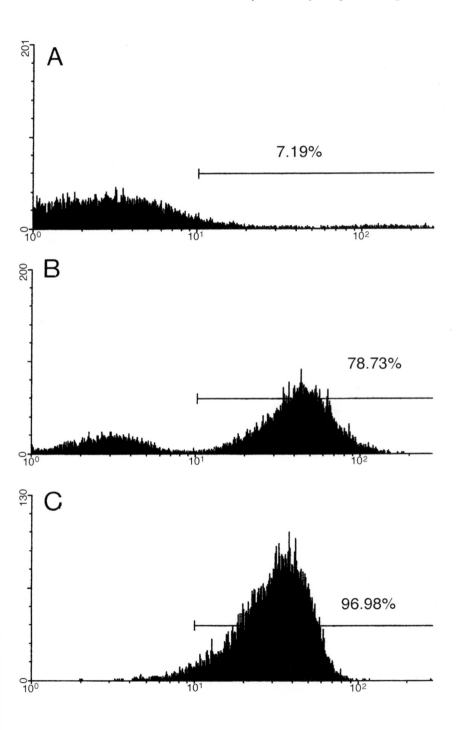

Treatment options in PNH are limited. Bone marrow transplantation has been used successfully in a few cases but is itself a life-threatening procedure which can only be contemplated in the most severe of cases[1176]. Androgens enhance the rate of production of cells from the bone marrow and improve symptoms in many individuals; steroids are beneficial in some; transfusion therapy can be useful but is often complicated by increased haemolysis of the patient's own cells.

## Isolated deficiency of DAF

Isolated deficiency of DAF has been described in a total of four families[1178–1180]. All had an unusual blood group phenotype known as *Inab*, an inherited null allele in the Cromer blood group, and were ascertained in studies of this phenotype. It was subsequently recognised that all were deficient in DAF and that the Cromer antigens resided on DAF[1179]. None of the proposita had symptoms suggestive of PNH although DAF was completely absent on all circulating cells and all other tissues, indicating that absence of the regulators of the membrane attack pathway is of more relevance to C susceptibility in PNH[1181, 1182]. In three of the four proposita a history of a poorly defined chronic intestinal disorder was obtained, provoking the suggestion that DAF might be involved in homeostasis in the gut. No evidence in support of this suggestion has yet emerged. In one of the families a point mutation in the *daf* gene has been found which introduces a stop codon early in the coding sequence[1183]. A second Cromer phenotype termed Dr(a-) is associated with greatly reduced expression of DAF in the absence of symptoms. The underlying genetic defect in this phenotype has also been unravelled[1184].

A high frequency of 'acquired' deficiency of DAF on malignant cells in non-Hodgkin's lymphoma has been reported[1185]. The molecular basis and functional consequences of this acquired deficiency are unclear.

## Deficiency of MCP

No individuals deficient in MCP have been described. Again we face the conundrum expressed earlier – is this because MCP is essential and deficiencies are incompatible with survival or because MCP is unimportant and deficiencies go unnoticed? MCP has been implicated as important in the reproductive system and in placental function (Chapter 6) and these functions may be vital. Gene knockouts of MCP in rodents may provide an answer to the conundrum, but given the species differences that are apparent in membrane regulators of the activation pathways (Chapter 9), extrapolation from the rodent knockout to man may be misleading.

# Deficiency of CR1

Deficiency of CR1 is an area of controversy. There are no reports of total, global deficiency of CR1 but it has long been recognised that individuals with immune complex diseases such as systemic lupus erythematosus (SLE) have reduced expression of CR1 on erythrocytes [1186, 1187]. The controversy has centred upon whether this is cause or effect. The early literature made the case for a relative deficiency in CR1 (very low CR1 expression on erythrocytes) being a causative factor in SLE by reducing the capacity of the cells to handle immune complexes. Although there is some evidence for a genetic component in the expression of low numbers of CR1 [564], it is now clear that low erythrocyte CR1 in SLE is primarily a consequence of the greatly increased immune complex load, a proportion of the CR1 being lost in the process of immune complex carriage and transfer to cells of the reticuloendothelial system [565, 568]. Perhaps most convincing is the demonstration that erythrocytes bearing normal CR1 numbers will, when transfused into an SLE patient, gradually lose CR1 from the membrane [574, 1188]. Acquired loss of CR1 from erythrocytes and other cells may also occur in other autoimmune diseases and in unrelated disorders. In AIDS it has been reported that erythrocyte CR1 expression is reduced, a consequence of enhanced proteolytic cleavage of CR1 [1189]. In the myelodyspastic syndrome reduced expression of CR1 on neutrophils has been described, although the mechanism is unknown [1190].

The Knops system antigens are a group of erythrocyte antigens identified in blood transfusion laboratories because of the common occurrence of anti-Knops antibodies in cross-matching; these antigens are resident on erythrocyte CR1 [1191–1193]. Null phenotypes for these antigens had been described, suggesting that erythrocytes in these individuals were CR1-negative. However, it has been clearly demonstrated using monoclonal anti-CR1 antibodies that the Knops-null phenotype correlates with low CR1 number, usually less than 100 per erythrocyte, rather than complete deficiency [1194]. The antibodies in patient sera can be efficiently neutralized using sCR1, providing a rapid means of identifying anti-Knops antibodies [1195].

# Isolated deficiency of CD59

Isolated deficiency of CD59 has been described in a single individual. The case report from Japan described a 22-year-old male who had been given the presumptive diagnosis of PNH at the unusually young age of 13 years [1052]. He had continued to have episodes of haemolysis and severe thrombotic incidents, including cerebral infarction which had left him with permanent neurological sequelae. He was investigated further because of his young age and the severity of his symptoms and it was shown that his circulating cells expressed normal levels of GPI-anchored molecules (eliminating the diagnosis of PNH) with the sole exception of CD59. Examination of CD59 expression in other tissues confirmed that this was a global deficiency not restricted to haemopoietic cells. The deficiency was shown to be due to single nucleotide deletions in the *cd59* gene which placed the gene product out of frame and introduced a stop codon early in the coding sequence[1053]. Two deletions were present, the first in the triplet encoding amino acid 16 of the mature protein and the second in the triplet encoding amino acid 96. The first deletion introduced a stop codon at position 54. The parents were cousins, were heterozygous for this first mutation and expressed half-normal levels of CD59 on circulating cells. One sibling was also heterozygous for the deficiency. No protein product of the truncated mRNA was detected in the patient.

This single case convincingly shows that deficiency of CD59 on PNH cells is the cause of erythrocyte susceptibility to haemolysis and is likely also to be the major factor rendering PNH neutrophils and platelets susceptible to activation by C. A confounding factor is that endothelia in this individual are also deficient in CD59 and would thus be expected to be susceptible to C damage. The observed thromboses are thus likely to be due to a combination of neutrophil/platelet activation and endothelial damage/activation.

In some individuals with leukaemia or lymphoma, a deficiency of CD59 and/or DAF occurs on the tumour cells[1185, 1196]. Other GPI-anchored proteins are expressed on the diseased cells, indicating that the deficit is not in the anchoring machinery. A recent study has examined expression of CD59 and DAF on leukaemic cell lines *in vitro*; these lines show very variable patterns of expression of the two regulators and commonly lack expression of one or both of these molecules[1197]. In the majority of leukaemic lines examined, the lack of expression was found to be due to the absence of specific mRNA for the proteins. However, some leukaemic cell lines do have a primary defect in GPI anchoring, indicating that this may also occur in some leukaemias[1197, 1198]. Cell lines deficient in the GPI-anchored C regulators have greatly increased susceptibility to lysis *in vitro* when compared to similar GPI-sufficient lines[1198]. No assessment of the consequences for the individual patient of deficiency of C regulators on the leukaemic clone has been made.

## CONCLUDING REMARKS

Deficiencies of C regulators are, with a few notable exceptions, very rare. However, the symptoms occurring in these rare disorders provide clues to the roles and relative importance of the various inhibitors *in vivo*. A large investment is currently being made to generate 'knockout' mice deficient in the murine analogues of the various regulators of C. The phenotypes arising in these animals will provide additional information on the likely roles of the regulators in man and may reveal hitherto unsuspected roles for particular molecules. The results of these ongoing experiments are eagerly awaited.

# Chapter 6

# COMPLEMENT REGULATION IN THE REPRODUCTIVE SYSTEM

## INTRODUCTION

The reproductive system represents a unique situation in which cells expressing foreign antigens are routinely introduced and must survive the threat of immune attack in the recipient. Specifically, the presence of foreign antigens on the surface of sperm and on cells of fetal tissue poses a threat to successful fertilization and fetal development due to the risk of targeting by the female immune system. In order to prevent this, a multitude of defence strategies are implemented such that these cells are isolated or protected from immune attack. The first strategy for defence is the lack of expression of HLA class I or class II molecules on the surface of sperm or cells of the villous trophoblast, such that these allogeneic cells are not recognised by host T cells. Secondly, seminal fluid contains a number of factors that mediate non-specific immunosuppressive effects (reviewed in [1199]). These include prostaglandins (PGE and 19-hydroxy PGE) which result in suppression of both lymphocyte proliferation and activation and NK cell cytotoxicity. Further immunosuppressive effects are achieved by the presence in human seminal fluid of factors such as transforming growth factor-β (TGFβ) and soluble receptors for both IgG and tumour necrosis factor-α (TNFα), which are either directly antiinflammatory or bind and neutralize inflammatory agents. The following chapter reviews the roles played by C inhibitors in reproduction, including contributions to gamete survival, protection of fetal cells at the interface with the maternal immune system and in the fertilization process itself[1200, 1201].

## ROLE OF C INHIBITORS IN THE PROTECTION OF SPERM

Antibodies specific for sperm have been identified in some individuals in plasma, in seminal fluid and also in cervical and follicular fluid [1202–1205].

This, in combination with the presence of a functional C system in the female reproductive tract, puts sperm at risk of C-mediated damage, lysis or immobilization[1206]. The production of anti-sperm antibodies has been linked to infertility, although it is unclear whether this is a consequence of C activation, or due to other factors such as blockade of the sperm–oocyte interaction[1207–1210]. Protection of sperm from C in the female reproductive tract is provided by 'bathing' the sperm in C inhibitors, present at high concentration in seminal plasma, and by the expression on the sperm surface of multiple C regulators which prevent activation and functional membrane attack complex (MAC) formation (Table 6.1). C regulators are also produced by cells of the female reproductive tissues.

Early studies indicated that seminal plasma contained C-regulatory activity sufficient to protect erythrocytes and bacteria from serum C attack, and that the C-inhibitory activity was present in seminal plasma samples of prostatic, vesicular and epididymal/testicular origin[1211–1213]. The most abundant C inhibitor in seminal plasma is clusterin; concentrations of between 250 µg/ml and 15 mg/ml have been reported[982, 983, 1214]. S-protein is reported to be also present at concentrations of 1 mg/ml[1215]. Clusterin was initially described as the major secreted product of rat Sertoli cells, then termed sulphated glycoprotein-2 (SGP-2); and was also found in ram seminal plasma where it constituted 18% of the total protein[953, 1216]. The identity of this major component of rat and ram seminal plasma with the human protein clusterin involved in C regulation was realised in 1989 when the human clusterin cDNA sequence was reported and demonstrated to be highly homologous to rat SGP-2[982, 983]. The structure and function of clusterin (also known as SP-40,40) is described in detail in Chapter 4. It is a MAC-inhibitory protein which binds to C5b-7 and prevents insertion of MAC into membranes. Clusterin has powerful cell adhesion properties and has many important roles outside of the C system. Indeed, it has been suggested that clusterin may have roles in seminal plasma other than C inhibition and levels of this protein have been correlated with successful fertilization[1214].

Other C regulatory proteins (CRP) in seminal plasma include those normally associated with the plasma membrane: membrane cofactor protein (MCP), decay-accelerating factor (DAF) and CD59. MCP and DAF are regulators of the C activation pathways and function by dissociating the C3/C5 convertases (DAF) or by acting as a cofactor for the cleavage and inactivation of C3b (MCP). CD59 is an inhibitor of the terminal pathway which binds to C8 in the forming membrane attack complex and prevents insertion and polymerization of C9. These inhibitors are discussed in depth elsewhere (Chapters 3 and 4). The 'soluble' forms of these membrane CRP found in seminal plasma are predominantly

**Table 6.1.** *Expression of CRP in the reproductive system.*

| | Distribution | Levels of expression | Comments |
|---|---|---|---|
| *Spermatozoa* | | | |
| CD59 | Global | High | Identical size to erythrocyte CD59 |
| DAF | Global | Weak | 10 kDa smaller than leukocyte DAF due to decrease in O-glycosylation |
| MCP | Inner acrosomal membrane | High | 10–20 kDa smaller than leukocyte MCP due to lack of O-glycosylation; may be expression of atypical truncated isoforms |
| Clusterin ⎫<br>S-protein ⎬ | Surface | | Probably adsorbed from seminal plasma, although spermatozoa may express S-protein |
| *Placenta* | | | |
| CD59 | Syncytiotrophoblast, cytotrophoblast, extra-villous cytotrophoblast | High throughout pregnancy | |
| DAF | Syncytiotrophoblast, cytotrophoblast, extra-villous cytotrophoblast | Moderate, expression increases during pregnancy | |
| MCP | Syncytiotrophoblast, cytotrophoblast, extra-villous cytotrophoblast | High throughout pregnancy | |
| *Seminal plasma* | | | |
| CD59 | Prostasome associated, also found in 'hydrophobic aggregates' | 20–200 µg/ml | |
| DAF | Prostasome associated, also found in 'hydrophobic aggregates' | 1 µg/ml | 65 kDa; heavy O-glycosylation, contrasts with sperm DAF |
| MCP | Prostasome associated | 250–700 ng/ml | 60 kDa; heavy O-glycosylation, contrasts with sperm MCP |
| Clusterin | Soluble | 0.25–15 mg/ml | |
| S-protein | Soluble | Up to 1 mg/ml | |

associated with small vesicular organelles called prostasomes derived from prostatic epithelium, or with as yet undefined structures described as 'hydrophobic aggregates', containing a high protein/lipid ratio [736, 737, 759, 1217, 1218]. CD59 is present in a GPI-anchored, prostasome-associated form in cell-free seminal plasma at a concentration of between 20 and 200 µg/ml and appears biochemically identical to erythrocyte CD59 [759]. Binding of prostasomes to sperm can 'deliver' CD59 to the cell surface, thereby increasing protection of sperm from C attack [759]. MCP has also been detected in seminal plasma at concentrations of 250–700 ng/ml, and is also associated with prostasomes [736–738, 740, 1218, 1219]. Interestingly, incubation of prostasomes with rat erythrocytes results in uptake by the cells not only of the GPI-anchored proteins but also of transmembrane MCP, indicating that transfer is probably due to association or fusion of whole prostasomes with cells rather than the 'painting' phenomenon whereby GPI-anchored proteins readily transfer between membranes [1218]. Prostasome-associated MCP has a molecular weight of 60 kDa and is heavily glycosylated, in contrast to the predominant form found on the sperm surface (described below). DAF is also present in seminal plasma but only at a concentration of about 1 µg/ml [1218]. In common with CD59, it is located on prostasomes and in the 'hydrophobic aggregates'; seminal plasma DAF closely resembles the erythrocyte protein in terms of molecular weight. The cell sources of the seminal plasma CRP are not well understood. However, studies in vasectomized males have demonstrated that as much as 50% of the various CRP are derived from points distal to the point of division of the ductus.

All membrane-associated CRP (excluding CR1) can also be demonstrated on the surface of sperm by immunostaining or immunoprecipitation (Figure 6.1) [662, 760, 1220]. MCP was the first C inhibitor to be demonstrated on spermatozoa. Several monoclonal antibodies were described with specificity for a widely expressed antigen located not only on leukocytes, but also on cells of the human trophoblast at the fetomaternal interface and the acrosomal region of sperm, and hence initially termed trophoblast-leukocyte common antigen (TLX ag) [709–711, 714]. Its identity with the C regulator, MCP, was realised when cross-reactivity of antibodies between antigen systems was demonstrated and sequence data became available [147, 717, 718]. These and other studies demonstrated that MCP on spermatozoa is located only on the inner acrosomal membrane (Figure 6.2) [760]. In contrast, CD59 is highly expressed on sperm, being present on all portions of the cell [662, 760, 1221]. The distribution of DAF on sperm has attracted some discussion with some claiming expression only on the acrosome whereas others claimed global expression [662, 663, 760, 1222]. The controversy is not without importance because if DAF were absent from

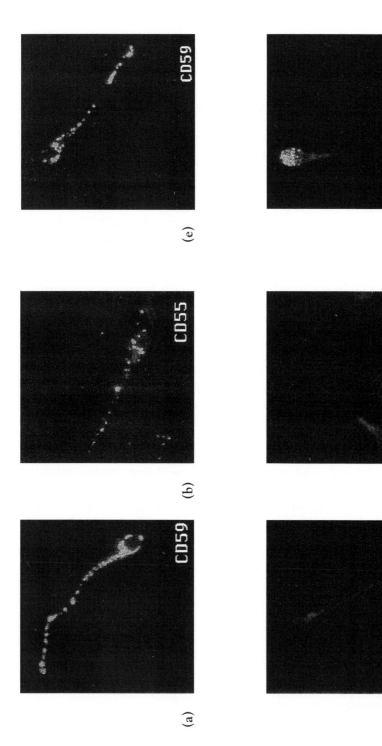

the membrane, sperm would have no endogenous protection against C activation. Our results indicate that DAF is expressed, though weakly in comparison with CD59, on all parts of the cell (Figure 6.1). Sperm MCP is exposed only following the acrosomal reaction, typically several hours following ejaculation. Sperm MCP can therefore play no role in protection of sperm in the female genital tract; expression of CD59 and DAF must be sufficient to provide protection from C attack. MCP may have an important role when exposed on the acrosome in proximity to the oocyte surface, either for C regulation or an as yet undefined role, possibly in the fertilization process itself (see below).

Sperm CD59 has a molecular weight identical to that on other cells and can protect sperm from the damaging effects of C. Incubation of sperm with neutralizing anti-CD59 antibodies and human serum results in immobilization and lysis, whereas sperm with functional CD59 are protected [662]. However, MCP and DAF on sperm are structurally different from their counterparts on cells elsewhere in the body. Spermatozoal MCP is 10–20 kDa smaller than leukocyte MCP due to an apparent lack of O-linked glycosylation [747, 748, 760, 1219]. A cDNA clone encoding MCP has been isolated from a testis cDNA library and shown to encode all four SCR domains, STP-C and CYT-2; all potential O- and N-linked glycosylation sites were normal [748]. Interestingly two unusual clones encoding MCP have been obtained either from a testis cDNA library or by RT-PCR from spermatozoan RNA. The two forms of MCP encoded by these clones were novel splice variants of MCP; both had a truncated transmembrane segment and a unique cytoplasmic tail [742, 749]. Due to differential use of splice sites in exon 12, their amino acid sequence varied slightly in that one isoform had an additional nine amino acids at the

---

**Figure 6.1. Location of complement regulatory proteins (CRP) on the surface of spermatozoa.**

CRP were located on the sperm surface using primary antibodies specific for the inhibitor and fluorescently-labelled secondary antibodies. The left-hand panel **(a–d)** illustrates staining of acrosome-intact spermatozoa, and the right-hand panel **(e, f)** illustrates staining of acrosome-reacted spermatozoa. CD59 is expressed in abundance on the surface of acrosome-intact spermatozoa **(a)**. DAF (CD55) is also expressed on the spermatozoal surface, although the level of expression is lower than that of CD59 **(b)**. MCP (CD46) is not found on the surface of acrosome-intact spermatozoa **(c)**. In contrast, high levels of MCP expression can be found on the inner acrosomal membrane which is exposed only following the acrosome reaction **(f)**. Levels of CD59 expression are similar on acrosome-intact and -reacted spermatozoa **(e)**. Panel **(d)**, secondary antibody only as control. Photographs courtesy of Rhian Morgan, Department of Medical Biochemistry, University of Wales College of Medicine.

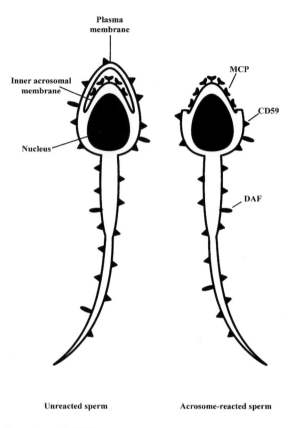

**Figure 6.2. Location of CRP on spermatozoa.**
High levels of CRP expression on the surface of spermatozoa protect them from C attack in the female genital tract. CD59 is abundantly expressed on all portions of the cell, including the acrosomal membrane (not shown here). DAF is also globally expressed, but at much lower levels than CD59. MCP is located only on the inner acrosomal membrane and is exposed following the acrosomal reaction.

carboxy-terminal end of the transmembrane segment. The possibility in both of these studies that the cDNA was derived from contaminating cells rather than spermatozoa is difficult to eliminate so this data should be treated with caution. Sperm DAF is also smaller than that on other cells, having a molecular weight of only 55 kDa, 10 kDa smaller than that on erythrocytes [663, 760]. The reason for the lower molecular weight of sperm DAF is unclear but may be due to reduced O-glycosylation [663]. Whilst both sperm MCP and DAF have little or no O-linked glycosylation, both inhibitors have been shown to be functional [663, 747].

# ROLE OF C REGULATORS IN THE PROTECTION OF THE FETUS

It is clear that regulation of C at the feto-maternal interface is of crucial importance for a successful pregnancy. Whilst C activation products can be detected in normal placenta, suggesting that low-level C activation occurs in all pregnancies, abnormally high levels of C activation have been associated with placental damage in conditions such as pre-eclampsia, although it is not proven that C is a primary cause of damage in these conditions [1223–1225]. Analysis of inhibitor expression on the oocyte and preimplantation embryo demonstrates that CRP are present on the surface. DAF and CD59 are both abundantly expressed on unfertilized oocytes but the presence of MCP is controversial [1222, 1226]. Several studies describe the presence of MCP on the unfertilized oocyte [280, 1222, 1227, 1228], whereas others report its absence [279, 1229] or its appearance only from the four/eight-cell stage, coinciding with gene expression [1226]. The discrepancies in the literature are unresolved but do not appear to relate to different fixation methodologies.

Whether or not MCP is expressed on the surface of the unfertilized oocyte, it is clear that CD59, DAF and MCP (but not CR1) are present early on in the developing embryo (Table 6.1). Differential expression appears with formation of the blastocyst, the embryonic stage at which implantation into the maternal tissues must occur. The outermost cells of the blastocyst form the trophoblast which has the capacity to erode and infiltrate the wall of the uterus. The trophoblast is in direct contact with maternal tissues and erodes into maternal blood vessels. Trophoblast must provide an immunoprotective barrier between mother and fetus, not least to protect against activation on fetal tissues of maternal C (Figure 6.3). In the first trimester, the trophoblastic villi consist of a bilayer of cells, the outer syncytiotrophoblast layer and the inner cytotrophoblast layer, surrounding a mesodermal core. The cytotrophoblast cells form a distinct layer between the syncytiotrophoblast and the villous mesenchyme. Early in pregnancy all three membrane CRP, DAF, MCP and CD59, are abundantly expressed on the trophoblast (Figure 6.4). MCP and CD59 are found on the apical and basal aspects of the chorionic villous syncytiotrophoblast and also on villous cytotrophoblast cells [733, 761].

The distribution of DAF on the trophoblast at the various stages of development is controversial. One group reports its presence only on the apical aspect of the syncytiotrophoblast layer during the first trimester and expression at much lower levels than CD59 and MCP [661, 761], whereas a second group locates high levels of DAF on both villous syncytiotrophoblast and cytotrophoblast cells [733]. Holmes and colleagues also report that at 6

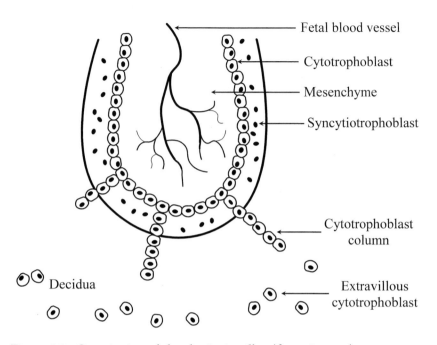

Fetal blood vessel

Cytotrophoblast

Mesenchyme

Syncytiotrophoblast

Cytotrophoblast
column

Decidua

Extravillous
cytotrophoblast

**Figure 6.3. Organization of the chorionic villus (first trimester).**
The trophoblast forms an immunoprotective barrier between mother and the
fetus. Early in pregnancy CRP (CD59, DAF and MCP) are expressed in abundance on the syncytiotrophoblast and also on villous cytotrophoblast cells and
extravillous cytotrophoblast cells. Whilst DAF is expressed at lower levels than
the other CRP during early pregnancy, its expression increases with advancing
gestation. CR1 is only expressed on villous mesenchymal cells.
Modified from Rooney *et al.* (1993), *Immunologic Research* **12**; 276.

weeks of gestation, immunostaining of both CD59 and MCP is evident on
human placenta whilst expression of DAF is 'patchy'; only after about 8
weeks of gestation do all chorionic villi stain positively for DAF[761]. It is
clear, however, that by the second trimester all CRP, with the exception of
CR1, are highly expressed on the syncytiotrophoblast cells of the chorionic villi (at this stage the villi contain few cytotrophoblast cells) and that
DAF expression increases with advancing gestational age[733, 761, 1022]. The
presence of all three membrane CRP on the trophoblast during the first
trimester and at term has been confirmed by western blot analysis of solubilized syncytiotrophoblast membranes[761].

Other trophoblast cell populations also express C inhibitors. The extravillous cytotrophoblast cell columns, which proliferate through the syncytiotrophoblast into the inter-villous space during early pregnancy, express

CD59, high levels of DAF and also MCP (the latter being at lower levels than in their villous counterparts). The cytotrophoblast cells forming part of the amniochorion which encloses the fetus also stain positively for MCP and CD59. CR1 is only present on the villous mesenchymal cells, and not on the trophoblast [733, 761].

We have described above the profusion of C regulatory molecules expressed on the interface between maternal and fetal tissue. However, amniotic fluid itself contains maternal IgG and IgM, as well as functional C activity, and thus poses a risk to fetal tissue exposed to the fluid [1230]. Examination of amniotic cells demonstrates that they too are adequately protected by C inhibitors and that the fluid itself contains soluble inhibitors capable of suppressing MAC formation. Cultured amniotic epithelial cells express CD59 [761, 1231], DAF [761, 1232] and MCP [733, 761]. Furthermore, CD59 is spontaneously shed from the epithelial cells and can be purified by affinity chromatography from amniotic fluid, with a yield of about 100 ng/ml. This purified CD59 can be incorporated into the surface of erythrocytes, indicating that it has a GPI anchor [1231]. The incorporated CD59 is functional and, after incorporation, protects erythrocytes from C-mediated lysis, suggesting that its presence in amniotic fluid may also protect the embryo.

## THE ROLE OF MCP IN FERTILIZATION

We have alluded to the possibility that sperm MCP may have a role other than protection from the female C system. Its absence from the exposed surface of the sperm indicates that it is not utilized to protect sperm from C in the lower reaches of the female reproductive tract. Its presence on the inner acrosomal membrane implies that it may have other roles subsequent to the acrosome reaction. Evidence implicating MCP in the process of fertilization is provided by the demonstration that monoclonal antibodies specific for MCP inhibit the sperm–oocyte interaction [279, 280]. It can be argued that this inhibition is due to the steric effects of the large antibody molecule binding to the acrosome, rather than indicating a primary role for MCP in fertilization and it remains possible that the abundant binding of antibody in the acrosomal region will inhibit fertilization whatever the antigen targeted. However, one recent study has analysed the effects of different anti-MCP monoclonal antibodies on sperm–oocyte interaction and demonstrated that a monoclonal antibody directed to the amino-terminal tip of the MCP molecule (SCR1) blocks fertilization, whilst a monoclonal to the C-regulatory domains (located in SCRs 2-4) binds sperm MCP but does not block

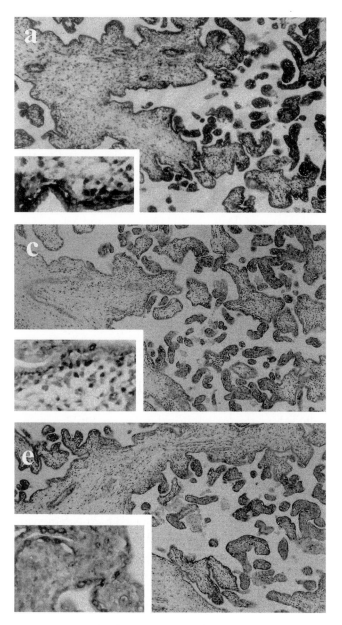

**Figure 6.4. Expression of CRP mRNAs in placenta, analysed by *in situ* hybridization.**

Sections from a 15-week placenta were analysed by *in situ* hybridization using ribo-probes specific for MCP (a, b), DAF (c, d) or CD59 (e, f). **(a)** Expression of MCP assessed using antisense probe; note strong expression both in syncytiotrophoblast and cytotrophoblast cells. **(b)** Negative control (MCP sense probe).

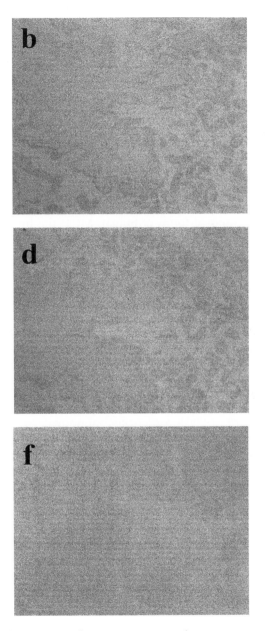

(c) Expression of DAF assessed using antisense probe; note expression is stronger in cytotrophoblast cells than in syncytiotrophoblast. (d) Negative control (DAF sense probe). (e) Expression of CD59 assessed using antisense probe; note strong expression in cytotrophoblast cells; (f) Negative control (CD59 sense probe). Magnification × 75, inserts × 262.5. Photographs courtesy of Sim Singhrao, Department of Medical Biochemistry, University of Wales College of Medicine.

fertilization [280]. This study supports the hypothesis that MCP has a direct role in sperm–oocyte interaction at the level of the oocyte plasma membrane. In this context it is interesting to note that MCP on spermatozoa has been suggested to be truncated and non-glycosylated [747, 748, 760]. In addition two novel cDNA clones encoding MCP have been detected in the testis; both isoforms were predicted to contain a truncated transmembrane domain and unique cytoplasmic tail [742, 749]. One of these atypical isoforms has been expressed on CHO cells, and indeed it lacked any O-linked carbohydrate [749]. Interestingly, the transfected cells formed rosettes with hamster eggs, lending support to a potential role in the fertilization process. It has been suggested that these 'short', non-glycosylated isoforms might have a distinct role on sperm membranes. In addition to these unusual forms of human MCP, the limited distribution of both murine and guinea pig MCP, and their predominant expression in the testis of both these animals, supports the suggestion that MCP fulfils a unique role in the reproductive system [1233, 1234].

A recent study has analysed in depth the interactions between C activation products and C3b-binding proteins on the sperm and oocyte surfaces [279]. These authors advance the theory that C3b is generated via C activation or by a C3-cleaving acrosomal enzyme (acrosin) following the acrosomal reaction. The C3 metabolites thus formed bind to MCP on sperm and C receptors (CR1 and CR3) on the oocyte and can form a physical link, or bridge, between the sperm and oocyte, facilitating fertilization (Figure 6.5). In support of the proposed role of MCP in the fertilization process, one recent study has identified abnormalities in spermatozoal MCP in infertile men [762]. Individuals were selected on the basis that their acrosome-reacted sperm did not agglutinate with anti-MCP-coated beads. Out of seven donors who demonstrated no agglutination (from a total of 63 infertile individuals), abnormalities in spermatozoal MCP were shown in three of them by immunoblotting using three monoclonal antibodies with specificities for different SCR. In two of these cases there was no detectable MCP on spermatozoa, and in the third case only one of the three antibodies detected MCP protein. Whilst MCP abnormalities were only detected in a limited number of cases, these data indicate that there may indeed be a link between abnormalities of sperm MCP and some forms of male infertility.

## CONCLUDING REMARKS

The emergence during the last 5 years of the CRP as important players in the protection of germ cells and the developing fetus has surprised those in

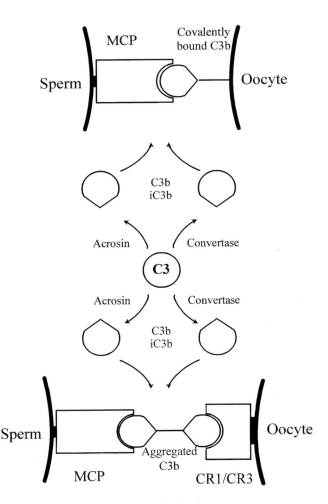

**Figure 6.5. Potential role of MCP in the fertilization process.**
Following the acrosomal reaction, C3b may be formed either through C activation or by the action of the acrosomal enzyme, acrosin. The C3b thus formed may form a physical link or 'bridge' between the spermatozoa and oocyte, possibly by binding to MCP on the sperm surface and receptors for C3 fragments on the oocyte surface. Covalent binding of C3b to either the spermatozoa or oocyte, and subsequent ligation of MCP or C receptors may also provide the link between the two cells.

C research and has been largely ignored by reproductive biologists outside of the field. However, the abundant expression of CRP on germ cells and in the placenta dictates that the involvement of the CRP in reproduction be seriously evaluated. The substantial body of evidence summarized above, implicating CRP not only in defence against C attack in the genital tract and at the materno-fetal interface but also in sperm–egg interaction, is clearly just the beginning. This is an area of C research which is still in its infancy, and will over the next few years generate yet more surprises which may lead to novel approaches to the treatment of reproductive problems such as infertility and placental failure.

# Chapter 7

# COMPLEMENT REGULATORS AND MICRO-ORGANISMS

## INTRODUCTION

Infectious agents, such as viruses, bacteria and parasites, have evolved many different strategies to evade clearance and destruction by the host's immune system [1235–1240]. C is a central component of the host immune system and evasion of C is vital to the survival of an infecting agent (reviewed in [426, 1235, 1241–1243]). The strategies used are as numerous and diverse as the organisms themselves but some general patterns emerge. Many bacteria are surrounded by a capsule which is refractory to C attack and reduces adherence to phagocytic cells. Some viruses utilize antigenic shift and antigenic drift to alter surface proteins, resulting in evasion of the host's acquired immunity by presenting an ever-changing facade of antigens to the immune system. Other viruses acquire host proteins during the budding process which may 'camouflage' a virus such that it looks more like the host; during this process there may even be sequestration of specific host proteins which provide protection from the immune system. Molecular mimicry is used by both viruses and parasites, resulting in production of proteins similar to host proteins which protect from immune surveillance and C attack. These are but a few of the many strategies employed by infecting agents to further infection (Figure 7.1). In this chapter we discuss the various ways in which invading organisms manipulate the host complement regulatory proteins (CRP) and C receptors, often subverting them to aid immune evasion and infection. We begin by discussing the subversion by micro-organisms of CRP and C receptors to aid infection of the cell. We then describe the appropriation of host CRP, both fluid-phase and membrane-bound, by various micro-organisms. Finally, we describe the use of molecular mimicry to generate proteins, endogenous to the organism, which in function and structure closely resemble the host's own CRP.

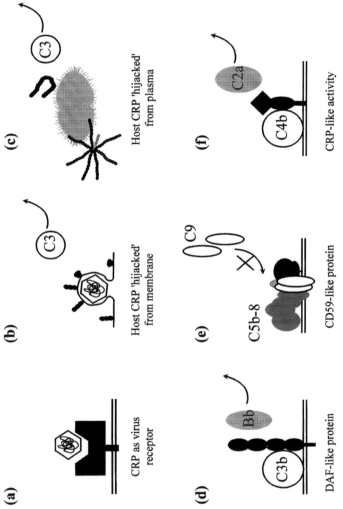

# COMPLEMENT REGULATORS AS RECEPTORS FOR MICRO-ORGANISMS

As described elsewhere in this volume the various CRP are abundantly expressed on the majority of cell types and in most tissues. As such, they offer particularly attractive targets for micro-organisms which must bind and infect cells. Decay-accelerating factor (DAF), membrane cofactor protein (MCP), complement receptor 1 (CR1) and CR2 are all utilized as receptors by different micro-organisms, in some instances providing the only portal through which that organism can enter host cells and establish an infection (Figure 7.2).

## DAF

DAF has recently been identified as a receptor for echovirus7 and other echoviruses [282, 283]. The first evidence for this was provided by the demonstration that incubation of infectable cells with monoclonal anti-bodies specific for DAF blocked binding of virus to the cell surface and inhibited infection. Furthermore, transfection of otherwise non-infectable hamster cell lines with human DAF cDNA rendered them susceptible to infection by echovirus7. Subsequent studies utilized short consensus repeat (SCR)-deletion mutants or DAF-MCP recombinant proteins to identify the regions in DAF responsible for binding virus [1244]. These studies demonstrated that SCRs 2–4 were required for virus-binding activity (see Chapter 3 for domain structure of DAF). The N- and O-linked carbohydrate moieties were not necessary for virus binding

---

**Figure 7.1. Strategies adopted by micro-organisms to evade or subvert the host's C system.**

(a) Complement regulatory proteins (CRP) and C receptors are widely distributed on many cell types and in many tissues. As a consequence they are utilized by various viruses as receptors, enabling attachment to the cell surface and establishment of infection. In some cases the binding between CRP and virus is direct, in other cases opsonization of virus with C3 fragments is thought to contribute to receptor binding.

(b–f) A multitude of different strategies have been adopted by micro-organisms which enable them to evade attack by host C. These evasion mechanisms include: (b) acquisition of host CRP from the plasma membrane of infected cells during the budding process; (c) acquisition of host CRP from plasma (particularly fH and C4bp); (d, e) generation of CRP, remarkably similar to the host's own CRP, which control C activation on the surface of the micro-organism, or in the vicinity of infected cells ('molecular mimicry'); (f) generation of C inhibitory proteins unrelated in structure to the host's own CRP.

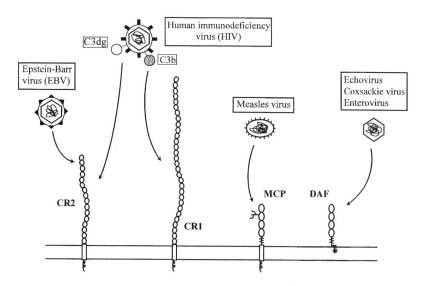

**Figure 7.2. CRP or C receptors as virus receptors.**

CR2 is the cellular receptor for Epstein–Barr virus (EBV); binding is mediated by the amino-terminal two short consensus repeats (SCRs). Opsonized HIV may also use C receptors (CR1, CR2 or CR3) as a means of enhancing infection. Binding and internalization may proceed in the absence of, or in concert with, CD4. Binding of opsonized HIV to C receptors on the surface of follicular dendritic cells may establish a 'reservoir' of virions in lymph nodes and contribute to infection of T cells. Membrane cofactor protein (MCP) is a cellular receptor for measles virus. Binding is mediated by the amino-terminal two SCRs, although the carbohydrate moiety on SCR2 is also implicated in the binding reaction. Decay-accelerating factor (DAF) has recently been identified as a virus receptor for a number of different viruses, including echovirus7, enterovirus 70 and coxsackieviruses A21, B1, B3 and B5. Binding sites are located throughout the SCR domains.

and DAF expressed on the native GPI anchor or on a transmembrane anchor was equally effective in mediating infection. DAF has also been demonstrated to be a cellular receptor for enterovirus 70 and coxsackieviruses A21, B1, B3 and B5 [284, 285, 287, 288]. In all these cases, it appears that a second factor may be required for effective cell entry and viral replication. In the case of echovirus7, this factor has not yet been identified [1245]. However, intercellular adhesion molecule 1 (ICAM-1) is required for cell entry of coxsackievirus A21, and indeed a close cell surface association of DAF and ICAM-1 has been proposed, based on chemical cross-linking studies [288]. In the case of coxsackievirus B3, the second attachment is to coxsackie-adenovirus receptor (CAR) [1246, 1247]. It is possible that interaction with DAF is of low affinity, and functions to 'tether' these viruses to the cell surface, and enhance their interaction with high-affinity receptors such as ICAM-1 and CAR [284, 285, 287]. Antibody blocking experiments have shown that the binding site for enterovirus 70 was present in SCR1 and/or SCR3, although effects of steric hindrance as a result of antibody binding could not be ruled out in these studies [284]. The binding site for coxsackievirus B3 (adapted to grow in RD rhabdomyosarcoma cells) has been firmly located in SCR 2 by transfection experiments with DAF SCR-deletion mutants [285], and binding to SCR1 has been proposed for coxsackievirus A21 [288]. DAF has also been identified as a ligand for the Dr family of adhesins expressed by *Escherichia coli*, thereby permitting the binding of pathogenic strains of *E. coli* to host cells [701, 702]. Binding sites for *E. coli* on DAF have been localized to either SCR 3 or 4 using SCR-deletion mutants and by blocking studies with antibodies [703].

## MCP

MCP has been identified as a cellular receptor for measles virus [289-291]. Following measles virus infection, virus disseminates throughout the body of an infected individual via the lymphoid system, resulting in suppression of cell-mediated immunity which, in extreme cases, can cause death. The mechanisms leading to immunosuppression are unclear, but downregulation of IL-12 production by monocytes, an effect that can be mimicked by cross-linking MCP, may play a role [1248]. Initially it was thought that MCP formed a complex with another membrane molecule, moesin, permitting binding and entry of the virus [1249]. However, recent publications suggest that moesin is not directly involved in measles virus binding, but rather has an effect on the shape and spreading of measles virus-mediated syncytia [1250, 1251]. Other evidence also implicates a second cellular receptor for measles virus [1252-1254]. The virus engages MCP through its envelope protein, haemagglutinin. Possession of a Tyr at position 481 of the haemagglutinin protein, as seen in laboratory-adapted strains of the virus such as Edmonston, is essential for binding to MCP. Wild-type strains of measles virus have Asn at this position; they do not bind MCP but rather an alternate as yet unidentified receptor which seems capable of binding both wild-type and laboratory-adapted strains of the virus.

The measles virus-binding domain has been located at the amino-terminal end of the MCP molecule, with binding sites identified in both SCR1 and SCR2 (see Chapter 3 for domain structure of MCP) [763-765]. It is interesting to note that MCP from New World monkeys consists only of three SCRs, equivalent to SCR2-4 in human MCP, and that cells expressing this truncated MCP do not bind the Edmonston measles virus [1255]. This strongly suggests that the location of the virus binding site is situated predominantly in SCR1 at the amino-terminal end of MCP. Other domains within MCP may also be involved in determining the susceptibility of cells to measles virus infection. For example, the N-linked carbohydrate moieties in MCP appear necessary for receptor function, particularly the N-linked sugar located on SCR2 [766–768]. MCP consists of multiple isoforms differing in the length of the O-glycosylated STP-rich region and the cytoplasmic tail (see Chapter 3). All isoforms of the protein permit infection, although the smaller isoforms with a shorter O-glycosylated region are more permissive [757, 1256, 1257]. It has been suggested that the MCP isoform with the largest glycosylated region is most active at binding the virus, whilst the smaller MCP isoform is more efficient at mediating subsequent membrane fusion, a process which is essential for infection and appears to be inhibited by an abundance of O-linked oligosaccharides [732, 1258]. It is unclear whether the transmembrane domain of MCP is required for measles virus infection; different groups working with recombinant, mutant forms of MCP have reported results which are conflicting. It has been claimed that MCP in which the extracellular portion of the protein is attached to the GPI anchor of DAF (both the transmembrane domain and cytoplasmic tail are missing), still functioned as a receptor rendering cells susceptible to measles virus infection [763]. However, Seya *et al.* have also expressed recombinant forms of the molecule and demonstrated that, whilst GPI-anchored MCP bound virus, virus entry and replication was impaired, resulting in low levels of cell fusion and syncytium formation [1259]. The differing results obtained by the two groups might have been attributable to variation in experimental conditions which utilized different doses of virus and incubation periods. An interesting recent observation is that cells persistently infected with measles virus downregulate expression of MCP, perhaps to limit the cytopathic effects of infection [1260]. This virus-mediated downregulation of MCP requires the amino acids Tyr-X-X-Leu in a region just proximal to the membrane in the cytoplasmic tail [1261].

## CR2

The first C receptor to be ascribed a role as a receptor for a virus was CR2. Although not a regulatory protein, CR2 is an important C receptor, binding the C3dg fragment on opsonized particles. CR2 is abundantly expressed on B cells and plays an important role in regulating the B cell response to antigen [274, 607]. Epstein–Barr virus (EBV), the causative agent of infectious mononucleosis, is a B cell-tropic virus. The identity of the B-cell receptor for EBV with that for the C fragment C3dg was realised in 1984 [281, 640, 641]. CR2 is thus often referred to as the EBV receptor. The EBV major surface glycoprotein, gp350/220, is the viral ligand for CR2. The binding sites in CR2 for both C3dg and EBV appear to be identical or very close and have been located in the two amino-terminal SCRs [634, 635]. A nine-amino acid sequence in gp350/220 homologous to the CR2-binding region in C3dg has been implicated in binding [1262].

## Complement receptors and HIV

HIV-1 and HIV-1-infected cells activate human C, resulting in the deposition of C3 fragments on the virus or cell surface. It has been suggested that anti-HIV antibodies and C may be in part responsible for the initial rapid clearance of viral load in the peripheral blood [1263-1265]. However, it is clear that C activation, occurring even in the absence of antibody, and deposition of C activation fragments on virus particles can enhance infectivity rather than promote viral lysis [426, 1266-1268]. The HIV envelope proteins, gp41 and gp120, bind human C1q (or mannan-binding protein) and activate C via the classical pathway (CP) but virions are not lysed [1269-1274]. The exceptional resistance of HIV to C-mediated lysis, very different from that of other retroviruses which are efficiently killed by C, may be due to the presence of C regulators on the virion surface, obtained as described below. Enhanced infection with the virus is due to the ability of HIV virions, coated with C3b, or with the degradation products iC3b and C3dg, to bind to and infect cells expressing C receptors (CR1, CR2 and CR3) [299, 585, 586]. Thus, HIV infection can proceed via interaction with C receptors acting in concert with, or even in the absence of, the primary HIV receptor CD4 [1275-1278]. The enhancement of infectivity seen following C activation may be most relevant at low levels of virus [298, 1279-1282]. Follicular dendritic cells (FDC) in the germinal centre of lymph nodes are a major reservoir of virions in an infected individual. It has been proposed that the binding of HIV onto FDC may be due to the abundant expression of three C receptors (CR1, CR2 and CR3) on these cells and thus dependent on the prior activation of C and deposition of C3 fragments on the virus surface [547, 1283, 1284]. Although debate continues as to whether FDC are infected with HIV or merely bind virus, localization of HIV on FDC may be an important mechanism for infection of T cells [587-589]. FDC binding of HIV may also explain the rapid accumulation of virus in lymphoid organs in early-stage disease despite a low viral burden in the peripheral blood [1285, 1286].

# 'HIJACKING' OF COMPLEMENT REGULATORS BY MICRO-ORGANISMS

## HIV

HIV virions bud from the surface of infected cells taking with them a portion of the host cell membrane. This results in the presence on the virion of host proteins including the GPI-anchored regulators, CD59 and DAF and the transmembrane regulator, MCP (Figure 7.3) [296, 297, 1287, 1288]. These CRP may be partly responsible for the resistance of HIV-1 to C attack; indeed, antibodies which neutralize DAF have been shown to sensitize the virus to C attack and virolysis [296, 1288]. Resistance of viral preparations to C attack can also be correlated with the levels of expression of the CRP on the cell from which the virions budded [1288]. Evidence suggests that the HIV envelope contains proportionately more of the GPI-anchored CRP than would be predicted from expression on the host cell; the cause of this might be the selective budding of virions through the GPI-rich 'patches' present on most cell types.

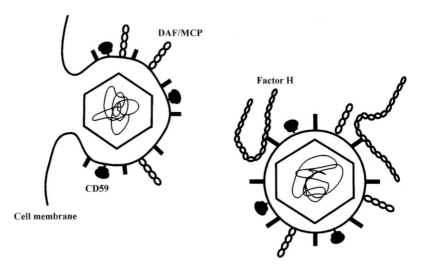

**Figure 7.3. Appropriation of host CRP by HIV.**

HIV acquires host CRP from the plasma membrane during the budding process. The GPI-anchored proteins CD59 and DAF are particularly highly 'expressed' on the virion surface, although recent evidence indicates that the transmembrane CRP, MCP, may also be acquired in this way. Envelope proteins on the virus may also bind fluid-phase CRP, such as C4bp and fH, thereby protecting free virus in the circulation from host C attack.

Another CRP, the fluid-phase regulator fH, is also 'hijacked' by HIV, conferring C resistance to the virion. The envelope proteins of HIV, gp41 and gp120, capture fH from serum and utilize the bound protein to protect the virus from C attack [292, 294, 295]. Virion-bound fH acts as a cofactor for fI and causes the inactivation of C3b by formation of iC3b, thus limiting C activation on the virus. The iC3b so formed can in turn interact with CR3 on cells (such as FDCs) enhancing HIV infection, as described above. When the protective effects of both fH and DAF are abrogated, for example by blocking DAF function on the viral surface using antibody and incubating in serum deficient in fH, efficient virolysis can be achieved [295].

CRP are also captured during the budding process by several other enveloped viruses, including human cytomegalovirus (HCMV) and the human T-cell leukaemia/lymphoma virus type 1 (HTLV-I), and contribute to C resistance of these viruses [1289, 1290]. HCMV has further refined this strategy by causing upregulation of DAF and MCP on the membranes of infected cells, thereby enhancing the C resistance of the host cell and of the virus particles budded from the cell [1290, 1291].

## Streptococcus pyogenes

M protein, a major component of the capsule of the pathogenic Gram-positive group A streptococcus *Streptococcus pyogenes*, binds several CRP. The fluid-phase regulator fH binds the M protein of group A streptococci, thereby localizing functionally active fH at the bacterial surface, resulting in the rapid inactivation of bound C3 convertases and protection from phagocytosis [300, 453]. A site on fH between SCR6 and 10 has been implicated in binding to *S. pyogenes* [427-429]. Streptococci also bind C4bp through the surface proteins Arp and Sir, members of the M protein family, and utilize the bound C4bp to protect against C [301, 507]. It has recently been suggested that these M proteins on *S. pyogenes* mimic C4b epitopes in order to capture C4bp on the bacterial surface [509]. Because of the multivalent nature of C4bp, capture through one 'arm' would still permit function. Together, bound fH and C4bp on *S. pyogenes* will provide effective protection against C activation through both pathways. A third RCA family member which may interact with streptococcal M proteins is MCP. Recent data indicate that MCP may function as the keratinocyte receptor for *S. pyogenes*; this is based in part on the observation that soluble MCP can inhibit adherence of the bacterium to keratinocytes [302]. The relevance of this event to penetration of the skin by the bacterium remains to be proven. Following entry into cutaneous tissue, *S. pyogenes* will bind to MCP on epidermal keratinocytes and utilize binding to establish an infection. The local inflammatory response which follows surface infection and tissue damage will result in an increased blood supply to the skin and influx of soluble C components and CRP, including high concentrations of fH. The bacterium will bind fH and C4bp via its M proteins which not only protect from C attack and subsequent phagocytosis, but also result in release of the bacterium from MCP due to the competition between fH and MCP for binding the same M proteins. Release of fH/C4bp coated bacteria will thus facilitate spread of *S. pyogenes* through the tissue, providing a partial explanation for its invasive nature. Taken together, all the above data imply that interactions with CRP play a major role in mediating infection with *S. pyogenes*.

### Sialylated micro-organisms

The presence of sialic acid on a cell surface is known to interfere with the activation of the alternative pathway (AP) by increasing the affinity of fH for cell-bound C3b (see Chapter 3). Many micro-organisms have adopted the strategy of coating themselves with sialic acid in order to present a non-activating surface to the host immune system. For example, capsular sialic acid on type III, group B streptococci protects them from lysis via the AP [1292]. Similarly, *Trypanosoma cruzi* trypomastigotes possess a surface-associated trans-sialidase, which enables the parasite to coat itself with sialic acid obtained from the host's own glycoconjugates present in serum or on cell surfaces [1293, 1294]. A recent study has analysed in depth the role of sialic acid in the protection of *Neisseria gonorrhoeae* from host C [1295]. In this case, it was demonstrated that sialic acid on the surface of the micro-organism enabled 'capture' of plasma fH, resulting in surface localization of the regulator where it conferred C resistance to the gonococci. Factor H possesses several putative polyanion binding sites (discussed in Chapter 3) and indeed involvement of one these sites, incorporating the carboxy-terminal SCR of fH, has been implicated in binding to sialylated *Neisseria gonorrhoeae*. This important finding is likely also to apply to other sialylated organisms.

## MOLECULAR MIMICRY

Many micro-organisms produce their own C regulators which protect against attack by host C, thus aiding survival. In many cases, these regulators, though clearly derived from the genome of the infecting agent, bear a structure remarkably similar to the host's own CRP. For example, a number of different regulators are produced which contain SCR-like domains and which can efficiently interfere with the C activation pathways. In other cases, the micro-organism produces regulatory proteins which are structurally distinct from the human CRP, but which nevertheless have the ability to bind human C components and inhibit activation. The wide range of proteins produced by micro-organisms are discussed in the following sections and summarized in Table 7.1.

## Poxvirus

Poxviruses are a large group of complex viruses, many of which are highly successful pathogens in man and animals. At the root of this success is a concerted strategy to overcome the effects of the host immune system. These strategies are best studied in the closely related vaccinia, variola and cowpox viruses [1236, 1296]. These viruses encode protective proteins acting as cytokine binders, serine protease inhibitors and C inhibitors. Cells infected with vaccinia virus secrete a 35-kDa protein, a product of the C21L open reading frame in the viral genome, consisting of four regions with features characteristic of the SCR domains found in the regulators of complement activation (RCA) proteins (see Chapter 3) (Figure 7.4). The protein has greatest homology (38%) with the amino-terminal half of the α chain of C4bp and can indeed bind C4b and inhibit the CP of C [1297, 1298]. Further analyses indicated that the protein also inhibited the AP and was a cofactor for fI in the cleavage and inactivation of both C3b and C4b [1299]. Functional analysis of a recombinant form of the protein, indicated that it could act as a cofactor for the first fI-mediated cleavage of C3b, but not for the second cleavage (forming iC3b) or the third (forming C3dg and C3c [1300]). However, the first cleavage of C3b in the α' chain is sufficient to render it non-functional in the formation of an AP C3 convertase. The recombinant protein also rendered C4b non-functional in the formation of the CP C3 convertase; in this case it supported both fI-mediated cleavages in the α' chain. In deference to these activities the protein has been termed vaccinia virus C-control protein (VCP). Secretion of VCP from infected cells inhibits C activation in the vicinity of the cell and contributes to the virulence of the virus [1236, 1301]. The vaccinia virus genome also encodes a second protein of molecular mass 42 kDa, the product of the B5R open reading frame, with sequence similarities to VCP. This protein also contains four SCR-like domains but differs from VCP in that it is not secreted; it has a transmembrane domain and forms part of the extracellular envelope of the virus (Figure 7.4) [1302, 1303]. The protein is 30% identical with VCP at the amino acid level and is most closely related to human fH. However, it remains to be determined whether C-regulatory activity is expressed in this protein. Since the

**Table 7.1.** *Production of complement regulatory proteins (CRP) by micro-organisms.*

A wide range of micro-organisms produce CRP which are either structurally related to host CRP ('molecular mimicry') or structurally distinct but with the ability to inhibit host C. These endogenous CRP protect the micro-organism, or infected cells, from host C attack. Abbreviations: VCP, vaccinia virus C-control protein; SCR, short consensus repeat; ORF, open reading frame; CCPH, C-control protein homologue; s, secreted; m, membrane; DAF, decay-accelerating factor; SCIP, schistosome complement inhibitory protein.

| Micro-organism | Protein | Surface bound or secreted | Comments |
|---|---|---|---|
| Vaccinia virus | VCP | Secreted | Contains SCR-like domains, homology with human C4bp. Inhibits activation pathways. |
| Vaccinia virus | Product of B5R ORF | Transmembrane | Contains SCR-like domains. Homology with VCP. |
| Herpesvirus saimiri | Product of HVS-15 ORF | GPI-anchored | CD59-like. Inhibits C. |
| Herpesvirus saimiri | mCCPH | Transmembane | Contains SCR-like domains. Homology with VCP and human C4bp. Inhibits activation pathways. |
| | sCCPH | Secreted | |
| Herpes simplex virus-1 and -2 | gC-1 and gC-2 | Membrane-anchored | gC-1 regulates C on infected cells and on virion; gC-2 regulates C on virion. gC-1 prevents properdin binding to convertase. |
| Epstein–Barr virus | Unidentified | Membrane-anchored | Decay-accelerating activity and fI cofactor activity. |
| Trypanosoma cruzi | T-DAF | Secreted | Decay-accelerating activity. Weak SCR consensus sequence. |
| Trypanosoma cruzi | T-CRP (gp160) | GPI-anchored, also cleaved and 'secreted' | Decay-accelerating activity. |
| Trypanosoma cruzi | gp58/68 | Membrane-anchored | Inhibits convertase formation. |
| Schistosoma mansoni | SCIP-1 | GPI-anchored | CD59-like. |
| Entamoeba histolytica | Galactose-specific lectin | Membrane-anchored | CD59-like |

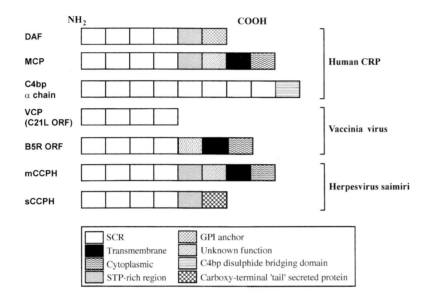

**Figure 7.4. Mimicry of the regulators of complement activation (RCA) proteins.**

Cells infected with vaccinia virus secrete a protein (VCP) containing four SCR-like domains with homology to human C4bp. A second protein is also encoded by the B5R open reading frame (ORF) in the vaccinia virus genome which forms part of the extracellular envelope of the virus; this too contains SCR-like domains. Complement control protein homologue (CCPH) is encoded by the herpesvirus saimiri genome and consists of four SCRs with homology to VCP and human C4bp. The overall structure of CCPH is remarkably similar to human MCP. A secreted form of CCPH (sCCPH) is also released from infected cells (mCCPH, membrane CCPH).

initial identification of VCP, various homologues of this inhibitory protein have been found in the genomes of other pox viruses, notably cowpox virus[1304] and smallpox virus[1305].

## Herpesvirus saimiri

Herpesvirus saimiri (HVS) is a T-cell lymphotropic virus of New World monkeys; in some species of monkey HVS causes malignant lymphomas and leukaemias. The HVS genome encodes several proteins with homology to the human CRP. The product of the HVS-15 open reading frame encodes a protein of 121 amino acids, with 48% amino acid identity to human CD59 and 69% amino acid identity with CD59 of the natural host, the squirrel monkey[1306–1308]. Expression of this protein in BALB3/3T3 cells demonstrated that it was attached to the cell via a GPI anchor and could indeed protect from C-mediated lysis[1308].

A second open reading frame in the HVS genome encoded a protein predicted to consist of 360 amino acids. It had four SCR-like repeats at the amino terminus with an inter-domain homology of between 33 and 37%, a carboxy-terminal domain predicted to encode a transmembrane segment and seven potential N-glycosylation sites[1306, 1309]. This putative C regulator, termed (membrane) C control protein homologue (mCCPH), bore highest amino acid identity with human C4bp and VCP (the CRP secreted by vaccinia virus), although the structural organization of the protein was more similar to that of the human regulators MCP and DAF (Figure 7.4). A splicing alternative was described in which deletion of a 193-bp intron removed the transmembrane domain and resulted in a predicted secreted form of the molecule (sCCPH) with a different carboxy terminus[1309]. The mechanism of splicing closely resembled that previously observed in human DAF[678]. The sCCPH protein has been further characterized using a virus-neutralizing monoclonal antibody, SE, which was selected from a panel of antibodies that had been raised against HVS[1310]. The SE antibody precipitated proteins of molecular weight 47–53 kDa from the supernatants of HVS-infected cells. Furthermore, a similarly sized protein was precipitated from the supernatant of COS-7 cells transfected with cDNA encoding the secreted form of CCPH, demonstrating that this antibody did indeed recognise CCPH. The same antibody immunoprecipitated proteins of molecular weight 65–75 kDa from purified virion particles but not from the membranes of infected cells. The data indicated that the two forms of CCPH were indeed membrane-associated (viral) and secreted (from infected cells). Transfection of mammalian cells with mCCPH has been demonstrated to confer upon them resistance to human C attack through interference with C3 convertase activity and C3 deposition[1311].

## Herpes simplex virus 1

Endothelial cells infected with herpes simplex virus 1 (HSV-1) express virally encoded receptors for IgFc and C3b on the cell surface[1312]. Infected cells form rosettes with C3b-coated erythrocytes, a phenomenon which can be inhibited by purified C3b, but not C3[1313]. The C3b receptor of HSV-1 has been identified as glycoprotein C (gC-1) and has been isolated by iC3 affinity chromatography from extracts of HSV-1 infected cells. Interestingly, an equivalent protein in HSV-2 infected cells, gC-2, can also be isolated from cell extracts by iC3 affinity chromatography, although infected cells do not form rosettes with C3b-coated erythrocytes[1314]. Both gC-1 and gC-2 have been shown to regulate C on the virion surface; gC-1 also regulates C on the surface of infected cells[1314–1319]. The C-inhibitory activity of gC-1 is thought to reside in its ability to interfere with properdin-mediated stabilization of the AP C3 convertase; separate putative functional domains capable of binding C3b and properdin have been identified on the protein[1320–1322]. The genes for gC-1 and gC-2 have been cloned and expressed in mammalian cells[1314, 1323, 1324]. Both gC proteins were expressed on the cell surface; gC-1 had a molecular weight of approximately 120 kDa and gC-2 of 68–78 kDa. Both recombinant proteins bound C3b and iC3b, indicating that other factors were responsible for the inability of HSV-2 infected cells to rosette with C3b-coated erythrocytes. In a screen of potential blocking monoclonal antibodies, one raised against human CR1, 5C11, had the ability to block rosetting of HSV-1 infected cells with EAiC3b, indicating that 5C11 has cross-reactivity with gC-1 and suggesting that this viral glycoprotein and CR1 share a common epitope[1325].

## Epstein–Barr virus (EBV)

An as yet unidentified protein (or proteins) present on EBV has potent C-regulating ability. Purified EBV virions provide efficient cofactor activity for fI in the cleavage and inactivation of both C3b and C4b (and also iC3b and iC4b), and also have decay-accelerating ability for the AP C3 convertase[1326]. It is intriguing that, despite these powerful C3-regulatory properties, binding of C3b to EBV has not been demonstrated. No SCR-like sequences have been identified in the EBV genome. Identification and characterization of the EBV-encoded C regulator(s) is now long overdue.

### *Trypanosoma cruzi*

The amistigote and trypomastigote stages of the protozoan parasite *Trypanosoma cruzi* are resistant to C. This may be due in part to the secretion of a molecule with C3-convertase decay-accelerating activity from the parasite[1327, 1328]. This C-regulatory protein has a molecular weight of 87–93 kDa and has been termed trypanosome decay accelerating factor (T-DAF) due to its functional activity. Cloning of a partial DNA sequence encoding T-DAF demonstrated that it had 40% nucleotide identity to human DAF through the sequence encoding the second half of SCR1 and most of SCR2, although the four cysteine residues seen in all SCRs found in members of the RCA gene cluster were not present, suggesting that the SCR structural unit was not conserved [1329]. A second trypanosome CRP has been identified which also has decay-accelerating activity [1330]. This second CRP (T-CRP) is shed by trypomastigotes and has a molecular weight of 160 kDa (hence originally termed gp160). The molecule is also expressed on the surface of the parasite where it is attached by a GPI anchor and has a molecular weight of 185 kDa [1330, 1331]. The protein is probably shed from the parasite surface and simultaneously converted from the 185 kDa form to the 160 kDa form by the action of an endogenous phospholipase C [1331]. The cDNA encoding T-CRP has been cloned and shown to encode a protein of 979 amino acids with a stretch of hydrophobic residues at the carboxy terminus preceded by a predicted GPI anchor addition site. Transfection of epimastigotes with this cDNA has been shown to confer resistance to C-mediated lysis and confirms T-CRP as a virulence factor of *Trypanosoma cruzi* [1332]. The gene encoding the protein is a member of a large family of trypomastigote-specific genes known as the FL-160-CEA gene family. These genes encode highly homologous proteins of unknown function [1333]. Whilst initial analyses indicated that T-CRP and human DAF shared some sequence homology (a human DAF cDNA probe hybridized with T-CRP DNA), no homology has been demonstrated between the human DAF coding sequence and the cDNA for T-CRP [1330, 1334].

*T. cruzi* possess several other strategies for evasion of host C attack, although these mechanisms are less well characterized than those described above. The first is the production of a protein which inhibits C attack on the trypomastigote stage parasite; this glycoprotein has been termed gp58/68 due to its electrophoretic mobility (58 kDa non-reduced and 68 kDa reduced) [1335]. It is a fibronectin/collagen receptor of *T. cruzi* and can be purified from parasite detergent lysate by affinity chromatography on Sepharose-wheatgerm agglutinin. The purified protein does not act as a convertase decay-accelerating factor, but rather it functions to inhibit the formation of the AP convertase by preventing the initial association of fB with C3b. A different protection mechanism is associated with the amistigote stage of *T. cruzi* which is also resistant to C attack. Whilst the parasite activates C and activation fragments are deposited on the parasite surface, no functional membrane attack complexes (MACs) are formed and the C9 which does become bound to the amastigote surface is unusually sensitive to trypsin, suggesting that it is not fully inserted into the membrane [1336]. Further, although deposition of C3 and C5b-7 on the surface of the amastigote is comparable to that on the epimastigote (C-sensitive insect stage), deposition of C9 is much lower on amastigotes, again suggesting the presence of an inhibitor that prevents C9 insertion into the membrane.

## Schistosoma mansoni

The parasite *Schistosoma mansoni* exhibits different degrees of C sensitivity at different stages. The free-swimming cercaria are C-sensitive whereas the host-dwelling schistosomula and adult worm are C-resistant[1337]. Schistosomula and adult worms of S. *mansoni* express several C-regulatory molecules which protect them from C attack. One of these molecules regulates the terminal pathway of human C and has been termed SCIP-1 (schistosome complement inhibitory protein type 1)[1338]. It shares antigenic determinants with the human CRP CD59; indeed polyclonal and monoclonal anti-human CD59 antibodies have been used to detect SCIP-1 by immunoprecipitation and western blotting. These studies indicated that the protein was much larger than CD59, with an apparent molecular weight of 94 kDa, although in common with CD59 it was attached to the parasite surface through a GPI anchor. SCIP-1 is suggested to function in a manner similarly to CD59 in that it binds constituents of the forming MAC (C8 and C9) and prevents formation of a lytic lesion. SCIP-1 has yet to be cloned; therefore the exact relationship with CD59 remains to be established. There are also reports that S. *mansoni* 'hijacks' human DAF, probably from erythrocytes, and uses this regulator on its own surface to diminish C attack[1339, 1340]. It is unclear whether human CD59 can be 'hijacked' in a similar manner.

## Entamoeba histolytica

A CD59-like molecule on the surface of the protozoan parasite E. *histolytica* has been identified by screening for antiparasite antibodies that augmented killing of the parasite by human C and identifying the antigens recognised by such antibodies[1341]. The monoclonal antibody, 3D12, markedly enhanced serum killing; this antibody immunoprecipitated a protein of apparent molecular weight 170 kDa from detergent lysates of amoebic trophozoites. The protein was identified as the 170-kDa subunit of the galactose-specific lectin or 'adhesin', a major cell surface heterodimeric glycoprotein also containing a second subunit of molecular weight 35kDa. Inhibition of C activation was shown to occur in the terminal pathway, and ligand blotting studies demonstrated that the 170-kDa adhesin bound $C8\alpha\gamma$, $C8\beta$ and C9. Comparison of the cDNA sequence of the adhesin and human CD59 revealed a limited sequence homology in the central third of CD59, and indeed a weak cross-reactivity of a polyclonal anti-CD59 with the adhesin has been demonstrated. The specific role of this adhesin in C inhibition awaits further analysis.

## C5a-PEPTIDASE/C5a-ASE

Virulent group A and group B streptococci have the capacity to suppress the host inflammatory response. This is achieved by the production of a factor which inactivates the chemotactic peptide C5a produced during C activation. The activity is particularly striking with some group B streptococci, which invoke a very much depressed inflammatory response due to the failure of neutrophils to localize at the site of invasion. The ability of virulent streptococci to inactivate the major C chemoattractant molecule, C5a, was recognized in the mid-1980s [1342]. The factor responsible was purified from group A streptococci and shown to be a bacterial membrane-associated enzyme which cleaved the carboxy-terminal six amino acids from C5a or C5adesArg, thereby inactivating these potent chemoattractants [1343-1345]. The nucleotide sequence of the group A streptococcal C5a peptidase (SCP-A) has been obtained and shown to be homologous to the serine protease subtilisin with 53% similarity in the region containing the reactive serine residue [1346]. Soon after the identification of the C5a peptidase in group A streptococci, it was demonstrated that group B organisms also possessed a similar C5a- and C5a-desarg degrading activity [1347, 1348]. The activity was associated with a 120-kDa enzyme present on the surface of the encapsulated organism which shared high homology with the C5a-peptidase from group A organisms [1349-1352]. Evidence suggests that similar anti-inflammatory strategies, aimed at inactivating C chemoattractants, are also adopted by other bacteria such as *Serratia marcescens* [1353].

## CONCLUSIONS

This chapter summarizes current knowledge on the subversion of CRP by micro-organisms to aid infection or the survival of the organism. This is one of the fastest developing areas of C research and many new surprises will emerge in the near future. Examples of recently described interactions between micro-organisms and the host CRP and C receptors include reports of a role of CR1 in malarial infection [592]. Studies in an individual deficient in CR1 on erythrocytes demonstrated that the malarial parasite *P. falcipurum* binds erythrocytes, in part, through an interaction with CR1. Individuals with certain polymorphisms in CR1, common in Africa, have a reduced affinity for binding *P. falcipurum* through CR1 and it has been proposed that this represents a protective mutation arising under selective pressure from the parasite. It has also recently been reported that CR1 may play a role in permitting entry of *Mycobacterium tuberculosis* into macrophages through interaction with C receptors [590, 591]. The process requires the prior opsonization of the bacillus and this is brought about by a novel C activating mechanism whereby the organism binds C2a to an as yet undefined receptor and utilizes bound C2a to activate C3, efficiently opsonizing the surface with C3b (C2a invasion pathway).

The identification and characterization of novel CRP from micro-organisms will also provide clues to alternative mechanisms and strategies for controlling activation of C, which may prove valuable in the design of C regulators for use in therapy of disease. Several of the viral CRP already described are good candidates for this approach, being powerful inhibitors of human C and having multiple modes of action, inhibiting at several points in the C pathway.

Finally, a clear understanding of C evasion strategies evolved by micro-organisms will better enable the design of counter-strategies to increase the efficiency of C-mediated killing *in vivo*.

# Chapter 8

# COMPLEMENT REGULATORY PROTEINS IN OTHER SPECIES

## INTRODUCTION

Although this book concerns primarily the human complement regulatory proteins (CRP), many important lessons have been learned from studies of CRP in other species. CRP acting at various stages of the C pathway have been described in many species, including primates, pigs, sheep, rabbits and rodents, and are summarized in Table 8.1. Information on the evolution of the CRP, location of active sites and species selectivity of C inhibition has been gleaned from these studies and the availability of reagents to identify and block CRP in other species has allowed the roles of these molecules to be examined in animal models of human disease. In this brief chapter we will summarize the current knowledge concerning CRP in other species, focusing on the membrane CRP acting in the activation and terminal pathways of C.

## REGULATORS OF THE ACTIVATION PATHWAYS

### Decay-accelerating factor (DAF; CD55)

Hoffmann in his classic studies of C-inhibiting activities in erythrocyte stromata described a C3 convertase decay accelerating activity in erythrocytes from human [142, 143], guinea pig and rabbit [141]. The active principle from guinea pig erythrocytes, guinea pig DAF, was purified by multi-step chromatography in 1981 [144], predating the isolation of human DAF by a year [145]. Okada and co-workers have recently further characterized guinea pig DAF using a haemolysis-enhancing monoclonal antibody raised against neuraminidase-treated guinea pig erythrocytes [1354]. Immunoaffinity chromatography using this antibody identified three different species of DAF in guinea pig erythrocyte membranes with molecular weights of 55 kDa, 70 kDa and 88 kDa, all of which had identical amino-terminal sequences (55% identical to human DAF). The molecular differ-

**Table 8.1. *Complement regulatory proteins (CRP) in other species.***
The table presents an incomplete list of the many species in which analogues of the various human CRP have been identified, isolated and/or cloned. Abbreviations: C1inh, C1 inhibitor; fH, factor H; fI, factor I; DAF, decay-accelerating factor; MCP, membrane cofactor protein; CR1, complement receptor 1.

*Fluid-phase regulators of activation pathways*

| | | |
|---|---|---|
| C1inh | Rabbit | Purified only |
| | Guinea pig | Purified only |
| | Rat | Purified only |
| | Mouse | Purified and cloned |
| | Cow | Purified only |
| fH | Many | Cloned in mouse, rat, pig |
| fI | Many | Cloned in mouse, xenopus |
| C4bp | Many | Cloned in mouse, rat, some others |

*Membrane regulators of activation pathways*

| | | |
|---|---|---|
| DAF | Guinea pig | Cloned and purified; multiple isoforms |
| | Rabbit | Purified only; GPI-anchored |
| | Orangutan | Purified; doublet predominates |
| | Mouse | Cloned; multiple isoforms (GPI and TM) |
| | Rat | Cloned and purified |
| MCP | Orangutan | Purified, on erythrocytes |
| | Rabbit | Purified from platelets |
| | Guinea pig | Purified from neutrophils, cloned |
| | Pig | Purified from erythrocytes, cloned |
| | Mouse | Cloned |
| CR1 | Primates | Cloned and purified, multiple forms |
| | Mouse | Cloned and purified |
| Crry | Mouse | Cloned and purified |
| | Rat | Cloned and purified |

ences between these three species were not resolved. DAF analogues have also been isolated from rabbit [1355] and mouse [1356] erythrocytes, using chromatography protocols similar to those used in the original DAF preparations. In each case the protein had a molecular mass of 65–70 kDa, identical to that of human DAF. Rabbit DAF was glycosyl-phosphatidyl-inositol (GPI)-anchored but the evidence suggested that mouse DAF contained both GPI- and transmembrane-anchored species. Rabbit and mouse DAF both caused the decay of C3 convertase on haemolytic intermediate target cells. For rabbit DAF the amino acid composition closely resembled that of human DAF [1355]. A DAF analogue has also been isolated from orangutan erythrocytes and partially sequenced [689, 1357]. Of interest, orangutan DAF, though 98% identical to human DAF at the protein level, exists primarily as a covalently associated dimer.

Within the last 3 years, cloning of the cDNAs for both murine and guinea pig DAF has been described, producing several surprises. Multiple cDNAs encoding guinea pig DAF were isolated from a spleen library[1358]. All encoded the four-SCR structure which typifies DAF and were identical with one another and 58% identical to human DAF through this region. However, alternative splicing generated several different isoforms. The various isoforms differed in the length of the STP-rich region, a feature also found in human MCP where the STP region is encoded on several exons which can be alternatively spliced. A similar mechanism operates in guinea pig DAF, STP variability being generated by differential usage of splice sites in the three exons encoding this region. Alternative splicing of two exons in the region encoding the carboxy teminus of the protein generates isoforms differing in the nature of the membrane anchor, with cDNAs encoding transmembrane, GPI-anchored and secreted forms of DAF all being found. The various isoforms were expressed in all tissues examined. These data provide an explanation for the multiple molecular weights of guinea pig DAF and the inefficiency of cleavage by phosphatidylinositol-specific phospholipase C (PIPLC) noted by Okada and co-workers[1354]. Very recently, Okada and co-workers have analysed the function of the various isoforms of guinea pig DAF expressed in CHO cells[1359]. The GPI-anchored and transmembrane forms of the protein were equally effective at C inhibition and isoforms with long STP regions were better at inhibiting C than those expressing short STP regions. The transmembrane form with the shortest STP region was a poor inhibitor of C.

Mouse DAF was independently cloned in three laboratories and provides an even more complex story[1360–1362]. Mouse DAF is encoded on two separate genes in the murine RCA cluster on chromosome 1 in close proximity to the gene encoding C4bp[1360]. The two genes are 85% identical at the nucleotide level and 78% identical at the amino acid level but encode respectively GPI-anchored (*Daf-gpi*) and transmembrane (*Daf-tm*) forms of the protein. By northern analysis, mRNA encoding the GPI-linked form of DAF was present in relative abundance in all tissues whereas mRNA encoding the transmembrane form was abundant only in testis and weakly detected in lymphoid tissue. The *Daf-gpi* gene is greatly upregulated in mouse uterus in response to oestrogen[1362]. Mouse DAF-GPI is 47% identical to both human and guinea pig DAF at the amino acid level[1361]. Our unpublished data indicates that the situation is even more complicated than suggested above. Alternative splicing occurs in both genes such that the *Daf-tm* gene generates mRNA species containing several distinct STP-rich regions and mRNAs encoding both transmembrane and GPI-linked forms of DAF, and the *Daf-gpi* gene can also generate mRNA encoding a predicted soluble form of DAF (Figure 8.1).

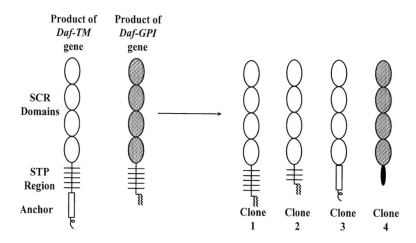

**Figure 8.1. Multiple isoforms of murine decay-accelerating factor (DAF).**

Murine DAF is encoded by two separate genes, one encoding a glycosyl phosphatidylinositol (GPI)-anchored form of the molecule and the other a transmembrane form. The products of the two genes are highly homologous (95%) through the SCR regions. Our unpublished data indicate that each of the genes can yield multiple products through alternative splicing. The *Daf-tm* gene can also encode GPI-anchored DAF with long or short STP regions (clones 1 and 2) and a transmembrane form completely lacking the Ser/Thr/Pro (STP) region (clone 3). The only alternative product of the *Daf-gpi* gene identified so far encodes an unusual carboxy terminus which is predicted to generate a secreted product (clone 4). Southern analysis of mouse tissues suggest that there are more products of these two genes still to be characterized.

We have recently identified and cloned cDNAs encoding GPI-anchored and secreted forms of rat DAF (Hinchliffe *et al.*, in press). These two forms are generated by alternative splicing of the same gene and there is no evidence of a second gene encoding DAF in the rat. The protein expressed in DAF-negative cells protects against C activation and antibody generated against the expressed protein identifies a 60–65-kDa band in rat erythrocyte membranes.

## Membrane cofactor protein (MCP; CD46)

Primate MCP was identified from a search for C3-binding proteins in primate erythrocytes and blood mononuclear cells. By passing erythrocyte membrane extracts over a column of immobilized human C3i, a protein doublet with molecular weights of 59 and 65 kDa, a pattern identical to that of human MCP, was isolated from orangutan erythrocytes [1357]. No such doublet was isolated from human or gorilla erythrocytes, the former known to be MCP-negative. The doublet was confirmed to be orangutan MCP by immunoprecipitation with a monoclonal anti-human MCP antibody and was shown to have fI cofactor activity. A C3i-binding protein which resembled MCP on SDS-PAGE had earlier been isolated from rabbit platelets but was not well characterized [1363]. A similar approach, utilizing immobilized guinea pig C3i, was adopted to isolate C3-binding proteins from guinea pig granulocytes [1364]. One of the identified proteins was a fI cofactor, migrated on SDS-PAGE as a doublet of 42 and 55 kDa species and was tentatively identified as guinea pig MCP.

MCP has been cloned from numerous primates. Of particular interest is the finding that several species of New World monkeys express a form of MCP in which the first short consensus repeat (SCR) is deleted [755, 1252]. It is this region in human MCP which is responsible for binding the measles virus (see Chapter 7). One might speculate that the loss of this SCR has occurred in these primates under selective pressure to provide protection from infection with the measles virus.

Multiple cDNAs encoding guinea pig MCP have recently been cloned from testis mRNA [1234]. Two species predominated; one encoded a form closely resembling human MCP, that is four SCR domains, an STP-rich domain corresponding to the human STP-C domain, a transmembrane region and a cytoplasmic tail (GM1); the other species encoded a form which was identical except that the fourth SCR was absent (GM2). Northern blotting analysis of various tissues indicated that the message was expressed to a significant degree only in testis. This surprising finding provoked the suggestion that MCP in the guinea pig is not an important regulator of C in host tissues but plays a much more restricted role, perhaps a specific role in reproduction.

The cloning of mouse MCP from a murine testis library has also recently been reported [1233]. The derived amino acid sequence was 45% identical to human MCP, and predicted a type I transmembrane protein with a short cytoplasmic tail. The protein expressed in CHO cells was a cofactor for fI cleavage of C3b in mouse or human C. Of note, northern analysis indicated that the mRNA was expressed predominantly or solely in testis.

Pig MCP has been purified, characterized and cloned during 1997 [1365, 1366]. A panel of monoclonal antibodies generated against pig lymphocytes all recognised several bands in the molecular mass range 50–60 kDa in western blots of pig erythrocytes and other pig cells. One of the antibodies was used to immunopurify the antigen. The complex banding pattern was retained in the purified protein and sequencing confirmed that all bands had the same amino terminus [1365]. The amino-terminal 28 amino acids were 43% identical with human MCP, provoking the conclusion that the protein was pig MCP. Purified pig MCP had fI cofactor activity using either human or pig fI and could catalyse the cleavage of both human and pig C3b. Pig MCP cDNA was identified by transfecting human cells with a pig cDNA library and selecting for resistance to lysis by pig C [1366]. Plasmids were rescued from surviving cells and cloned. The cDNA thus obtained encoded a protein with a structure identical to that of human MCP: four SCR domains, an STP-rich region, transmembrane and cytoplasmic domains. Overall identity at the amino acid level was 42%. Alternatively spliced forms have not yet been identified. Pig MCP further resembles human MCP in that it is very widely and abundantly distributed – on all circulating cells, endothelia and epithelia [1365]. An important difference is that pig erythrocytes abundantly express MCP whereas human erythrocytes are completely negative.

## Complement receptor 1 (CR1; CD35)

Erythrocytes from non-human primates express CR1-like proteins which serve as receptors for C3b-opsonized particles. Passage of primate erythrocyte membrane extracts over columns of Sepharose-immobilized human C3b identified multiple CR1-like proteins. Proteins with molecular weights in the ranges 55–75 kDa and 130–165 kDa predominated and only in a minority of primate species was a protein with molecular weight similar to that of human CR1 (approx. 200 kDa) found in erythrocytes [1367]. Both the 200-kDa and the 75-kDa chimpanzee CR1-like molecules have been cloned [1368]. The former was a 30-SCR molecule 99% identical to human CR1 (termed chimpanzee CR1) and the latter, termed CR1a, consisted of sequence identical to SCRs 1–6, 28–30 and the transmembrane and cytoplasmic portions of chimpanzee CR1. A third CR1-like molecule, an eight-SCR-containing protein termed CR1b, was shown to be derived from a separate gene related to the human *cr1-like* genomic sequence. CR1b has been further characterized in the baboon where it encodes a 65-kDa GPI-anchored molecule which is the major C receptor on baboon erythrocytes (Figure 8.2) [1369].

An excellent example of the way in which study of CRP in other species can contribute new data on the human analogues is provided from functional and structural comparisons of human and chimpanzee CR1 [582]. Although chimpanzee and human CR1 differ at the C4b binding site by only two amino acids (Tyr37>Ser; Asp79>Gly), chimpanzee CR1 binds both C3b and C4b at this site. Single substitutions at either of these sites in human CR1 fails to confer C3b binding whereas the double substitution does. This information has helped to guide the engineering of human CR1 to generate more active C inhibitors for use in therapy.

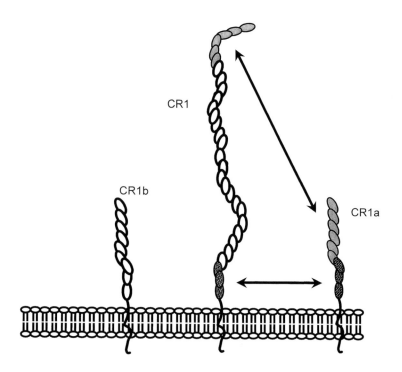

**Figure 8.2. Complement receptor 1 (CR1) analogues in primates.**
Chimpanzees and baboons express several different CR1 analogues which are likely to fulfil specific roles. In chimpanzees, the largest form, termed CR1, is a 30-short consensus repeat (SCR) molecule, 99% identical to human CR1 at the amino acid level. Despite this high degree of identity, there are functional differences in that the site in LHR1which in human CR1 binds C4b, in the chimpanzee binds both C4b and C3b. CR1a is a truncated form derived from alternative splicing in the chimpanzee *cr1* gene and comprising SCRs 1–6, 28–30 and the transmembrane and cytoplasmic portions (filled SCRs in figure). CR1b is derived from a separate gene homologous to the human *cr1-like* pseudogene and encodes a transmembrane protein consisting of eight SCRs which, in baboons, is the major erythrocyte C receptor.

The presence of C receptors on murine spleen cells and macrophages was recognised more than two decades ago [1370, 1371]. C receptors involved in immune complex clearance have also been identified on mouse platelets but not erythrocytes [1372]. Passage of murine spleen cell membrane extracts over immobilized C3i identified three distinct C3-binding proteins with molecular weights of 190 kDa, 150 kDa and 50 kDa, a pattern similar to that obtained with human spleen cells where the three C3-binding proteins are, respectively, CR1, CR2 and MCP [1373]. It was anticipated that the situation in the mouse would be comparable; however, antibodies raised against the 190-kDa protein (termed murine CR1) also recognised the 150-kDa protein (termed murine CR2), suggesting that these two proteins were much more similar in the mouse than in man. The cloning of murine CR1 and CR2 has resolved this surprising situation and explained the differences between the murine CR1/CR2 analogues and their human counterparts [1374]. C receptors 1 and 2 (CR1 and CR2), encoded on separate genes in man, are derived in the mouse from the same gene, termed *cr1/cr2*, on chromosome 1. Alternative splicing of this gene gives rise to a 15-SCR-containing transmembrane-anchored molecule (molecular weight 155 kDa) which is approximately 70% identical to human CR2 (murine CR2) and a 21-SCR-containing molecule, identical to murine CR2 except for the presence of six additional SCRs at the amino terminus (molecular weight 190 kDa; murine CR1) (Figure 8.3) [1375, 1376]. The six

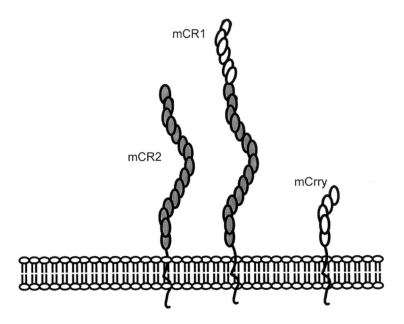

**Figure 8.3. CR1/CR2/Crry in the mouse.**
Murine CR1 and CR2 are derived from alternative splicing of the same gene (termed *cr1/cr2*), in contrast to the situation in man where these proteins are encoded on separate genes. mCR2 consists of 15 SCRs, transmembrane and cytoplasmic domains and is homologous with human CR2. mCR1 is identical to mCR2 except for the addition of six SCRs (unfilled) at the amino terminus which are closely related to the same region in human CR1. Murine Crry is a five-SCR-containing transmembrane molecule encoded on yet another gene but exhibiting high homology with the five N-terminal SCRs of mCR1.

additional SCRs in murine CR1 are homologous to human CR1 and contain binding sites for both C3b and C4b and cofactor activity for fI cleavage of these substrates [1377, 1378].

Recent data, particularly the development of knockout mice deficient in CR1 and CR2, have implicated murine CR1 and CR2 in B-cell signalling and the development of the humoral immune response [1379–1381].

In the rat a 200-kDa molecular weight C3b-binding protein has been identified on platelets, neutrophils and splenocytes [1382–1384]. It remains unclear whether this represents rat CR1.

## Crry/5I2 antigen

Regulation of the C activation pathways in rats and mice is further complicated by the presence of yet another CRP, termed Crry or 5I2 antigen. When, during the search for murine CR1, a polyclonal anti-human CR1 antiserum was used to immunoprecipitate mouse spleen cell membranes, the only protein identified had a molecular weight of 65 kDa and was initially designated p65 [1385]. Around the same time, other workers had used a probe derived from the C3b/C4b-binding region of human CR1 to screen a murine cDNA library; the screen identified a cDNA encoding a five-SCR-containing transmembrane protein which was dubbed Crry [1376, 1386]. Crry expressed in the human K562 cell line resembled p65 in many respects, including cross-reactivity with anti-human CR1 antiserum, provoking the suggestion that the proteins were identical [1376]. Crry is a 65–70-kDa molecule widely expressed on cells and tissues in the mouse [1387]. The fact that the majority of cells which expressed Crry had no surface C receptor activity, assessed by binding of C3b-coated erythrocytes, indicated that Crry did not function as a receptor. Crry was then shown to inhibit C activation in both the classical and alternative pathways, leading to the conclusion that Crry was the functional analogue of human MCP and/or DAF [1387, 1388]. Further functional analyses revealed that Crry did indeed contain both fI cofactor activity and decay-accelerating activity for the C3 convertases of the activation pathways, strengthening the conviction that it was the murine functional analogue of both MCP and DAF [1389].

Mouse Crry consists of five SCR domains, a transmembrane domain and a cytoplasmic domain [1390, 1391]. The gene is located in the murine regulators of complement activation (RCA) cluster on chromosome 1 (an area syntenous with the location of the human RCA cluster on chromosome 1q) in close proximity to the gene encoding murine CR1/CR2. The *crry* gene spans more than 25 kb of DNA and is composed of ten exons [1391]. As predicted from antibody cross-reactivity and the use of a probe derived from the sequence of human CR1 in the original cloning studies, the sequence of Crry most closely resembles that of human CR1, specifically the five amino-terminal SCRs of CR1 [1376]. Homology with human MCP and DAF is weak but SCRs 2 and 3 of Crry and SCRs 3 and 4 of mouse DAF are 50% identical at the amino acid level [1361].

A Crry-like molecule was identified in the rat by two groups and was termed respectively rat Crry or 512 antigen [1383, 1392]. Okada and co-workers had generated a monoclonal antibody, 512, which enhanced the susceptibility of rat erythrocytes to lysis by homologous C and identified the ligand for this antibody as a transmembrane-anchored molecule of molecular mass 70-kDa in rat erythrocyte membranes, hence the term 512 antigen. The first cloning of rat Crry (512 antigen) identified a six-SCR-containing molecule [1393]. The first five SCRs were highly homologous with mouse Crry while the extra SCR bore strong homology to the sixth SCR of human CR1. Shortly afterwards, others identified two forms of rat Crry cDNA expressed in all tissues and encoding respectively the six-SCR form described above and a seven-SCR form in which an exact duplication of a single SCR has occurred [1394]. The expression of these two mRNAs explains the two-band pattern (65–70 kDa and 75–80 kDa) observed for Crry in most tissues. In short, rat Crry has a structure which closely resembles that of the murine molecule (85–90% identity through the shared regions) but contains either one or two additional SCRs. Whether there are functional differences between the two forms of rat Crry has not yet been tested. As in the mouse, rat Crry is very widely and abundantly distributed, on all circulating cells, epithelia, endothelia and in most tissues examined [1395].

In summary, rats and mice abundantly express an efficient and intrinsically active CRP, undetected as yet in other species, which contains within a single molecule activities associated with two separate CRP (DAF and MCP) in other species. The central importance of this molecule in C homeostasis is vividly demonstrated by *in vivo* blocking of Crry using a monoclonal antibody in the rat [877]. Such treatment induced systemic C activation and profound hypotension, increased vascular permeability, leukocytopenia and thrombocytopenia. What then is the role of DAF and MCP in rats and mice? Evidence summarized above indicates that although mice and rats express DAF, the tissue distribution may be restricted – primarily in the testis for the mouse. Mice also express MCP, though again it appears that expression may be restricted to testis [1233]. Our interpretation of this complicated data is that the major CRP inhibiting the activation pathways on most tissues in rats and mice is Crry. DAF and MCP are expressed in specific sites, perhaps primarily for functions unrelated to regulation of C, and play little role in the regulation of C activation in these species. The pressures which have caused this surprising situation to arise in rats and mice and apparently not in other species remain undetermined. One clue is provided by the demonstration in man that both MCP and DAF have been hijacked for use by viruses and other micro-organisms as gateways into cells (see Chapter 7). It is possible that such piracy has provided a potent negative selection against the abundant and widespread expression of these molecules on murine cells. The possibility that a Crry analogue exists in other species (including man), perhaps with restricted tissue distribution and roles, also deserves further investigation.

## Fluid-phase regulators of the activation pathways

C1 esterase activity has been detected in serum from many species and analogues of C1inh have been isolated from rabbit, guinea pig, rat, mouse and bovine plasma [1396–1400]. Murine C1inh was recently cloned and shown to have a structure very similar to that of human C1inh, although overall sequence conservation was only 39% [1401]. The activity of C1inh is not species-restricted, human C1inh being effective at inhibiting C1 activation in many different species [1402].

Species analogues of fI, fH and C4bp are known from functional studies to be present in all higher mammals and in some more primitive species. Human C3b/C4b-cleaving activities have been demonstrated in the plasma of reptiles, amphibians and fish. The fragments obtained in these cleavages are identical to those generated using human serum, indicating that fI-like activity and the appropriate cofactors might be particularly ancient [1403, 1404].

FI has been cloned from *Xenopus* and shown to cleave human C3b and C4b and also to function with *Xenopus* C3 and C4 to regulate C activation in serum [1405, 1406]. The cDNA encoding murine fI was recently cloned and sequences compared between man, mouse, chicken and toad [1407]. The sequences were markedly conserved between species with the exception of a single region the 'divergent' or D region, which was poorly conserved. The biological significance of this observation is unknown.

Factor H analogues have been isolated in all species from fish to primates. Structural conservation between species is remarkable – A 24 kDa protein isolated from sand bass plasma consists of three SCRs and bears strong homology to human fH and fH-related protein 3 [1408]. Murine fH displays multiple allotypes which vary in cofactor activity, and also comprises several distinct molecular forms comparable to the large family of fH-like and fH-related proteins described in man [159, 430]. Sequence comparison between human, mouse and bovine fH has been used to identify conserved residues involved in binding of C3 and mediation of fI cofactor activity[1409]. Porcine fH is worthy of specific mention. A sub-strain of Yorkshire pigs was discovered in Norway which had a high incidence of spontaneous glomerulonephritis. Examination of the C system in this strain identified the underlying deficit as a deficiency of porcine fH which permitted uncontrolled C activation and glomerular injury[1152].

An analogue of C4bp has been described in sand bass plasma[1404] and a C4bp-like molecule has even been identified as a major haemolymph protein in insect larvae[1410]. Murine C4bp is closely related to human C4bp, consisting of six or seven α chains, each comprising multiple SCR units, joined at their carboxy terminus[1411]. Murine and human C4bp are 51% identical at the amino acid level; however, murine C4bp differs from the human protein in several respects. The murine α chains consist of only six SCRs, homologous with SCRs 1–4, 7 and 8 of the eight-SCR human α chain, and the chains are non-covalently associated at their carboxy termini (Figure 3.20). Even more striking is the observation that murine C4bp does not contain the β chain which in human C4bp interacts with protein S. It has been shown by genomic analysis that the β-chain gene has become a pseudogene in the murine genome[493]. Bovine C4bp also fails to bind protein S despite the presence of a β chain. Molecular cloning of the bovine β chain has revealed that the chain contains only two SCRs (three in human). The missing SCR is SCR1, the site of protein S binding in human C4bp[506]. It has recently been shown that C4bp in rat serum is complexed with protein S, the first non-primate example of this interaction[1412]. Both the α and β chains of rat C4bp were cloned and shown to be highly homologous to the human C4bp chains (α, 60%; β, 68%).

# REGULATORS OF THE TERMINAL PATHWAY

## CD59

The majority of the species analogues of CD59 identified to date have been isolated, characterized and cloned in our laboratory. The first to be identified was rat CD59. Rat CD59 was purified from detergent extracts of rat erythrocyte membranes, initially by classical chromatography and later, after the development of specific monoclonal antibodies, by immunoaffinity methods[1413]. The purified protein ran as a broad band of molecular mass 18–22 kDa on SDS-PAGE, was GPI-anchored, spontaneously incorporated into guinea pig erythrocytes and rendered these cells resistant to lysis by human or rat C. The protein was, like human CD59, broadly and abundantly distributed[1395, 1414]. Amino-terminal sequencing revealed high homology with human CD59, strengthening the conviction that this was the rat analogue of CD59. Molecular cloning of the cDNA confirmed this conviction [1415]. The full-length cDNA isolated from a rat kidney library encoded a protein sequence comprising a 22-amino acid signal peptide and a 104-amino acid coding region. The full sequence was 44% identical with that of human CD59; all ten of the Cys residues were conserved and there were several highly conserved sequence stretches. The carboxy-terminal portion of the protein was compatible with a GPI-anchor addition signal and the predicted site of anchor addition was at Asn-79. Within the coding region were two potential sites for N-glycosylation (Asn-16 and Asn-70) (Figure 8.4). Importantly, rat CD59 whether purified from erythrocytes or expressed as a recombinant protein inhibited C from many different species, including human [1413, 1415]. CD59 analogues purified from sheep and pig erythrocytes showed a similar lack of species selectivity, inhibiting C from most but not all species tested [1416–1418].

CD59 analogues from several non-human primates have been cloned and shown to be highly homologous to human CD59 and to inhibit effectively human C [1308, 1419]. We have gone on to clone the CD59 analogues from mouse [1420] and pig [1421]. In both of these species CD59 analogues are widely and abundantly distributed. Cloning of rabbit CD59 has also recently been reported [1069]. Comparison of the sequences between the various CD59 analogues revealed a surprisingly low degree of overall homology – only 48% identity at the amino acid level between human and pig, 44% between human and rat, and even between rat and mouse, only 65% (Figure 8.4). The reasons for this low inter-species conservation of sequence are unknown but within this rather weak overall conservation there are several regions of high homology. All, with the sole exception of rabbit CD59, have an amino-terminal Leu residue, all are N-glycosylated at or near residue 18, all are between 70 and 80 amino acids in length and are GPI-anchored, and the five intra-molecular disulphide bonds are completely conserved.

Because of the lack of species selectivity of C-inhibiting activity described above, we have predicted that the regions concerned with function will be well conserved between species. We have therefore undertaken mutagenesis studies in human and rat CD59 targeting the highly conserved regions, and have identified an active site in the molecule [1057]. Others have taken advantage of the gaps in cross-species activity to generate chimeric CD59 molecules in order to identify sites conferring species selectivity [1067, 1068].

Mouse CD59 is encoded by a gene on murine chromosome 2 in a region syntenous with the location of human CD59 on chromosome 11 [1420]. The

```
           -20        -10         1          10         20         30

HUM:  MGIQGGSVLFGLLLVLAVFCHSGHSLQCYNCPNP-TADCKTAVNCSSDFDACLITKAGLQVYN-K

RAT:  MRARRGFIL--LLL-LAVLCSTGVSLRCYNCLDP-VSSCKTNSTCSPNLDACLVAVSGKQVYQ-Q

MUR:  MRAQRGLIL--LLLLLAVFCSTAVSLTCYHCFQPVVSSCNMNSTCSPDQDSCLYAVAGMQVYQ-R

PIG:  MGSKGGFILLWLLSILAVLCHLGHSLQCYNCINP-AGSCTTAMNCSHNQDACIFVEAVPPKTYYQ


           40         50         60         70         80         90

HUM:  CWKFEHCNFNDVTTRLRENELTYYCCKKDLCNFNEQLEN--GGTSLSEKTVLLLVTPFLAAAWSLHP.

RAT:  CWRFSDCNAKFILSRLEIANVQYRCCQADLCNKSFEDKPNNGAISLLGKTALL-VTSVLAAILKPCF.

MUR:  CWKQSDCHGEIIMDQLEETKLKFRCCQFNLCNKSD------GSL-GKTPLLGTSVLVAILNLCFLSHL.

PIG:  CWRFDECNFDFISRNLAEKKLKYNCCRKDLCNKSD-----ATIS-SGKTALLVILLVATWHFCL
```

**Figure 8.4. Comparison of sequences of CD59 analogues.**

Comparison of the protein sequences of CD59 analogues from human, rat, mouse and pig CD59. Numeration refers to the predicted human CD59 sequence, with the first residue of the mature protein known from protein sequencing to be L. Vertical lines ( | ) indicate identity of conserved residues between human CD59 and CD59 in other species. Gaps inserted in the sequence to maintain alignments are indicated as (-).

gene structure closely resembles that of human CD59 (unpublished data). Current research efforts are focused on the targeted deletion of the CD59 gene in the mouse.

## Fluid-phase regulators of the terminal pathway

Both clusterin and S-protein are highly conserved proteins which have been identified and characterized in many species. As clarified in Chapter 4, these proteins have roles outside of the C system which are of key importance and probably explain their conservation through evolution. Clusterin was first identified as a major protein constituent of ram seminal plasma and was similarly identified as a seminal plasma protein in the rat [953, 962, 1422]. Although it is likely that clusterin and S-protein contribute to the regulation of the terminal pathway in other species there are, to our knowledge, no studies which have formally addressed this issue. However, data from studies of the human proteins make it unlikely that clusterin and S-protein will be of major importance as C regulators in other mammals. Indeed, mice in which the gene encoding murine S-protein has been deleted appear completely normal and show no evidence of excessive activation of C [933].

## MANIPULATION OF MEMBRANE C REGULATORS *IN VIVO*

The characterization of the rat and murine analogues of the human CRP and the generation of blocking antibodies against some of the rodent CRP has enabled studies to be undertaken in which specific CRP have been neutralized *in vivo*. In the first of these studies [1423], Fab fragments of a function-blocking anti-rat CD59 MoAb were infused into one kidney of a rat in order to block CD59 on endothelium. A glomerulonephritis was then induced in the animal by delivery of immune complexes. Inflammation and injury were much more severe in the kidney in which CD59 had been blocked than in the contralateral control kidney, indicating that CD59 on the endothelium played a role in the protection of the kidney from C-induced injury in this model. A dramatic demonstration of the role played by the CRP in homeostasis in the normal animal was provided by a study in which an identical approach was used to block CD59 and/or Crry on the endothelium (and circulating cells) throughout the animal [877]. Large amounts of the appropriate antibody (as Fab fragments) were infused into normal rats and the effects of the infusion carefully monitored. Within minutes of infusing the neutralizing anti-Crry antibody, animals developed a transient increase in blood pressure followed by a prolonged period of hypotension accompanied by a rapid fall in circulating leukocytes and platelets. C activation products were found in plasma and tissues of the animals treated with blocking anti-Crry antibody but not in those treated with anti-CD59 (Figure 8.5).

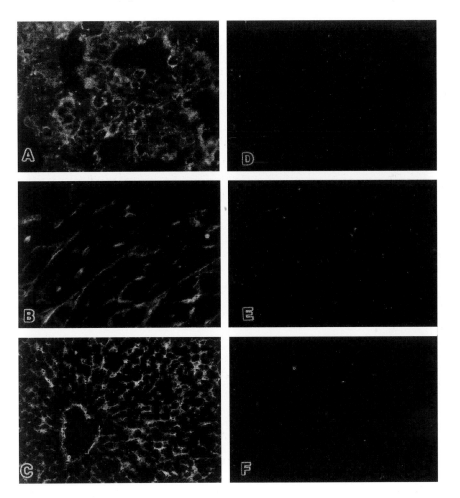

**Figure 8.5. Neutralization of Crry.**
Neutralization of Crry *in vivo* in rats causes deposition of C in the tissues. F(ab')$_2$ monoclonal antibodies capable of blocking the activities of Crry (5I2) or CD59 (6D1) were administered intravenously at a dose of 1 mg/kg. Physiological and haematological parameters were monitored. At sacrifice, binding of the antibody and C fragments in tissues was assessed by immunofluorescence. The figure shows immunofluorescence staining for C3 in animals given anti-Crry (A–C) or anti-CD59 (D–F) in lung (A, D), heart (B, E) and liver (C, F).

The combination of the approach described above to neutralize *in vivo* the rodent CRP and the imminent availability of mice in which individual CRP have been deleted will permit a detailed examination of the roles and interactions of the various CRP *in vivo* in health and disease.

# Chapter 9

# COMPLEMENT REGULATORS IN THERAPY

## INTRODUCTION

The C system developed as an efficient means of rapidly destroying invading micro-organisms, either directly by lysis or indirectly by recruiting and activating phagocytes. However, C also carries a considerable potential for self-injury. As a reflection of this fact C is often referred to as a 'double-edged sword'. If C is activated to an excessive degree or in an inappropriate site the proinflammatory products generated will contribute to pathology. In many inflammatory and infectious diseases C, although not the primary cause of the disease, is a major factor in exacerbating and perpetuating injury. Therapies which inhibit C activation can therefore break the proinflammatory cycle and reduce inflammation. Application of this attractive strategy has until recently been limited by the lack of specific and effective pharmacological inhibitors of C. Numerous agents have been proposed as C-inhibitory therapies based upon their anti-C activities *in vitro* but all have proved disappointing when tested *in vivo*. The naturally occurring C regulators described in this volume are excellent inhibitors of C. Both fluid-phase and appropriately modified membrane-associated C regulators have potential as therapeutic agents and should provide models to permit the design of even better anti-C therapeutics. The first steps in this process have begun and are discussed in this chapter. While the results obtained are encouraging, it should be emphasized that this is a field in its infancy. Developments over the next decade are likely to revolutionize the clinical use of anti-C therapies.

## THERAPY WITH SOLUBLE C REGULATORS

### Therapy with C1 inhibitor (C1inh)

C1inh was the first of the naturally occurring C regulators to be used therapeutically, as a replacement therapy in hereditary angiedema (HAE) [152].

In this context, C1inh purified from plasma proved safe and effective in acute attacks of HAE, rapidly inhibiting the activity of C1 and other proteolytic cascades and resolving the oedema. The role of C1inh in therapy of HAE is described in detail in Chapter 5.

Little attention has been given to the potential role of C1inh in other therapeutic contexts. C1inh has been shown to be effective in therapy of several experimental models of shock, including septic shock [1424] and in rodent models of acute pancreatitis [1425, 1426]. A beneficial effect of C1inh therapy has also been demonstrated in feline and porcine models of myocardial infarction [1427, 1428]. In man, high dose therapy with C1inh has been shown to be of benefit in septic shock and the related vascular leak syndrome [1429] and has been used to reduce the toxic effects of therapy with high dose interleukin-2 [1430]. It is surprising that this inexpensive and effective C regulator, with a 35-year history of safe use, has yet to find wider applications. To the best of our knowledge large-scale human trials of the efficacy of C1inh in the various shock syndromes and reperfusion injuries (including myocardial infarction) have yet to take place.

## Therapy with other fluid-phase regulators of C

The C3 convertases are inhibited from the fluid phase by fH and C4bp which act as essential cofactors for fI cleavage of C3b or C4b in the convertases of the alternative and classical pathways respectively. These cofactors are present in plasma at relatively high concentrations (fH, 400 mg/l; C4bp, 200 mg/l) and together are responsible for restricting C activation in the fluid phase. This essential role is graphically illustrated in individuals deficient in fH or fI in whom uncontrolled activation of C rapidly consumes all the plasma C activity. However, despite their high concentration in plasma, fH and C4bp do not prevent activation of C on activating surfaces such as bacterial capsules. Indeed, the inability of the fluid-phase C3 convertase inhibitors to effectively block C activation on surfaces is an essential prerequisite for an efficient C system capable of activating rapidly and in an unimpeded manner on foreign surfaces, and is the reason for the existence on host cells of additional defence in the form of the membrane CRP [1431]. The fact that the fluid-phase C3 convertase inhibitors fail to inhibit C activation on surfaces despite their high concentration in plasma makes it unlikely that they would prove effective as therapeutics. To bring about even a 50% increase in the plasma concentration of fH would require administration of several grams of protein! Nevertheless, modified forms of C4bp and/or fH with improved efficiencies of inhibition might yield agents of sufficient activity to be used as therapeutics. However, this potential direction has been neglected in favour of developments described below involving the membrane regulators of the C3 convertases.

A similar argument can be made with regard to the fluid-phase inhibitors of the terminal pathway, S-protein and clusterin. Already present in plasma at high concentrations (S-protein, 500 mg/l; clusterin, 250 mg/l) and acting to mop up C5b-7 complexes in a rather non-specific manner, it is unlikely that the modest increases in plasma concentration which could be achieved by administration of recombinant forms of these agents could effectively inhibit C.

## Therapy with recombinant soluble forms of the membrane regulators

Unlike the fluid-phase regulators, the membrane regulators have evolved to inhibit C activation on surfaces in order to protect host tissue from damage in the face of ongoing C activation. Their roles in protecting the host are graphically illustrated by experimental blockade of one or several of these regulators in experimental animals. Systemic blockade of the rat C3 convertase inhibitor 512 Ag (Crry) permitted unregulated C activation on endothelia and precipitated profound shock, further exacerbated if the rat analogue of CD59 was simultaneously blocked[877].

In the late 1980s, Fearon advanced the concept that soluble forms of the membrane regulators might retain their capacity to inhibit efficiently C activation on surfaces. To test this concept he engineered a recombinant soluble form of the membrane regulator complement receptor 1 (sCR1)[1432–1434]. The resultant molecule, comprising the 30 short consensus repeats (SCRs) (the 60-amino acid repeating units which are found in all C3 convertase regulators) of the most common isotype of CR1 (see chapter 3), but lacking the transmembrane and cytoplasmic portions of the molecule, was expressed as a soluble protein, initially in insect cells and later in mammalian cells (Figure 9.1). Recombinant sCR1 was shown to be a very potent inhibitor of C activation when present in the fluid-phase with a molar inhibitory activity some 20-fold greater than that of fH. The recombinant molecule also inhibited C in serum from numerous other species, allowing the *in vivo* testing of sCR1 in animal models of disease[1435].

The original studies reported the protective effects of sCR1 in a rat model of myocardial infarction, cardiac ischaemia–reperfusion injury[1434]. In the myocardial ischaemia-reperfusion model an area of myocardium is made ischaemic by clamping off a coronary artery in the anaesthetized rat. After an ischaemic interval of several hours, the clamp is released. Severe

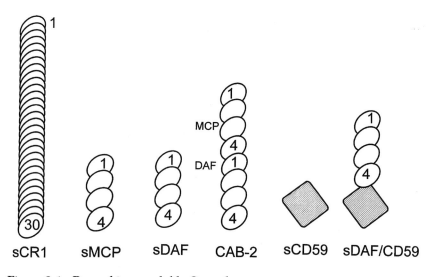

**Figure 9.1. Recombinant soluble C regulators.**

Recombinant soluble forms of each of the major membrane complement regulatory proteins (CRPs) have been generated. The figure illustrates the structures of these molecules. Recombinant soluble CR1 (sCR1) consists of the 30 short consensus repeats (SCRs, represented as overlapping ovals) of CR1 which represent almost all of the extracellular portion of this molecule (1 is N-terminal). Recombinant soluble membrane cofactor protein (sMCP) and decay-accelerating factor (sDAF) consist of the four SCRs which contain the C regulatory activities. CAB-2 is a chimaeric molecule in which sMCP is attached through the C-terminus to the N-terminus of sDAF. CD59 is a globular molecule with no SCR structures. DAF/CD59 chimaeras have been expressed on cells but not, to date, as the soluble molecules depicted here.

inflammation and necrosis of the myocardium accompany reperfusion. Intravenous administration of a single dose of sCR1 at the time of reperfusion markedly decreased the amount of myocardial damage, confirming that a substantial proportion of the damage was mediated by C.

Subsequent work has shown that sCR1 is effective at preventing or reducing injury in numerous animal models of disease, many but not all of which involve ischaemia–reperfusion injury of an organ or tissue (Table 9.1). We have used sCR1 in therapy of experimental demyelination in the rat[1436]. In this model, administration of sCR1 daily for several days from just prior to the onset of clinical disease almost completely prevented demyelination and markedly reduced inflammation and disease (Figure 9.2). A therapeutic effect of sCR1 on established disease, an effect more relevant to therapy of human demyelinating disease, has proved difficult to

**Table 9.1.** *Soluble complement receptor 1 (sCR1) in animal models of disease.*

| Model disease | Effect of sCR1 | Reference |
|---|---|---|
| Myocardial ischaemia–reperfusion (rat) | Reduced infarct size | (1432, 1433) |
| Intestinal ischaemia–reperfusion (rat) | Reduced local (gut) and remote (lung) inflammation and tissue destruction | (1513, 1514) |
| Skeletal muscle ischaemia–reperfusion (mouse) | Improved muscle reperfusion and muscle viability, reduced lung injury | (1515) |
| Liver ischaemia–reperfusion (rat) | Improved liver function, reduced release of liver enzymes | (1516) |
| Thermal injury (rat) | Reduced local (skin) and remote (lung) inflammation and tissue destruction | (1517) |
| Xenograft rejection (guinea pig to rat; pig to primate) | Increased survival time of xenograft | (1498, 1499) |
| Allograft rejection (rat) | Reduced inflammation in renal allograft | (1518) |
| Experimental demyelination (rat) | Reduced brain inflammation and myelin loss | (1436) |
| Traumatic brain injury (rat) | Reduced neutrophil accumulation in brain | (1519) |
| Experimental glomerulonephritis (rat) | Improved renal function, reduced glomerular damage and proteinuria | (1520) |
| Haemodialysis (*ex vivo*) | Reduced neutrophil activation, reduced lung injury | (1521, 1522) |
| Acid instillation in lung (rat) | Decreased neutrophil accumulation in lung, decreased plasma tumour necrosis factor (TNF) | (1523) |
| Experimental myaesthenia gravis (rat) | Reduced loss of acetylcholine receptors; improved muscle function | (1524) |
| Experimental autoimmune neuritis (rat) | Improved nerve conduction, reduced incidence of paralysis | (1525) |

demonstrate in this fulminant model. These experiments proved that it was possible to inhibit C in rats for several days by daily or twice-daily infusion of sCR1. Beyond day 6, C inhibition and the therapeutic effect were lost as the rats developed serum antibodies against the human CR1. In almost all the applications so far described, sCR1 has been given systemically, usually intravenously. However, we have recently shown that local therapy with sCR1 can also be effective. Antigen arthritis is

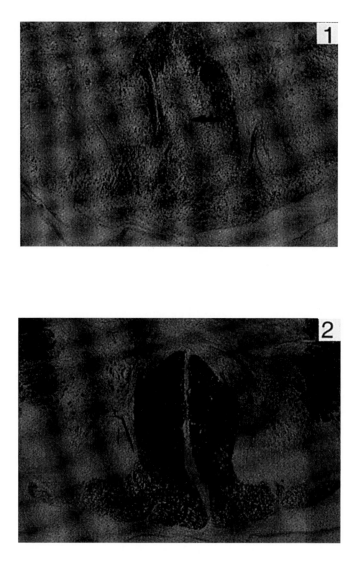

**Figure 9.2. Therapy with sCR1 in experimental demyelination.**

Sections from cervical spinal cords from rats at day 12 in the course of induction of experimental demyelination. Myelin is stained dark (blue) using the Luxol fast blue myelin stain. Magnification × 60. **1.** No therapy; severe loss of myelin is apparent, particularly around vessels (arrowed). **2.** Treated with sCR1, 20 mg/kg/day by intraperitoneal injection from day 8; relative sparing of myelin with demyelination restricted to the areas immediately surrounding blood vessels. Modified from Piddlesden *et al.* (1994), *Journal of Immunology* **152**; 5477.

induced in rats by the intra-articular injection of antigen into a primed recipient; intra-articular therapy with sCR1 at the time of disease onset markedly inhibited arthritis as assessed by joint swelling and joint histology (Figure 9.3)[(1437)].

One problem with sCR1 as a potential therapeutic agent is its short plasma half-life, particularly in experimental animals. The β-phase half-life has been estimated at 8 hours in man but only 1.7 hours in rodents. A variety of approaches have been adopted in order to increase the plasma half-life, permitting lower doses and less frequent administration of agent in therapy.

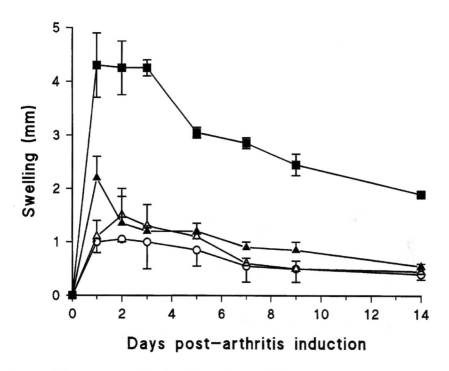

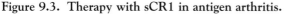

**Days post–arthritis induction**

**Figure 9.3. Therapy with sCR1 in antigen arthritis.**

A uniarticular antigen arthritis was induced in Lewis rats by intra-articular (i.a.) injection of methylated BSA (m-BSA) in animals previously immunized with m-BSA. Disease was either sham-treated (closed squares), treated with a single i.a. injection of sCR1 (200 μg; closed triangles), with a combination of intravenous (i.v.; 20 mg/kg/day) and i.a. sCR1 (open circles), or a combination of i.v. CVF (100 I.U.) and i.a. sCR1 (open triangles). Disease was assessed by measurement of knee swelling using digital calipers and comparison with the uninjected knee. All treatment regimens significantly inhibited disease and i.a. sCR1 alone was as effective as combinations of i.a. and i.v. therapy. Modified from Goodfellow *et al.* (1997), *Clinical and Experimental Immunology* **110**; 45.

Apart from its short half-life, the use of the human protein sCR1 in animal models is also limited by its antigenicity. In rats given anti-sCR1 daily at doses required for effective C inhibition, sCR1 antibodies are detectable within 4 days and by day 6 are present in sufficient quantities to abrogate the anti-C effect of sCR1 (our unpublished data). Longer term anti-C therapy in rodents and meaningful comparisons with the likely effects of inhibitors in human disease will require therapy with recombinant soluble forms of the rodent membrane inhibitors; such reagents are currently under development in several laboratories.

The Boston-based company T Cell Sciences Inc. have driven the development of sCR1 (termed TP10 by T Cell Sciences) for use in therapy of human disease. Safety studies have been completed; the agent has proven itself to be well-tolerated and, so far, non-antigenic. However, it remains a possibility that sCR1 will prove antigenic in some individuals because of the polymorphic nature of the CR1 molecule. Clinical trials using sCR1 have begun in the adult respiratory distress syndrome, myocardial infarction and lung transplantation (T Cell Sciences; personal communication). All of these initial disease targets can be considered as shock and/or ischaemia–reperfusion injuries and all are acute. These are the appropriate first targets for this new agent but applications may eventually extend to chronic autoimmune disorders such as multiple sclerosis and rheumatoid arthritis.

Recombinant soluble forms of other membrane regulators have also been engineered and tested for their C-inhibiting activity. Soluble forms of decay-accelerating factor (DAF) and membrane cofactor protein (MCP) have been engineered and expressed using strategies similar to that described for sCR1 [1438, 1439]. For both, the expressed molecules comprised the four SCRs together with the Ser/Thr/Pro (STP) region but without the transmembrane and cytoplasmic portions in the case of MCP, and without the GPI-anchoring signal in the case of DAF (Figure 9.1). In the membrane-associated molecules the STP region acts as a rigid 'spacer' placing the active site-containing SCRs in the appropriate position relative to the membrane. It is thus unlikely to be required for function in the soluble molecule. Both sDAF and sMCP have been shown to inhibit C activation *in vitro* and *in vivo* in experimental animals [1438-1440].

A recent report provided a partial comparison of the relative C inhibitory efficiencies of the three recombinant soluble convertase inhibitors, sDAF, sMCP and sCR1 [1440]. As expected, sCR1 was the most efficient inhibitor on a molar basis for the classical pathway of C. A combination of sDAF and sMCP was more effective than either alone and approached the efficacy of sCR1. In the alternative pathway, sMCP and sCR1 were almost equally effective whereas sDAF was a poor inhibitor. Soluble CR1 contains cofactor and decay-accelerating activities for the classical and alternative pathway convertases in a single molecule and it seems likely that this unique combination of activities confers significant advantages over other recombinant CRP. Neither sDAF nor sMCP has been widely tested in animal models of disease, although the latter has recently been shown to be effective in a mouse heart to rat xenograft model and in the reverse passive Arthus reaction in rabbits [1439, 1440]. To our knowledge, neither has been tested in humans.

It is clear from the above studies that an ideal inhibitor of the activation pathways should combine decay-accelerating and cofactor activities in a single molecule; indeed, this logic has recently been exploited with the development of a hybrid recombinant molecule in which DAF and MCP have been combined (Figure 9.1) [1441, 1442]. The soluble form of this chimaera, termed CAB-2, combines decay-accelerating and cofactor activities in a single molecule and has proven more effective as a C inhibitor *in vitro* and *in vivo* than either parent molecule or even a combination of the two parent molecules. Whether this agent, currently in the early stages of testing in humans, offers any significant advantage over sCR1 is unclear at present.

CD59 inhibits at the stage of assembly of the membrane attack complex (MAC) [1443]. There are some theoretical advantages inherent in the use in therapy of agents inhibiting only this lytic pathway. Many of the pathological effects of C are mediated by the MAC whereas the bulk of the physiological effects are mediated by products of the activation pathways [114, 879]. Therefore inhibitors acting early in the pathway and switching off activation are more likely to predispose to bacterial infections and immune complex disease, particularly if used for long-term therapy. In contrast, a specific inhibitor of the MAC is unlikely to cause such problems. Indeed, deficiencies of components of the MAC have remarkably few consequences, the sole associated clinical problem being an increased susceptibility to infections with organisms of the genus *Neisseria* [73]. Soluble forms of CD59 lacking the GPI anchor have been extracted from urine and generated by recombinant techniques [1033, 1444, 1445]. Soluble CD59 (sCD59) inhibits MAC formation and target lysis when tested in a serum-free reactive lysis system using purified C components, demonstrating that the soluble molecule can interact with and inhibit the forming MAC (Figure 9.4). However, sCD59 loses its inhibitory activity in the presence of serum, a consequence of its association with serum lipoproteins [1444]. Recombinant sCD59 thus cannot achieve significant inhibition of MAC generation *in vivo* even at very high doses (our unpublished data).

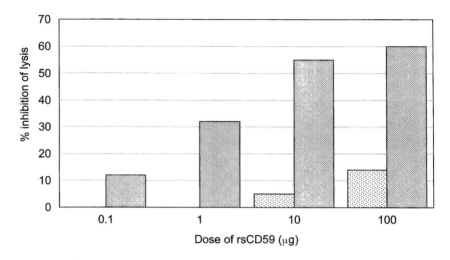

**Figure 9.4. C inhibition by recombinant soluble CD59.**
The membrane attack complex (MAC)-inhibitory effects of different doses of recombinant soluble (rs) CD59 on lysis of indicator guinea pig erythrocytes bearing preformed human C5b-7 sites. Lysis was developed either by adding EDTA-serum (1:10) as a source of C8 and C9 (dotted bars) or by adding purified human C8 and C9 (0.5 μg C8, 5 μg C9; solid bars). Data are presented as percentage inhibition of lysis. In each case, lysis in the absence of rsCD59 was 80–95%. Whereas rsCD59 was effective at inhibiting lysis when purified components were used, inhibition was only seen at the highest doses in the presence of serum.

Modification of sCD59 to retain inhibitory activity in the presence of serum presents a difficult challenge. Strong clues to the active site of CD59 have emerged in 1997–98 (see Chapter 4 for details), and it is possible that knowledge of the active site will permit the truncation of CD59 or the generation of active site peptides to reduce the binding to lipoproteins while retaining MAC-inhibitory activity. It may also be possible to target sCD59 to specific tissue sites by generating chimaeras of CD59 and antibody fragments. Perhaps simplest of all, it might be possible to deliver CD59, either GPI-anchored or soluble, directly to tissue sites such as the joint in arthritis or the brain in demyelination.

A molecule which inhibited C both in the activation pathways and in the terminal pathway might also be of use in therapy. The first step towards the generation of such an agent has been the engineering of a DAF-CD59 chimaeric molecule which, when expressed on the cell membrane, retained the activities of both parent molecules (Figure 9.1) [1446]. The activities of the component molecules were only retained when the chimaera was assembled such that DAF, truncated in the STP region, was attached through its C-terminus to the N-terminus of CD59 which was anchored normally to the membrane. Assembly in the reverse orientation resulted in expression of an inactive chimaera, indicating that the spatial relationships between the inhibitors and their respective ligands are very important. Whether a soluble form of this DAF-CD59 chimaera would be capable of inhibiting C from the fluid phase remains to be demonstrated.

It should be clear from the above account that the current 'first wave' of C regulators developed from the membrane CRP are far from ideal. We are now reaching a stage where the properties of these first-generation regulators are being optimized for use in therapy. To take sCR1 as an example, at least three complementary approaches to optimization are underway. First, molecules are being developed which contain only the active sites by deleting the portions of the molecule redundant for activity in the plasma. The SCRs of sCR1 are arranged in four sets of seven, termed long homologous repeats (Figure 9.1). Only LHRs 1, 2 and 3 are involved in ligand binding and only the first three SCRs in each LHR are required for this activity, although the spacing of the LHRs is likely to be essential for the multivalency intrinsic to CR1. Molecules containing only these active portions or multiple copies in tandem of the active portions are currently under development and may provide more effective C inhibitors. A recent report described the generation of a truncated form of sCR1, lacking LHR-A within which resides the C4b binding site [1447, 1448]. This modified sCR1 was a specific inhibitor of the alternative pathway, active at inhibiting C *in vivo*. Secondly, strategies are being developed for selective targeting to areas of C activation. Modification of sCR1 to express multiple copies of the sialyl-Lewis-X carbohydrate epitope yields an agent which will bind P-selectin expressed by activated endothelia at inflammatory sites [1449, 1450]. Thirdly, modifications aimed at extending the half-life *in vivo* are being explored. Fusion of the albumin-binding domains from streptococcal protein G to the carboxy terminus of sCR1 caused a doubling of the *in vivo* half-life in the rat without adversely affecting the activity of the protein [1451]. Others have expressed a truncated form of sCR1 as a fusion protein with Ig F(ab')$_2$ [1452–1454]. The resultant (CR1)$_2$- F(ab')$_2$ chimaeras were effective inhibitors of C activation. Expression of sCR1 or truncated forms of the molecule as chimaeras with Ig Fc or intact IgG might be more effective at increasing the plasma half-life because of the very long half-life of IgG in plasma. Other strategies for modifying sCR1 and improvements in the other soluble recombinant C regulators will undoubtedly follow.

## ANTIBODIES AGAINST C COMPONENTS AS THERAPEUTICS

A very different strategy for inhibition of C has been adopted by some research groups. They have exploited the observation that some mono-clonal antibodies against key components of C can effectively remove the component, thereby preventing activation of C beyond that point. Most attention has focused on antibody against C5. A murine monoclonal antibody against mouse C5 generated in C5-deficient mice was shown to have the unique property of binding C5 in a manner which prevented its cleavage by the C5-convertases of the activation pathways [1455–1459]. Like sCD59, this agent inhibits MAC while permitting opsonization. A poten-tial advantage over sCD59 is that the agent also inhibits formation of the powerful anaphylactic and chemotactic peptide C5a. Therapy with this antibody inhibited generation of MAC and C5a in mice and rats. This antibody has been shown to be very effective in prophylaxis of model dis-eases in which C activation is known to contribute [1458]. An anti-human C5 monoclonal antibody with similar blocking properties, N19/8, has been generated [1460]. This murine monoclonal antibody has been engin-eered to produce a recombinant single-chain Fv (scFv) antibody and a humanized antibody, both of which retain the capacity to bind C5 with high affinity and inhibit cleavage, have reasonably long half-lives and are well-tolerated *in vivo* [1461, 1462]. These agents are in the early stages of testing in humans.

Antibodies targeting C5a or the C5a receptor and peptide blockers of the C5a receptor have also been developed and proposed as therapeutic agents [1463–1465]. Comparison of the therapeutic effects of anti-C5 (blocking C5a and MAC) and anti-C5a or anti-C5a receptor (blocking C5a alone) in experimental disease will provide important information on the relative roles of these two effectors in pathology.

## COMPLEMENT REGULATORS IN XENOTRANSPLANTATION

Xenotransplantation refers to the transplantation of organs, tissues or cells across species barriers. There is a long and almost universally disastrous history of attempts at xenotransplantation. Perhaps the first to try were the Ancient Greeks, who attempted transfusions of animal blood into people with predictably dire consequences. During the 1960s and 1970s, as human organ transplantation became routine, several groups experimented with limited success with organs harvested from primates, initially in renal transplantation and later in cardiac transplantation, the latter in a neonate [1466–1468]. The problem facing transplant surgeons is that there is an enormous shortfall in the numbers of human donor organs available for transplant. As a consequence, the majority of patients who require a heart or liver transplant will die before an organ becomes available.

The ideal scenario would be to have an abundant source of organs which could be harvested as needed and transplanted safely and successfully. The prospect of using non-human primates as a source of organs is superficially attractive because the immune response to transplantation between such closely related (*concordant*) species is relatively mild [1469]. However, the use of primate organs for a large-scale transplant programme is not feasible because of problems of supply, cost, infection and ethical considerations [1470]. Attention has thus switched to domestic animals as donors, specifically the pig [1471]. Pigs can be bred in large numbers and in germ-free environments. The major target organs – heart, lung, liver, kidney – are similar in size and function to human organs. However, a major hurdle stands in the way of pig-to-human transplants, that of hyperacute rejection [1472, 1473].

Transplants between distantly related or *discordant* species (e.g. pig–man; mouse–guinea pig) result in the rapid destruction of the donor organ, as natural antibody and C from the recipient serum attack the endothelium [1472, 1474, 1475]. With the realization that hyperacute rejection was mediated by C attack [1476] came the suggestion that the human C regulators might be utilized to overcome this hurdle [133, 1473].

The favoured approach has been to utilize molecular engineering techniques to generate animals modified in a manner which renders the organs resistant to hyperacute rejection when transplanted into humans. This has involved the introduction into the donor animal of the genes encoding the human CRP. The validity of the concept was first demonstrated by transfection of cultured cells [721, 1477] and later confirmed by creating transgenic mice, a relatively straightforward procedure [1478–1480]. The crucial site at which protection must be provided is the endothelium, the site of C attack (Figure 9.5). Mice abundantly expressing human DAF, CD59 or MCP on endothelium and other sites have been created [1481, 1482]. Hearts harvested from each of these transgenic strains could be perfused *ex vivo* with human blood or serum for prolonged periods whereas non-transgenic hearts were rapidly destroyed by C activation [1480]. Encouraged by these results, several groups have gone on to generate by transgenesis pigs expressing human C regulators on endothelium and other tissues [1483, 1484]. Organs from pigs transgenic for human DAF were resistant to C damage when perfused with human blood *ex vivo* and when transplanted into

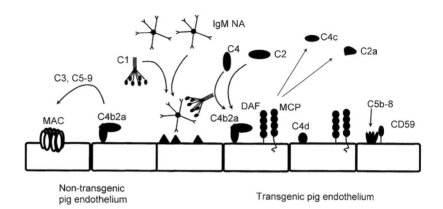

**Figure 9.5. Modification of pig endothelium to express human CRP.**

On pig endothelium IgM natural antibody binds carbohydrate epitopes (▲) and subsequently binds and activates C1. On unmodified endothelium (left of centre) C activation proceeds through cleavage of C4 and C2, formation of the C4b2a convertase, cleavage of C3 and C5 and initiation of the membrane attack pathway. On transgenic endothelium expressing the human CRP, the C4b2a convertase is rapidly inactivated by the expressed human DAF (causing dissociation of C2a) and MCP (catalysing cleavage of C4b). Any membrane attack pathway activation occurring despite this control will be intercepted by human CD59 before formation of the lytic MAC.

primates [1483, 1485–1487]. High levels of DAF expression were obtained in the pig organs, levels greater than the natural expression in the equivalent human organ, and it is likely but unproven that hyperexpression of the human C regulatory protein is essential to survival of the organ. Others have generated pigs expressing human MCP and/or CD59 and have shown similar protection against human C in *ex vivo* perfused hearts [1488–1490].

A second molecular approach to avoiding hyperacute rejection has involved the engineering of pigs such that the endothelium is no longer a target for natural antibody. The major target on pig endothelium for human IgM natural antibody is the carbohydrate epitope Galα1-3Gal, present in pigs and other non-primate mammals but absent in man and other primates [1491, 1492]. A clever tactic has been adopted to eliminate this epitope. The enzyme fucosyl transferase, hyperexpressed in transgenic pigs, competes with the natural glycosylation system to replace the Galα1-3Gal epitope with a fucosylated sugar not recognised by natural antibody (Figure 9.6) [1493, 1494]. The success of this strategy is dependent on the assumption that classical pathway activation triggered by natural antibody is the primary route for C activation. However, there is considerable

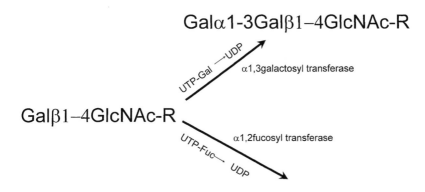

**Figure 9.6. Modification of pig endothelium to inhibit expression of galα1-3gal epitope.**

Pig tissues express the enzyme α1,3galactosyl transferase, absent in man and higher primates, which generates the α galactosyl epitope, the major target on pig cells for human natural antibody. Hyperexpression by transgenesis in pig endothelium of the enzyme α1,2fucosyl transferase will cause the bulk of the substrate, N-acetylactosamine, to become fucosylated, greatly reducing the expression of the α-galactosyl epitope on the endothelium.

evidence that the porcine endothelium also triggers activation of the alternative pathway, even in the absence of natural antibody[1495–1497]. Nevertheless, a combination of these two complementary approaches, that is hyperexpressing human CRP and eliminating binding of natural antibody, holds enormous promise for the future.

It is not at all clear that the engineering approaches described above are essential to permit survival of a xenograft. There is abundant evidence that the systemic inhibition of C, either using sCR1 or the C-depleting agent CVF, markedly prolongs the survival of a discordant xenograft[1498–1501]. Indeed, a number of other agents which cause systemic inhibition of C, including the small molecule C inhibitors FK506[1502] and K76-COOH[1503], the monoclonal anti-C5 antibody described earlier in this chapter[1456] and intravenous immunoglobulin[1504] also significantly extend the lifetime of the xenotransplanted organ. In many cases survival extends beyond the period of C inhibition, suggesting that the vigorous attack characteristic of hyperacute rejection does not occur if the organ has been present in the C-inhibited host for several days, even when host C levels return to normal. This phenomenon, known as *accommodation* remains somewhat mysterious but probably relates to changes in the endothelium of the organ which render it unreactive with natural antibody and much less C-activating [1505, 1506]. It is possible that relatively short-term inhibition of C using sCR1 or other inhibitors would be sufficient to circumvent hyperacute rejection in pig-human transplants. Even when engineered organs are used, there is certainly a case for the concomitant administration of systemic inhibitors of C to further reduce the risk of damage to the endothelium. The part played by the endogenous pig C inhibitors in protection against C has also received little attention. The pig analogues of CD59 and MCP have been isolated and characterized and both have been shown to have the capacity to inhibit human C [1365, 1418]. These findings further indicate that protection of the xenograft is more a factor of the amount of the CRP expressed on the endothelium than of the species source of the CRP.

## ANTI-CRP ANTIBODIES AND THERAPY

The above sections all describe the application of modified forms of the naturally occurring CRP to the inhibition of C *in vivo*. However, there are situations where it might be advantageous to 'switch off' CRP at a particular site in order to increase susceptibility of a cell or tissue to C lysis. The obvious example is in cancer. Many tumours abundantly express CRP, often at levels much greater than in the equivalent non-neoplastic tissue [1029, 1507–1509]. As a consequence, the tumour cells are resistant to killing by C even when C-fixing anti-tumour antibodies are present. *In vitro*, antibodies blocking the function of the various CRP can markedly increase the sensitivity of tumour cells to C killing. In order to utilize this finding *in vivo* to enhance tumour clearance, it is necessary to develop strategies which will target the anti-CRP to the tumour. We and others have generated chimaeric antibodies with one 'arm' specific for the tumour cell and the other capable of blocking a CRP on the tumour cell [1510, 1526] (Figure 9.7). By judicious selection of the two constituent antibodies it may be possible to deliver the CRP-blocking activity specifically to the tumour, even when the CRP is broadly expressed in other cells and tissues. The constructs can themselves be made C-activating by engineering on one or more Fc, thereby creating a potent anti-tumour agent (Figure 9.7). Molecular approaches to the generation of these chimaeric immuno-toxins, now being developed here and elsewhere, will provide a far greater range of reagents for testing as therapeutics.

A second situation in which such immunotoxins might be of benefit is in the therapy of certain infections. Viruses and other pathogens frequently subvert CRP to protect infected cells and shed virus (see chapter

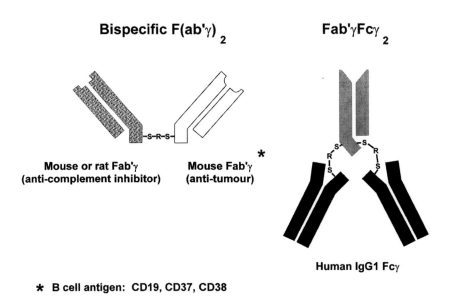

**Bispecific F(ab'γ)₂**

**Fab'γFcγ₂**

**Mouse or rat Fab'γ**
**(anti-complement inhibitor)**

**Mouse Fab'γ**
**(anti-tumour)**

**Human IgG1 Fcγ**

**\* B cell antigen: CD19, CD37, CD38**

**Figure 9.7. Anti-CRP antibodies in immunotherapy.**

Bispecific F(ab')₂ containing determinants recognising the tumour cell target and a membrane CRP (neutralizing antibody) are generated chemically from the individual F(ab') fragments. Lysis can be instigated by using an anti-tumour cell F(ab') to which are coupled two or more Fc portions. Combinations of these molecules containing anti-inhibitor/anti-tumour reactivity with C-activating capacity can also be engineered. The focus so far has been on targeting B-cell epitopes with a view to therapy in leukaemia and lymphoma.

7). HIV provides a good example of a virus which utilizes CRP to protect itself from C attack[426]. Immunotoxins which neutralize CRP and fix C on virally infected cells and shed virus might dramatically increase the efficiency with which serum kills these agents. In this context it is interesting to note that viruses utilize not only the membrane CRP but also the fluid-phase CRP, notably fH. Neutralizing antibodies against fH, appropriately targeted, might thus also find a place in the therapy of infection[426, 1511, 1512].

## CONCLUDING REMARKS

The concept of using the naturally occurring membrane CRP as models for the design of therapeutic agents is in its infancy. The enormous interest which the first generation of agents has generated among clinicians is not surprising, given that they offer for the first time the prospect of specific and effective anti-C therapies. These agents are far from ideal as therapeutics, but the current wealth of research aimed at a better understanding of the C system and its control will provide the information necessary for the development of better agents in the near future. It is anticipated that within a few years clinicians will have at their disposal a number of anti-C therapies based upon CRP and anti-C antibodies.

# REFERENCES

1. Buchner, H. 1889. Uber die nahere natur der bakterientoden substanz in blutserum. *Zentralblatt für Bakteriologie* **6**:561.
2. Ferrata, A. 1907. Die univerksamkeit der komplex haemolysise in salzfrein losungen und ihre ursache. *Berliner Klinische Wochenschrift* **44**:366.
3. Mayer, M. M. 1984. Complement. Historical perspectives and some current issues. *Complement* **1**:2.
4. Turner, M. W. 1991. Deficiency of mannan binding protein—a new complement deficiency syndrome. *Clinical and Experimental Immunology* **86**:53.
5. Reid, K. B. and M. W. Turner. 1994. Mammalian lectins in activation and clearance mechanisms involving the complement system. *Springer Seminars in Immunopathology* **15**:307.
6. Ehrlich, P., and J. Morgenroth. 1899. Zur theorie der lysinwirkung. *Berliner Klinische Wochenschrift* **36**:6.
7. Ehrlich, P., and J. Morgenroth. 1900. Ueber haemolysin dritte Mitteilung. *Berliner Klinische Wochenschrift* **37**:453.
8. Taylor, P. W. 1993. Non-immunoglobulin activators of the complement system. In *Activators and inhibitors of the complement system*. R. B. Sim, ed. Kluwer, Amsterdam, p. 37.
9. Hughes-Jones, N. C., B. D. Gorick, N. G. Miller, and J. C. Howard. 1984. IgG pair formation on one antigenic molecule is the main mechanism of synergy between antibodies in complement-mediated lysis. *European Journal of Immunology* **14**:974.
10. Loos, M. 1985. The complement system: activation and control. *Current Topics in Microbiology and Immunology* **121**:7.
11. Loos, M. 1988. 'Classical' pathway of activation. In *The complement system*. K. Rother, Till, G.O., eds. Springer, Berlin, p. 136.
12. Ziccardi, R. J., and N. R. Cooper. 1977. The subunit composition and sedimentation properties of human C1. *Journal of Immunology* **118**:2047.
13. Ziccardi, R. J. 1983. The first component of human complement (C1): activation and control. *Springer Seminars Immunopathology* **6**:213.
14. Porter, R. R., and K. B. Reid. 1979. Activation of the complement system by antibody–antigen complexes: the classical pathway. *Advances in Protein Chemistry* **33**:1.
15. Reid, K. B. 1983. Proteins involved in the activation and control of the two pathways of human complement. *Biochemical Society Transactions* **11**:1.
16. Loos, M., and M. Colomb. 1993. C1, the first component of complement:

structure-function-relationship of C1q and collectins (MBP, SP-A, SP-D, conglutinin), C1-esterases (C1r and C1s), and C1-inhibitor in health and disease. *Behring Institut Mitteilungen* **93**:1.

17. Sellar, G. C., D. J. Blake, and Reid, K. B. 1991. Characterization of the genes encoding the A-, B- and C- chains of the human complement subcomponent C1q. *Biochemical Journal* **274**:481.

18. Perkins, S. J., A. S. Nealis, B. J. Sutton, and A. Feinstein. 1991. Solution structure of human and mouse immunoglobulin M by synchrotron X-ray scattering and molecular graphics modelling. A possible mechanism for complement activation. *Journal of Molecular Biology* **221**:1345.

19. Daha, M. R., A. Gorter, M. Rits, L. A. van Es, and P. S. Hiemstra. 1989. Interaction of immunoglobulin A with complement and phagocytic cells. *Progress in Clinical and Biological Research* **297**:247.

20. Rits, M., P. S. Hiemstra, H. Bazin, L. A. Van Es, J. P. Vaerman, and M. R. Daha. 1988. Activation of rat complement by soluble and insoluble rat IgA immune complexes. *European Journal of Immunology* **18**:1873.

21. Lucisano Valim, Y. M., and P. J. Lachmann. 1991. The effect of antibody isotype and antigenic epitope density on the complement-fixing activity of immune complexes: a systematic study using chimaeric anti-NIP antibodies with human Fc regions. *Clinical and Experimental Immunology* **84**:1.

22. Reid, K. B. 1986. Activation and control of the complement system. *Essays in Biochemistry* **22**:27.

23. Reid, K. B., and A. J. Day. 1989. Structure function relationships of the complement components. *Immunology Today* **10**:177.

24. Schreiber, R. D., and H. J. Müller-Eberhard. 1974. Fourth component of human complement: description of a three polypeptide chain structure. *Journal of Experimental Medicine* **140**:1324.

25. Janatova, J., and B. F. Tack. 1981. Fourth component of human complement: studies of an amine-sensitive site comprised of a thiol component. *Biochemistry* **20**:2394.

26. Campbell, R. D., I. Dunham, and C. A. Sargent. 1988. Molecular mapping of the HLA-linked complement genes and the RCA linkage group. *Experimental and Clinical Immunogenetics* **5**:81.

27. Campbell, R. D. 1988. The molecular genetics of components of the complement system. *Baillières Clinical Rheumatology* **2**:547.

28. Law, S. K., and R. P. Levine. 1977. Interaction between the third complement protein and cell surface macromolecules. *Proceedings of the National Academy of Sciences of the United States of America* **74**:2701.

29. Law, S. K., and A. W. Dodds. 1990. C3, C4 and C5: the thioester site. *Biochemical Society Transactions* **18**:1155.

30. Dodds, A. W., X. D. Ren, A. C. Willis, and S. K. Law. 1996. The reaction mechanism of the internal thioester in the human complement component C4. *Nature* **379**:177.

31. Law, S. K., and A. W. Dodds. 1997. The internal thioester and the covalent binding properties of the complement proteins C3 and C4. *Protein Science* **6**:263.

32. Nagasawa, S., and R. M. Stroud. 1977. Cleavage of C2 by C1s into the antigenically distinct fragments C2a and C2b: demonstration of binding of C2b to C4b. *Proceedings of the National Academy of Sciences of the United States of America* **74**:2998.

33. Lambris, J. D. 1988. The multifunctional role of C3, the third component of complement. *Immunology Today* **9**:387.

34. Kozono, H., T. Kinoshita, Y. U. Kim, Y. Takata-Kozono, S. Tsunasawa, F. Sakiyama, J. Takeda, K. Hong, and K. Inoue. 1990. Localization of the covalent C3b-binding site on C4b within the complement classical pathway C5 convertase, C4b2a3b. *Journal of Biological Chemistry* **265**:14444.

35. Kinoshita, T., Y. Takata, H. Kozono, J. Takeda, K. S. Hong, and K. Inoue. 1988. C5 convertase of the alternative complement pathway: covalent linkage between two C3b molecules within the trimolecular complex enzyme. *Journal of Immunology* **141**:3895.

36. Takata, Y., T. Kinoshita, H. Kozono, J. Takeda, E. Tanaka, K. Hong, and K. Inoue. 1987. Covalent association of C3b with C4b within C5 convertase of the classical complement pathway. *Journal of Experimental Medicine* **165**:1494.

37. Ebanks, R. O., A. S. Jaikaran, M. C. Carroll, M. J. Anderson, R. D. Campbell, and D. E. Isenman. 1992. A single arginine to tryptophan interchange at beta-chain residue 458 of human complement component C4 accounts for the defect in classical pathway C5 convertase activity of allotype C4A6. Implications for the location of a C5 binding site in C4. *Journal of Immunology* **148**:2803.

38. Kitamura, H., M. Tsuboi, and K. Nagaki. 1986. C3-independent immune haemolysis: mechanism of membrane attack complex formation. *Immunology* **59**:147.

39. Masaki, T., M. Matsumoto, T. Hara, I. Nakanishi, H. Kitamura, and T. Seya. 1995. Covalently-bound human C4b dimers consisting of C4B isotype show higher hemolytic activity than those of C4A in the C3-bypass complement pathway. *Molecular Immunology* **32**:21.

40. Hammer, C. H., G. Hansch, H. D. Gresham, and M. L. Shin. 1983. Activation of the fifth and sixth components of the human complement system: C6-dependent cleavage of C5 in acid and the formation of a bimolecular lytic complex, C5b,6a. *Journal of Immunology* **131**:892.

41. Dessauer, A., U. Rother, and K. Rother. 1984. Freeze–thaw activation of the complement attack phase: II. Comparison of convertase generated C–56 with C–56 generated by freezing and thawing. *Acta Pathologica, Microbiologica, et Immunologica Scandinavica Supplement* **284**:83.

42. Ratnoff, W. D. 1980. A war with the molecules: Louis Pillemer and the history of properdin. *Perspectives in Biology and Medicine* **23**:638.

43. Pillemer, L., L. Blum, I. H. Lepow, O. A. Ross, E. W. Todd, and A. C. Wardlaw. 1954. The properdin system and immunity. I. Demonstration of a new serum protein, properdin, and its role in immune phenomena. *Science* **120**:279.

44. Nelson, R. A. 1959. An alternative mechanism for the properdin system. *Journal of Experimental Medicine* **108**:515.

45. Lepow, I. H. 1980. Presidential address to American Association of Immunologists in Anaheim, California, April 16, 1980. Louis Pillemer, properdin, and scientific controversy. *Journal of Immunology* **125**:471.

46. Chapitis, J., and I. H. Lepow. 1976. Multiple sedimenting species of properdin in human serum and interaction of purified properdin with the third component of complement. *Journal of Experimental Medicine* **143**:241.

47. Götze, O. 1986. The alternative pathway of activation. In *The complement system*. K. Rother, Till, G.O., eds. Springer, Berlin, p. 154.

48. Choy, L. N., B. S. Rosen, and B. M. Spiegelman. 1992. Adipsin and an endogenous pathway of complement from adipose cells. *Journal of Biological Chemistry* **267**:12736.

49. Weiler, J. M., M. R. Daha, K. F. Austen, and D. T. Fearon. 1976. Control of the amplification convertase of complement by the plasma protein beta1H. *Proceedings of the National Academy of Sciences of the United States of America* **73**:3268.

50. Fearon, D. T., M. R. Daha, J. M. Weiler, and K. F. Austen. 1976. The natural modulation of the amplification phase of complement activation. *Transplantation Reviews* **32**:12.

51. Minta, J. O., I. H., Lepow, 1974. Studies on the subunit structure of human properdin. *Immunochemistry* **11**:361.

52. Nolan, K. F., and K. B. Reid. 1993. Properdin. *Methods in Enzymology* **223**:35.

53. Lachmann, P. J., and N. C. Hughes-Jones. 1984. Initiation of complement activation. *Springer Seminars in Immunopathology* **7**:143.

54. Pangburn, M. K. 1983. Activation of complement via the alternative pathway. *Federation Proceedings* **42**:139.

55. Sahu, A., and M. K. Pangburn. 1993. Identification of multiple sites of interaction between heparin and the complement system. *Molecular Immunology* **30**:679.

56. Pangburn, M. K. 1986. The alternative pathway. In *Immunobiology of the complement system*. G. D. Ross, ed. Academic Press, New York, p. 45.

57. Super, M., S. Thiel, J. Lu, R.J. Levinsky, and M.W. Turner. 1989. Association of low levels of mannan-binding protein with a common defect of opsonization. *Lancet* **2(8674)**:1236.

58. Holmskov, U., Malhotra, R., Sim, R. B., Jensenius, J. C. 1994. Collectins: collagenous C-type lectins of the innate immune defense system. *Immunology Today*:67.

59. Matsuhita, M., and T. Fujita. 1992. Activation of the classical complement pathway by mannose-binding protein in association with a novel C1s-like serine protease. *Journal of Experimental Medicine* **176**:1497.

60. Thiel, S., T. Vorup-Jensen, C. M. Stover, W. Schwaeble, S. B. Laursen, K. Poulsen, A. C. Willis, P. Eggleton, S. Hansen, U. Holmskov, K. B. Reid, and J. C. Jensenius. 1997. A second serine protease associated with mannan-binding lectin that activates complement. *Nature* **386**:506.

61. Jiang, H., T. F. Lint, and H. Gewurz. 1991. Defined, chemically cross-linked oligomers of human C-reactive protein: characterization and reactivity with the complement system. *Immunology* **74**:725.

62. Podack, E. R., W. P. Kolb, and H. J. Müller-Eberhard. 1976. The C5b-9 complex: subunit composition of the classical and alternative pathway-generated complex. *Journal of Immunology* **116**:1431.

63. Podack, E. R., G. Biesecker, and H. J. Müller-Eberhard. 1979. Membrane attack complex of complement: generation of high-affinity phospholipid binding sites by fusion of five hydrophilic plasma proteins. *Proceedings of the National Academy of Sciences of the United States of America* **76**:897.

64. Tschopp, J., and E. R. Podack. 1981. Membranolysis by the ninth component of human complement. *Biochemical and Biophysical Research Communications* **100**:1409.

65. Tamura, N., A. Shimada, and S. Chang. 1972. Further evidence for immune cytolysis by antibody and the first eight components of complement. *Immunology* **22**:131.

66. Podack, E. R., J. Tschoop, and H. J. Müller-Eberhard. 1982. Molecular organization of C9 within the membrane attack complex of complement. Induction of circular C9 polymerization by the C5b-8 assembly. *Journal of Experimental Medicine* **156**:268.

67. Tschopp, J., E. R. Podack, and H. J. Müller-Eberhard. 1982. Ultrastructure of the membrane attack complex of complement: detection of the tetramolecular C9-polymerizing complex C5b-8. *Proceedings of the National Academy of Sciences of the United States of America* **79**:7474.

68. Dankert, J. R., J. W. Shiver, and A. F. Esser. 1985. Ninth component of complement: self-aggregation and interaction with lipids. *Biochemistry* **24**:2754.

69. Bhakdi, S., and J. Tranum-Jensen. 1991. Complement lysis: a hole is a hole. *Immunology Today* **12**:318.

70. Esser, A. F. 1991. Big MAC attack: complement proteins cause leaky patches. *Immunology Today* **12**:316.

71. Sauer, H., L. Pratsch, J. Tschopp, S. Bhakdi, and R. Peters. 1991. Functional size of complement and perforin pores compared by confocal laser scanning microscopy and fluorescence microphotolysis. *Biochimica et Biophysica Acta* **1063**:137.

72. Rogne, S., O. Mykelbost, J. H. Olving, H. T. Kyrkjebo, R. Jonasen, B. Olaisen, and T. Gedde-Dahl. 1991. The gene for human complement component C9 is on chromosome 9. *Journal of Medical Genetics* **28**:587.

73. Morgan, B. P., and M. J. Walport. 1991. Complement deficiency and disease. *Immunology Today* **12**:301.

74. Figueroa, J., J. Andreoni, and P. Densen. 1993. Complement deficiency states and meningococcal disease. *Immunologic Research* **12**:295.

75. Pepys, M. B., D. D. Mirjah, A. C. Dash, and M. H. Wansbrough-Jones. 1976. Immunosuppression by cobra factor: distribution, antigen-induced blast transformation and trapping of lymphocytes during in vivo complement depletion. *Cellular Immunology* **21**:327.

76. Pepys, M. B. 1976. Role of complement in the induction of immunological responses. *Transplantation Reviews* **32**:93.

77. Klaus, G. G., and J. H. Humphrey. 1977. The generation of memory cells. I. The role of C3 in the generation of B memory cells. *Immunology* **33**:31.

78. Dickler, H. B., and H. G. Kunkel. 1972. Interaction of aggregated gamma-globulin with B lymphocytes. *Journal of Experimental Medicine* **136**:191.

79. Tenner, A. J., and N. R. Cooper. 1981. Identification of types of cells in human peripheral blood that bind C1q. *Journal of Immunology* **126**:1174.

80. Ghebrehiwet, B., and M. Hamburger. 1982. Purification and partial characterization of a C1q inhibitor from the membranes of human peripheral blood lymphocytes. *Journal of Immunology* **129**:157.

81. Andrews, B. S., M. Shadforth, P. Cunningham, and J. S. Davis. 1981. Demonstration of a C1q receptor on the surface of human endothelial cells. *Journal of Immunology* **127**:1075.

82. Arvieux, J., A. Reboul, J. C. Bensa, and M. G. Colomb. 1984. Characterization of the C1q receptor on a human macrophage cell line U937. *Biochemical Journal* **218**:547.

83. Ghebrehiwet, B., L. Silvestri, and C. McDevitt. 1984. Identification of the Raji cell membrane-derived C1q inhibitor as a receptor for human C1q: Purification and immunochemical characterization. *Journal of Experimental Medicine* **160**:1375.

84. Ghebrehiwet, B. 1989. Functions associated with the C1q receptor. *Behring Institut Mitteilungen* **84**:204.

85. Sim, R. B., and R. Malhotra. 1994. Interactions of carbohydrates and lectins with complement. *Biochemical Society Transactions* **22**:106.

86. Malhotra, R. 1993. Collectin receptor (C1q receptor): structure and function. *Behring Institute Mitteilungen* **93**:254.

87. Malhotra, R., J. Lu, U. Holmskov, and R. B. Sim. 1994. Collectins, collectin receptors and the lectin pathway of complement activation. *Clinical and Experimental Immunology* **97**:4.

88. Guan, E., W. H. Burgess, S. L. Robinson, E. B. Goodman, K. J. McTigue, and A. J. Tenner. 1991. Phagocytic cell molecules that bind the collagen-like region of C1q. Involvement in the C1q-mediated enhancement of phagocytosis. *Journal of Biological Chemistry* **266**:20345.

89. Peerschke, E. I., K. B. Reid, and B. Ghebrehiwet. 1994. Identification of a novel 33kDa C1q-binding site on human platelets. *Journal of Immunology* **152**:5896.

90. Eggleton, P., B. Ghebrehiwet, K. N. Sastry, J. P. Coburn, K. S. Zaner, K. B. Reid, and A. I. Tauber. 1995. Identification of a gC1q-binding protein (gC1q-R) on the surface of human neutrophils. Subcellular localisation and binding properties in comparison with the cC1q-R. *Journal of Clinical Investigation* **95**:1569.

91. Fearon, D. T., and W. W. Wong. 1983. Complement ligand–receptor interactions that mediate biological responses. *Annual Review of Immunology* **1**:243.

92. Fearon, D. T., L. B. Klickstein, W. W. Wong, J. G. Wilson, F. D. Moore, Jr., J. J. Weis, J. H. Weis, R. M. Jack, R. H. Carter, and J. A. Ahearn. 1989. Immunoregulatory functions of complement: structural and functional studies of complement receptor type 1 (CR1; CD35) and type 2 (CR2; CD21). *Progress in Clinical and Biological Research* **297**:211.

93. Fearon, D. T., and R. H. Carter. 1995. The CD19/CR2/TAPA-1 complex of

B lymphocytes: linking natural to acquired immunity. *Annual Review of Immunology* **13**:127.

94. Rothlein, R., and T. A. Springer. 1985. Complement receptor type three-dependent degradation of opsonized erythrocytes by mouse macrophages. *Journal of Immunology* **135**:2668.

95. Anderson, D. C., L. J. Miller, F. C. Schmalstieg, R. Rothlein, and T. A. Springer. 1986. Contributions of the Mac-1 glycoprotein family to adherence-dependent granulocyte functions: structure–function assessments employing subunit-specific monoclonal antibodies. *Journal of Immunology* **137**:15.

96. Larson, R. S., and T. A. Springer. 1990. Structure and function of leukocyte integrins. *Immunological Reviews* **114**:181.

97. Corbi, A. L., T. K. Kishimoto, L. J. Miller, and T. A. Springer. 1988. The human leukocyte adhesion glycoprotein Mac-1 (complement receptor type 3, CD11b) alpha subunit. Cloning, primary structure, and relation to the integrins, von Willebrand factor and factor B. *Journal of Biological Chemistry* **263**:12403.

98. Chenoweth, D. E., and M. G. Goodman. 1983. The C5a receptor of neutrophils and macrophages. *Agents and Actions Supplements* **12**:252.

99. van Epps, D. E., and D. E. Chenoweth. 1984. Analysis of the binding of fluorescent C5a and C3a to human peripheral blood leukocytes. *Journal of Immunology* **132**:2862.

100. Gerard, N. P., Gerard, C. 1991. The chemotactic receptor for human C5a anaphylatoxin. *Nature* **349**:614.

101. Wetsel, R. A. 1995. Expression of the complement C5a anaphylatoxin receptor (C5aR) on non-myeloid cells. *Immunology Letters* **44**:183.

102. Gasque, P., S. K. Singhrao, J. W. Neal, O. Gotze, and B. P. Morgan. 1997. Expression of the receptor for complement C5a (CD88) is up-regulated on reactive astrocytes, microglia, and endothelial cells in the inflamed human central nervous system. *American Journal of Pathology* **150**:31.

103. Hopken, U. E., Lu, B., Gerard, N. P., Gerard, C. 1996. The C5a chemoattractant receptor mediates mucosal defense to infection. *Nature* **383**:86.

104. Ames, R. S., Y. Li, H. M. Sarau, P. Nuthulaganti, J. J. Foley, C. Ellis, Z. Zeng, K. Su, A. J. Jurewicz, R. P. Hertzberg, D. J. Bergsma, and C. Kumar. 1996. Molecular cloning and characterization of the human anaphylatoxin C3a receptor. *Journal of Biological Chemistry* **271**:20231.

105. Gasque, P., S. K. Sihghrao, J. W. Neal, P. Wang, S. Sayah, M. Fontaine, and B. P. Morgan. 1998. The receptor for complement anaphylatoxin C3a is expressed by myeloid cells and non-myeloid cells in inflamed human CNS: analysis in multiple sclerosis and bacterial meningitis. *Journal of Immunology* **160**:3543.

106. Taylor, P. W. 1983. Bactericidal and bacteriolytic activity of serum against gram-negative bacteria. *Microbiological Reviews* **47**:46.

107. Schreiber, R. D. 1984. The chemistry and biology of complement receptors. *Springer Seminars in Immunopathology* **7**:221.

108. Hughli, T. E. 1981. The structural basis for anaphylatoxic and chemotactic functions of C3a, C4a and C5a. *Critical Reviews in Immunology* **1**:321.

109. Till, G. O. 1986. Chemotactic factors. In *The complement system*. K. Rother, Till, G.O., eds. Springer, Berlin, p. 354.

110. McCoy, R., D. L. Haviland, E. P. Molmenti, T. Ziambaras, R. A. Wetsel, and D. H. Perlmutter. 1995. N-formylpeptide and complement C5a receptors are expressed in liver cells and mediate hepatic acute phase gene regulation. *Journal of Experimental Medicine* **182**:207.

111. Schifferli, J. A., Y. C. Ng, and D. K. Peters. 1986. The role of complement and its receptor in the elimination of immune complexes. *New England Journal of Medicine* **315**:488.

112. Schifferli, J. A. 1996. Complement and immune complexes. *Research in Immunology* **147**:109.

113. Schifferli, J. A., and R. P. Taylor. 1989. Physiological and pathological aspects of circulating immune complexes. *Kidney International* **35**:993.

114. Morgan, B. P. 1989. Mechanisms of tissue damage by the membrane attack complex of complement. *Complement and Inflammation* **6**:104.

115. Morgan, B. P. 1994. Clinical complementology: recent progress and future trends. *European Journal of Clinical Investigation* **24**:219.

116. Kinoshita, T. 1991. Biology of complement: the overture. *Immunology Today* **12**:291.

117. Dierich, M. P., A. Erdei, H. Huemer, A. Petzer, R. Stauder, T. F. Schulz, and J. Gergely. 1987. Involvement of complement in B-cell, T-cell and monocyte/macrophage activation. *Immunology Letters* **14**:235.

118. Dempsey, P. W., and D. T. Fearon. 1996. Complement: instructing the acquired immune system through the CD21/CD19 complex. *Research in Immunology* **147**:71.

119. Dempsey, P. W., M. E. Allison, S. Akkaraju, C. C. Goodnow, and D. T. Fearon. 1996. C3d of complement as a molecular adjuvant: bridging innate and acquired immunity. *Science* **271**:348.

120. Weiler, J. M., and M. V. Hobbs. 1987. Role of the third component of complement in immunoregulation. *Concepts in Immunopathology* **4**:103.

121. Feldbush, T. L., M. V. Hobbs, C. D. Severson, Z. K. Ballas, and J. M. Weiler. 1984. Role of complement in the immune response. *Federation Proceedings* **43**:2548.

122. Colten, H. R., and F. S. Rosen. 1992. Complement deficiencies. *Annual Review of Immunology* **10**:809.

123. Fukumori, Y., K. Yoshimura, S. Ohnoki, H. Yamaguchi, Y. Akagaki, and S. Inai. 1989. A high incidence of C9 deficiency among healthy blood donors in Osaka, Japan. *International Immunology* **1**:85.

124. Whaley, K. 1989. Measurement of complement activation in clinical practice. *Complement and Inflammation* **6**:96.

125. Laurell, A.-B. 1986. Complement determinations in clinical diagnosis. In *The complement system*. K. Rother, G. O. Till, eds. Springer, Berlin, p. 272.

126. Rauterberg, E. W. 1986. Demonstration of complement deposits in tissues. In *The complement system*. K. Rother, G. O. Till, ed. Springer, Berlin, p. 287.

127. Mollnes, T. E., V. Videm, J. Riesenfeld, P. Garred, J. L. Svennevig, E. Fosse, K. Hogasen, and M. Harboe. 1991. Complement activation and bioincompatibility. *Clinical and Experimental Immunology* **86(Supp.1)**:21.

128. Kirklin, J. K., D. E. Chenoweth, D. C. Naftel, E. H. Blackstone, J. W. Kirklin, D. D. Bitran, J. G. Curd, J. G. Reves, and P. N. Samuelson. 1986. Effects of protamine administration after cardiopulmonary bypass on complement, blood elements, and the hemodynamic state. *Annals of Thoracic Surgery* **41**:193.

129. Cheung, A. K. 1994. Complement activation as index of haemodialysis membrane biocompatibility: the choice of methods and assays. *Nephrology, Dialysis, Transplantation* **9**:96.

130. Mellbye, O. J., S. S. Froland, P. Lilleaasen, J. L. Svennevig, and T. E. Mollnes. 1988. Complement activation during cardiopulmonary bypass: comparison between the use of large volumes of plasma and dextran 70. *European Surgical Research* **20**:101.

131. Ovrum, E., T. E. Mollnes, E. Fosse, E. A. Holen, G. Tangen, M. A. Ringdal, and V. Videm. 1995. High and low heparin dose with heparin-coated cardiopulmonary bypass: activation of complement and granulocytes [see comments]. *Annals of Thoracic Surgery* **60**:1755.

132. Mollnes, T. E. 1997. Biocompatibility: complement as mediator of tissue damage and as indicator of incompatibility. *Experimental and Clinical Immunogenetics* **14**:24.

133. White, D. J. 1992. Transplantation of organs between species. *International Archives of Allergy and Immunology* **98**:1.

134. Ziegler, J. B., C. A. Alper, R. S. Rosen, P. J. Lachmann, and L. Sherington. 1975. Restoration by purified C3b inactivator of complement-mediated function in vivo in a patient with C3b inactivator deficiency. *Journal of Clinical Investigation* **55**:668.

135. Bordet, J. 1900. Les sérums hémolytiques, leurs antitoxines et les théories des sérums cytolytiques. *Annales de l'Institut Pasteur* **15**:257.

136. Muir, R., C. H. Browning. 1904. On chemical combination and toxic action as exemplified in haemolytic sera. *Proceedings of the Royal Society, London* **74**:298.

137. Muir, R. 1911. On the relationships between the complements and immune bodies of different animals. *Journal of Pathology and Bacteriology* **16**:523.

138. Houle, J. J., and E. M. Hoffmann. 1984. Evidence for restriction of the ability of complement to lyse homologous erythrocytes. *Journal of Immunology* **133**:1444.

139. van Dijk, H., E. Heezius, P. J. van Kooten, P. M. Rademaker, R. van Dam, and J. M. Willers. 1983. A study of the sensitivity of erythrocytes to lysis by heterologous sera via the alternative complement pathway. *Veterinary Immunology and Immunopathology* **4**:469.

140. Hansch, G. M., C. H. Hammer, P. Vanguri, and M. L. Shin. 1981. Homologous species restriction in lysis of erythrocytes by terminal complement proteins. *Proceedings of the National Acadamy of Sciences of the United States of America* **78**:5118.

141. Hoffmann, E. M., and H. M. Etlinger. 1973. Extraction of complement inhibitory substances from the erythrocytes of non-human species. *Journal of Immunology* **111**:946.

142. Hoffmann, E. M. 1969. Inhibition of complement by a substance isolated from human erythrocytes. I. Extraction from human erythrocyte stroma. *Immunochemistry* **6**:391.

143. Hoffmann, E. M. 1969. Inhibition of complement by a substance isolated from human erythrocytes. II. Studies on the site and mechanism of action. *Immunochemistry* **6**:405.

144. Nicholson-Weller, A., J. Burge, and K. F. Austen. 1981. Purificaiton from guinea pig erythrocyte stroma of a decay-accelerating factor for the classical C3 convertase, C4b2a. *Journal of Immunology* **127**:2035.

145. Nicholson-Weller, A., J. Burge, and D. T. Fearon. 1982. Isolation of a human erythrocyte membrane glycoprotein with decay-accelerating activity for C3 convertases of the complement system. *Journal of Immunology* **129**:184.

146. Fearon, D. T. 1979. Regulation of the amplification C3 convertase of human complement by an inhibitory protein isolated from human erythrocyte membranes. *Proceedings of the National Acadamy of Sciences of the United States of America* **76**:5867.

147. Lublin, D. M., M. K. Liszewski, T. W. Post, M. A. Arce, M. M. Le Beau, M. B. Rebentisch, L. S. Lemons, T. Seya, and J. P. Atkinson. 1988. Molecular cloning and chromosomal localization of human membrane cofactor protein (MCP). Evidence for inclusion in the multigene family of complement-regulatory proteins. *Journal of Experimental Medicine* **168**:181.

148. Seya, T., and J. P. Atkinson. 1989. Functional properties of membrane cofactor protein of complement. *Biochemical Journal* **264**:581.

149. Schonermark, S., E. W. Rauterberg, M. L. Shin, S. Loke, D. Roelcke, and G. M. Hansch. 1986. Homologous species restriction in lysis of human erythrocytes: a membrane-derived protein with C8-binding capacity functions as an inhibitor. *Journal of Immunology* **136**:1772.

150. Zalman, L. S., L. M. Wood, and H. J. Muller-Eberhard. 1986. Isolation of a human erythrocyte membrane protein capable of inhibiting expression of homologous complement transmembrane channels. *Proceedings of the National Academy of Sciences of the United States of America* **83**:6975.

151. Pensky, J., Levy, L., Lepow, I. 1961. Partial purification of a serum inhibitor of C1′esterase. *Journal of Biological Chemistry* **236**:1674.

152. Donaldson, V. H., Evans, R. R. 1963. A biochemical abnormality in hereditary angioneurotic edema. Absence of serum inhibitor of C1′-esterase. *American Journal of Medicine* **35**:37.

153. Lachmann, P. J., and H. J. Müller-Eberhard. 1969. The demonstration in human serum of conglutinogen-activating factor and its effects on the third component of complement. *Journal of Immunology* **100**:691.

154. Fearon, D. T. 1977. Purification of C3b inactivator and demonstration of its two polypeptide chain structure. *Journal of Immunology* **119**:1248.

155. Whaley, K., and S. Ruddy. 1976. Modulation of the alternative complement pathways by beta 1 H globulin. *Journal of Experimental Medicine* **144**:1147.

156. Ferreira, A., M. Takahashi, and V. Nussenzweig. 1977. Purificaiton and characterization of mouse serum protein with specific binding affinity for C4 (Ss protein). *Journal of Experimental Medicine* **146**:1001.

157. Davis, A. E. 1988. C1 inhibitor and hereditary angioneurotic edema. *Annual Review of Immunology* **5**:595.

158. Davis, A. E. 1989. Hereditary and acquired deficiencies of C1 inhibitor. *Immunodeficiency Reviews* **1**:207.

159. Vik, D. P., P. Munoz-Canoves, D. D. Chaplin, and B. F. Tack. 1990. Factor H. *Current Topics in Microbiology and Immunology* **153**:147.

160. Hessing, M. 1991. The interaction between complement component C4b-binding protein and the vitamin K-dependent protein S forms a link between blood coagulation and the complement system. *Biochemical Journal* **277**:581.

161. Gigli, I., T. Fujita, and V. Nussenzweig. 1979. Modulation of the classical pathway C3 convertase by the plasma proteins C4 binding protein and C3b inactivator. *Proceedings of the National Academy of Sciences of the United States of America* **76**:6596.

162. Smith, C. A., M. K. Pangburn, C-W. Vogel, and H. J. Müller-Eberhard. 1984. Molecular architecture of human properdin, a positive regulator of the alternative pathway of human complement. *Journal of Biological Chemistry* **259**:4582.

163. Podack, E. R., K. T. Preissner, and H. J. Müller-Eberhard. 1984. Inhibition of C9 polymerization within the SC5b-9 complex of complement by S-protein. *Acta Pathologica, Microbiologica, et Immunologica Scandinavica Supplement* **284**:89.

164. Jenne, D. E., and J. Tschopp. 1992. Clusterin: the intriguing guises of a widely expressed glycoprotein. *Trends in Biochemical Sciences* **17**:154.

165. Lint, T. F., C. L. Behrends, and H. Gewurz. 1977. Serum lipoproteins and C567-INH activity. *Journal of Immunology* **119**:883.

166. Nemerow, G. R., K. I. Yamamoto, and T. F. Lint. 1979. Restriction of complement-mediated membrane damage by the eighth component of complement: a dual role for C8 in the complement attack sequence. *Journal of Immunology* **123**:1245.

167. Zalman, L. S. 1992. Homologous restriction factor. In *Membrane defenses against attack by complement and perforins*. C. J. Parker, ed. Springer, Berlin, p. 87.

168. Zalman, L. S., and H. Muller-Eberhard. 1994. Homologous restriction factor: effect on complement C8 and C9 uptake and lysis. *Molecular Immunology* **31**:301.

169. Lachmann, P. J. 1991. The control of homologous lysis. *Immunology Today* **12**:312.

170. Davies, A., Lachmann, P. J. 1993. Membrane defence against complement lysis: the structure and biological properties of CD59. *Immunologic Research* **12**:258.

171. Morgan, B. P., J. R. Dankert, and A. F. Esser. 1987. Recovery of human neutrophils from complement attack: removal of the membrane attack complex by endocytosis and exocytosis. *Journal of Immunology* **138**:246.

172. Burgi, B., T. Brunner, and C. A. Dahinden. 1994. The degradation product of the C5a anaphylatoxin C5adesArg retains basophil-activating properties. *European Journal of Immunology* **24**:1583.

173. Painter, R. H. 1993. The binding of C1q to immunoglobulins. *Behring Institute Mitteilungen* **93**:131.
174. Ratnoff, O. D., and I. H. Lepow. 1957. Some properties of an esterase derived from preparations of the first component of complement. *Journal of Experimental Medicine* **106**:327.
175. Pensky, J., L. R. Levy, and I. H. Lepow. 1961. Partial purification of a serum inhibitor of C1 esterase. *Journal of Biological Chemistry* **236**:1674.
176. Schultze, H. E., K. Heide, and H. Haupt. 1962. Uber ein sisher unbekanntes saures α2 Glykoprotein. *Naturwissenschaften* **49**:133.
177. Pensky, J., and H. G. Schwick. 1969. Human serum inhibitor of C'1 esterase: identity with alpha-2-neuraminoglycoprotein. *Science* **163**:698.
178. Engh, R. A., A. J. Schulze, R. Huber, and W. Bode. 1993. Serpin structures. *Behring Institute Mitteilungen* **93**:41.
179. Potempa, J., E. Korzus, and J. Travis. 1994. The serpin superfamily of proteinase inhibitors: structure, function, and regulation. *Journal of Biological Chemistry* **269**:15957.
180. Ratnoff, O. D., J. Pensky, D. Ogston, and G. B. Naff. 1969. The inhibition of plasmin, plasma kallikrein, plasma permeability factor, and the C'1r subcomponent of the first component of complement by serum C'1 esterase inhibitor. *Journal of Experimental Medicine* **129**:315.
181. Forbes, C. D., J. Pensky, and O. D. Ratnoff. 1970. Inhibition of activated Hageman factor and activated plasma thromboplastin antecedent by purified serum C1 inactivator. *Journal of Laboratory and Clinical Medicine* **76**:809.
182. Gigli, I., J. W. Mason, R. W. Colman, and K. F. Austen. 1970. Interaction of plasma kallikrein with the C1 inhibitor. *Journal of Immunology* **104**:574.
183. Curd, J. G., L. J. Prograis, Jr., and C. G. Cochrane. 1980. Detection of active kallikrein in induced blister fluids of hereditary angioedema patients. *Journal of Experimental Medicine* **152**:742.
184. Schapira, M., L. D. Silver, C. F. Scott, A. H. Schmaier, L. J. Prograis, Jr., J. G. Curd, and R. W. Colman. 1983. Prekallikrein activation and high-molecular-weight kininogen consumption in hereditary angioedema. *New England Journal of Medicine* **308**:1050.
185. Wuillemin, W. A., M. Minnema, J. C. Meijers, D. Roem, A. J. Eerenberg, J. H. Nuijens, H. ten Cate, and C. E. Hack. 1995. Inactivation of factor XIa in human plasma assessed by measuring factor XIa-protease inhibitor complexes: major role for C1-inhibitor. *Blood* **85**:1517.
186. Huisman, L. G., J. M. van Griensven, and C. Kluft. 1995. On the role of C1-inhibitor as inhibitor of tissue-type plasminogen activator in human plasma. *Thrombosis and Haemostasis* **73**:466.
187. Sakai, K., and R. M. Stroud. 1973. Purification, molecular properties, and activation of C1 proesterase, C1s. *Journal of Immunology* **110**:1010.
188. Ziccardi, R. J. 1982. Spontaneous activation of the first component of human complement (C1) by an intramolecular autocatalytic mechanism. *Journal of Immunology* **128**:2500.
189. Ziccardi, R. J. 1985. Demonstration of the interaction of native C1 with monomeric immunoglobulins and C1 inhibitor. *Journal of Immunology* **134**:2559.

190. Ziccardi, R. J. 1982. A new role for C-1-inhibitor in homeostasis: control of activation of the first component of human complement. *Journal of Immunology* **128**:2505.

191. Hosoi, S., M. G. Colomb, and T. Borsos. 1987. Activation of human C1: analysis with Western blotting reveals slow self-activation. *Journal of Immunology* **139**:1602.

192. Bianchino, A. C., P. H. Poon, and V. N. Schumaker. 1988. A mechanism for the spontaneous activation of the first component of complement, C1, and its regulation by C1-inhibitor. *Journal of Immunology* **141**:3930.

193. Ziccardi, R. J. 1981. Activation of the early components of the classical complement pathway under physiologic conditions. *Journal of Immunology* **126**:1769.

194. Tenner, A. J., and M. M. Frank. 1986. Activator-bound C1 is less susceptible to inactivation by C1 inhibition than is fluid-phase C1. *Journal of Immunology* **137**:625.

195. Sim, R. B., G. J. Arlaud, and M. G. Colomb. 1979. C1 inhibitor-dependent dissociation of human complement component C1 bound to immune complexes. *Biochemical Journal* **179**:449.

196. Chesne, S., C. L. Villiers, G. J. Arlaud, M. B. Lacroix, and M. G. Colomb. 1982. Fluid-phase interaction of C1 inhibitor (C1 Inh) and the subcomponents C1r and C1s of the first component of complement, C1. *Biochemical Journal* **201**:61.

197. Perkins, S. J. 1990. Hydrodynamic data show that C1- inhibitor of complement forms compact complexes with C1-r and C1-s. *FEBS Letters* **271**:89.

198. Ziccardi, R. J., and N. R. Cooper. 1979. Active disassembly of the first complement component, C-1, by C-1 inactivator. *Journal of Immunology* **123**:788.

199. Laurell, A. B., U. Martensson, and A. G. Sjoholm. 1976. C1 subcomponent conplexes in normal and pathological sera studied by crossed immunoelectrophoresis. *Acta Pathologica et Microbiologica Scandinavica, Section C, Immunology* **84C**:455.

200. Laurell, A. B., U. Johnson, U. Martensson, and A. G. Sjoholm. 1978. Formation of complexes composed of C1r, C1s, and C1 inactivator in human serum on activation of C1. *Acta Pathologica et Microbiologica Scandinavica, Section C, Immunology* **86C**:299.

201. Harrison, R. A. 1983. Human C1 inhibitor: improved isolation and preliminary structural characterization. *Biochemistry* **22**:5001.

202. Gregorek, H., M. Kokai, T. Hidvegi, G. Fust, K. Sabbouh, and K. Madalinski. 1991. Concentration of C1 inhibitor in sera of healthy blood donors as studied by immunoenzymatic assay. *Complement and Inflammation* **8**:310.

203. Johnson, A. M., C. A. Alper, F. S. Rosen, and J. M. Craig. 1971. C1 inhibitor: evidence for decreased hepatic synthesis in hereditary angioneurotic edema. *Science* **173**:553.

204. Colten, H. R. 1972. Ontogeny of the human complement system: in vitro biosynthesis of individual complement components by fetal tissue. *Journal of Clinical Investigation* **51**:725.

205. Morris, K. M., D. P. Aden, B. B. Knowles, and H. R. Colten. 1982. Complement biosynthesis by the human hepatoma-derived cell line HepG2. *Journal of Clinical Investigation* **70**:906.

206. Prandini, M. H., A. Reboul, and M. G. Colomb. 1986. Biosynthesis of complement C1 inhibitor by Hep G2 cells. Reactivity of different glycosylated forms of the inhibitor with C1s. *Biochemical Journal* **237**:93.

207. Gulati, P., C. Lemercier, D. Guc, D. Lappin, and K. Whaley. 1993. Regulation of the synthesis of C1 subcomponents and C1-inhibitor. *Behring Institute Mitteilungen* **93**:196.

208. Katz, Y., and R. C. Strunk. 1989. Synthesis and regulation of C1 inhibitor in human skin fibroblasts. *Journal of Immunology* **142**:2041.

209. Schmaier, A. H., S. C. Murray, G. D. Heda, A. Farber, A. Kuo, K. McCrae, and D. B. Cines. 1989. Synthesis and expression of C1 inhibitor by human umbilical vein endothelial cells. *Journal of Biological Chemistry* **264**:18173.

210. Bensa, J. C., A. Reboul, and M. G. Colomb. 1983. Biosynthesis in vitro of complement subcomponents C1q, C1s and C1 inhibitor by resting and stimulated human monocytes. *Biochemical Journal* **216**:385.

211. Laiwah, A. C. Y., L. Jones, A. O. Hamilton, and K. Whaley. 1985. Complement-subcomponent-C1-inhibitor synthesis by human monocytes. *Biochemical Journal* **226**:199.

212. Randazzo, B. P., R. J. Dattwyler, A. P. Kaplan, and B. Ghebrehiwet. 1985. Synthesis of C1 inhibitor (C1-INA) by a human monocyte-like cell line, U937. *Journal of Immunology* **135**:1313.

213. Schmaier, A. H., P. M. Smith, and R. W. Colman. 1985. Platelet C1-inhibitor. A secreted alpha-granule protein. *Journal of Clinical Investigation* **75**:242.

214. Schmaier, A. H., S. Amenta, T. Xiong, G. D. Heda, and A. M. Gewirtz. 1993. Expression of platelet C1 inhibitor. *Blood* **82**:465.

215. Gasque, P., A. Ischenko, J. Legoedec, C. Mauger, M. T. Schouft, and M. Fontaine. 1993. Expression of the complement classical pathway by human glioma in culture. A model for complement expression by nerve cells. *Journal of Biological Chemistry* **268**:25068.

216. Walker, D. G., O. Yasuhara, P. A. Patston, E. G. McGeer, and P. L. McGeer. 1995. Complement C1 inhibitor is produced by brain tissue and is cleaved in Alzheimer disease. *Brain Research* **675**:75.

217. Davis, A. E. 3rd, 1988. C1 inhibitor and hereditary angioneurotic edema. *Annual Review of Immunology* **6**:595.

218. Tosi, M. 1993. Molecular genetics of C1-inhibitor and hereditary angioedema. In *Complement in health and disease.* K. Whaley, ed. Kluwer Academic Publishers, Lancaster, UK, p. 245.

219. Lachmann, P. J., and F. S. Rosen. 1984. The catabolism of C1(-)-inhibitor and the pathogenesis of hereditary angio-edema. *Acta Pathologica, Microbiologica, et Immunologica Scandinavica, Supplement* **284**:35.

220. Cugno, M., J. Nuijens, E. Hack, A. Eerenberg, D. Frangi, A. Agostoni, and M. Cicardi. 1990. Plasma levels of C1- inhibitor complexes and cleaved C1- inhibitor in patients with hereditary angioneurotic edema. *Journal of Clinical Investigation* **85**:1215.

221. Donaldson, V. H., and J. J. Bissler. 1992. C1 inhibitors and their genes: an update. *Journal of Laboratory and Clinical Medicine* **119:**330.

222. Davis, A. E. 3rd, J. J. Bissler, and M. Cicardi. 1993. Mutations in the C1 inhibitor gene that result in hereditary angioneurotic edema. *Behring Institute Mitteilungen* **93:**313.

223. Caldwell, J. R., S. Ruddy, P. H. Schur, and K. F. Austen. 1972. Acquired C1-inhibitor deficiency in lymphosarcoma. *Clinical Immunology and Immunopathology* **1:**39.

224. Gelfand, J. A., G. R. Boss, C. L. Conley, R. Reinhart, and M. M. Frank. 1979. Acquired C1 esterase inhibitor deficiency and angioedema: a review. *Medicine* **58:**321.

225. Frank, M. 1989. Acquired C1-inhibitor deficiency. *Behring Institute Mitteilungen* **84:**161.

226. Haupt, H., N. Heimburger, T. Kranz, and H. G. Schwick. 1970. Ein beitrag zur isolierung und characterizierung des C1-inaktivators aus humanplasma. *European Journal of Biochemistry* **17:**254.

227. Harpel, P. C., and N. R. Cooper. 1975. Studies on human plasma C1 inactivator-enzyme interactions. I. Mechanisms of interaction with C1s, plasmin, and trypsin. *Journal of Clinical Investigation* **55:**593.

228. Bock, S. C., K. Skriver, E. Nielsen, H. C. Thogersen, B. Wiman, V. H. Donaldson, R. L. Eddy, J. Marrinan, E. Radziejewska, R. Huber, and et al. 1986. Human C1 inhibitor: primary structure, cDNA cloning, and chromosomal localization. *Biochemistry* **25:**4292.

229. Que, B. G., and P. H. Petra. 1986. Isolation and analysis of a cDNA coding for human C1 inhibitor. *Biochemical and Biophysical Research Communications* **137:**620.

230. Davis, A. E. 3rd, A. S. Whitehead, R. A. Harrison, A. Dauphinais, G. A. Bruns, M. Cicardi, and F. S. Rosen. 1986. Human inhibitor of the first component of complement, C1: characterization of cDNA clones and localization of the gene to chromosome 11. *Proceedings of the National Academy of Sciences of the United States of America* **83:**3161.

231. Tosi, M., C. Duponchel, P. Bourgarel, M. Colomb, and T. Meo. 1986. Molecular cloning of human C1 inhibitor: sequence homologies with alpha 1-antitrypsin and other members of the serpins superfamily. *Gene* **42:**265.

232. Rauth, G., G. Schumacher, P. Buckel, and W. Muller-Esterl. 1988. Molecular cloning of the cDNA coding for human C1 inhibitor. *Protein Sequences and Data Analysis* **1:**251.

233. Strecker, G., M. P. Ollier-Hartmann, H. van Halbeek, J. F. Vliegenthart, J. Montreuil, and L. Hartmann. 1985. Primary structure of the glycan chains of normal C 1 esterase inhibitor (C 1-INH) after NMR analysis at 400 MHz. *Comptes Rendus de l Académie des Science, Serie Iii, Sciences de la Vie* **301:**571.

234. Perkins, S. J., K. F. Smith, S. Amatayakul, D. Ashford, T. W. Rademacher, R. A. Dwek, P. J. Lachmann, and R. A. Harrison. 1990. Two-domain structure of the native and reactive centre cleaved forms of C1 inhibitor of human complement by neutron scattering. *Journal of Molecular Biology* **214:**751.

235. Reboul, A., M. H. Prandini, and M. G. Colomb. 1987. Proteolysis and deglycosylation of human C1 inhibitor. Effect on functional properties. *Biochemical Journal* **244**:117.

236. Coutinho, M., K. S. Aulak, and A. E. Davis, 3rd. 1994. Functional analysis of the serpin domain of C1 inhibitor. *Journal of Immunology* **153**:3648.

237. Odermatt, E., H. Berger, and Y. Sano. 1981. Size and shape of human C1-inhibitor. *FEBS Letters* **131**:283.

238. Schumaker, V. N., and M. L. Phillips. 1993. Electron microscope studies of C1s, C1s2, C1r2, C1r2C1s2 and C1-inhibitor. *Behring Institute Mitteilungen* **93**:17.

239. Loebermann, H., R. Tokuoka, J. Deisenhofer, and R. Huber. 1984. Human alpha 1-proteinase inhibitor. Crystal structure analysis of two crystal modifications, molecular model and preliminary analysis of the implications for function. *Journal of Molecular Biology* **177**:531.

240. Lennick, M., S. A. Brew, and K. C. Ingham. 1985. Changes in protein conformation and stability accompany complex formation between human C1 inhibitor and C1-s. *Biochemistry* **24**:2561.

241. Carter, P. E., B. Dunbar, and J. E. Fothergill. 1988. Genomic and cDNA cloning of the human C1 inhibitor. Intron–exon junctions and comparison with other serpins. *European Journal of Biochemistry* **173**:163.

242. Carter, P. E., C. Duponchel, M. Tosi, and J. E. Fothergill. 1991. Complete nucleotide sequence of the gene for human C1 inhibitor with an unusually high density of Alu elements. *European Journal of Biochemistry* **197**:301.

243. Zahedi, K., A. E. Prada, and A. E. Davis, 3rd. 1993. Structure and regulation of the C1 inhibitor gene. *Behring Institute Mitteilungen* **93**:115.

244. Stoppa-Lyonnet, D., P. E. Carter, T. Meo, and M. Tosi. 1990. Clusters of intragenic Alu repeats predispose the human C1 inhibitor locus to deleterious rearrangements. *Proceedings of the National Academy of Sciences of the United States of America* **87**:1551.

245. Ariga, T., P. E. Carter, and A. E. Davis. 1990. Recombinations between Alu repeat sequences that result in partial deletions within the C1 inhibitor gene. *Genomics* **8**:607.

246. Huber, R., and R. W. Carrell. 1989. Implications of the three-dimensional structure of $\alpha$1-antitrypsin for structure and function of serpins. *Biochemistry* **28**:8951.

247. Perkins, S. J. 1993. Three-dimensional structure and molecular modelling of C1-inhibitor. *Behring Institute Mitteilungen* **93**:63.

248. Stein, P., and C. Chothia. 1991. Serpin tertiary structure transformation. *Journal of Molecular Biology* **221**:615.

249. Stein, P. E., A. G. Leslie, J. T. Finch, W. G. Turnell, P. J. McLaughlin, and R. W. Carrell. 1990. Crystal structure of ovalbumin as a model for the reactive centre of serpins. *Nature* **347**:99.

250. Bode, W., and R. Huber. 1992. Natural protein proteinase inhibitors and their interaction with proteinases. *European Journal of Biochemistry* **204**:433.

251. Wei, A., H. Rubin, B. S. Cooperman, and D. W. Christianson. 1994. Crystal structure of an uncleaved serpin reveals the conformation of an inhibitory reactive loop. *Nature Structural Biology* **1**:251.

252. Haris, P. I., D. Chapman, R. A. Harrison, K. F. Smith, and S. J. Perkins. 1990. Conformational transition between native and reactive center cleaved forms of alpha 1-antitrypsin by Fourier transform infrared spectroscopy and small-angle neutron scattering. *Biochemistry* **29**:1377.

253. Perkins, S. J., K. F. Smith, A. S. Nealis, P. I. Haris, D. Chapman, C. J. Bauer, and R. A. Harrison. 1992. Secondary structure changes stabilize the reactive-centre cleaved form of SERPINs. A study by 1H nuclear magnetic resonance and Fourier transform infrared spectroscopy. *Journal of Molecular Biology* **228**:1235.

254. Gettins, P., and B. Harten. 1988. Properties of thrombin- and elastase-modified human antithrombin III. *Biochemistry* **27**:3634.

255. Pemberton, P. A., P. E. Stein, M. B. Pepys, J. M. Potter, and R. W. Carrell. 1988. Hormone binding globulins undergo serpin conformational change in inflammation. *Nature* **336**:257.

256. Bruch, M., V. Weiss, and J. Engel. 1988. Plasma serine proteinase inhibitors (serpins) exhibit major conformational changes and a large increase in conformational stability upon cleavage at their reactive sites. *Journal of Biological Chemistry* **263**:16626.

257. Carrell, R. W., and M. C. Owen. 1985. Plakalbumin, alpha 1-antitrypsin, antithrombin and the mechanism of inflammatory thrombosis. *Nature* **317**:730.

258. Salvesen, G. S., J. J. Catanese, L. F. Kress, and J. Travis. 1985. Primary structure of the reactive site of human C1-inhibitor. *Journal of Biological Chemistry* **260**:2432.

259. Pemberton, P. A., R. A. Harrison, P. J. Lachmann, and R. W. Carrell. 1989. The structural basis for neutrophil inactivation of C1 inhibitor. *Biochemical Journal* **258**:193.

260. Hekman, C. M., and D. J. Loskutoff. 1985. Endothelial cells produce a latent inhibitor of plasminogen activators that can be activated by denaturants. *Journal of Biological Chemistry* **260**:11581.

261. Mottonen, J., A. Strand, J. Symersky, R. M. Sweet, D. E. Danley, K. F. Geoghegan, R. D. Gerard, and E. J. Goldsmith. 1992. Structural basis of latency in plasminogen activator inhibitor-1. *Nature* **355**:270.

262. Smith, C. F., R. A. Harrison, and S. J. Perkins. 1990. Structural comparisons of the native and reactive-centre-cleaved forms of α1-antitrypsin by neutron- and X-ray-scattering in solution. *Biochemical Journal* **267**:203.

263. Sim, R. B., G. J. Arlaud, and M. G. Colomb. 1980. Kinetics of reaction of human C1-inhibitor with the human complement system proteases C1r and C1s. *Biochimica et Biophysica Acta* **612**:433.

264. Arlaud, G. J., A. Reboul, R. B. Sim, and M. G. Colomb. 1979. Interaction of C1-inhibitor with the C1r and C1s subcomponents in human C1. *Biochimica et Biophysica Acta* **576**:151.

265. Lawrence, D. A., D. Ginsburg, D. E. Day, M. B. Berkenpas, I. M. Verhamme, J. O. Kvassman, and J. D. Shore. 1995. Serpin–protease complexes are trapped as stable acyl-enzyme intermediates. *Journal of Biological Chemistry* **270**:25309.

266. Wilczynska, M., M. Fa, P. I. Ohlsson, and T. Ny. 1995. The inhibition mechanism of serpins. Evidence that the mobile reactive center loop is cleaved in the native protease-inhibitor complex. *Journal of Biological Chemistry* **270**:29652.

267. Egelund, R., K. W. Rodenburg, P. A. Andreasen, M. S. Rasmussen, R. E. Guldberg, and T. E. Petersen. 1998. An ester bond linking a fragment of a serine proteinase to its serpin inhibitor. *Biochemistry* **37**:6375.

268. Rodriguez de Cordoba, S., T. R. Dykman, F. Ginsberg-Fellner, G. Ercilla, M. Aqua, J. P. Atkinson, and P. Rubinstein. 1984. Evidence for linkage between the loci coding for the binding protein for the fourth component of human complement (C4BP) and for the C3b/C4b receptor. *Proceedings of the National Academy of Sciences of the United States of America* **81**:7890.

269. Holers, V. M., J. L. Cole, D. M. Lublin, T. Seya, and J. P. Atkinson. 1985. Human C3b- and C4b-regulatory proteins: a new multi-gene family. *Immunology Today* **6**:188.

270. Reid, K. B. M., D. R. Bentley, R. D. Campbell, L. P. Chung, R. B. Sim, T. Kristensen, and B. F. Tack. 1986. Complement proteins which interact with C3b or C4b. A superfamily of structurally related proteins. *Immunology Today* **7**:230.

271. Kristensen, T., P. D'Eustachio, R. T. Ogata, L. P. Chung, K. B. M. Reid, and B. F. Tack. 1987. The superfamily of C3b/C4b binding proteins. *Federation Proceedings* **46**:2463.

272. Hourcade, D., V. M. Holers, and J. P. Atkinson. 1989. The regulators of complement activation (RCA) gene cluster. *Advances in Immunology* **45**:381.

273. Tedder, T. F., L. J. Zhou, and P. Engel. 1994. The CD19/CD21 signal transduction complex of B lymphocytes. *Immunology Today* **15**:437.

274. Fearon, D. T., and R. H. Carter. 1995. The CD19/CR2/TAPA-1 complex of B lymphocytes: linking natural to acquired immunity. *Annual Review of Immunology* **13**:127.

275. Cornacoff, J. B., L. A. Hebert, W. L. Smead, M. E. VanAman, D. J. Birmingham, and F. J. Waxman. 1983. Primate erythrocyte-immune complex-clearing mechanism. *Journal of Clinical Investigation* **71**:236.

276. Davis, L. S., S. S. Patel, J. P. Atkinson, and P. E. Lipsky. 1988. Decay-accelerating factor functions as a signal transducing molecule for human T cells. *Journal of Immunology* **141**:2246.

277. Stefanova, I., and V. Horejsi. 1991. Association of the CD59 and CD55 cell surface glycoproteins with other membrane molecules. *Journal of Immunology* **147**:1587.

278. Shenoy-Scaria, A. M., J. Kwong, T. Fujita, M. W. Olszowy, A. S. Shaw, and D. M. Lublin. 1992. Signal transduction through decay-accelerating factor. Interaction of glycosyl-phosphatidylinositol anchor and protein tyrosine kinases p56lck and p59fyn 1. *Journal of Immunology* **149**:3535.

279. Anderson, D. J., A. F. Abbott, and R. M. Jack. 1993. The role of complement component C3b and its receptors in sperm-oocyte interaction. *Proceedings of the National Academy of Sciences of the United States of America* **90**:10051.

280. Taylor, C. T., M. M. Biljan, C. R. Kingsland, and P. M. Johnson. 1994. Inhibition of human spermatozoon-oocyte interaction in vitro by monoclonal antibodies to CD46 (membrane cofactor protein). *Human Reproduction* **9:**907.

281. Fingeroth, J. D., J. J. Weis, T. F. Tedder, J. L. Strominger, P. A. Biro, and D. T. Fearon. 1984. Epstein–Barr virus receptor of human B lymphocytes is the C3d receptor CR2. *Proceedings of the National Academy of Sciences of the United States of America* **81:**4510.

282. Bergelson, J. M., M. Chan, K. R. Solomon, N. F. St. John, H. Lin, and R. W. Finberg. 1994. Decay-accelerating factor (CD55), a glycosylphosphatidylinositol-anchored complement regulatory protein, is a receptor for several echoviruses. *Proceedings of the National Academy of Sciences of the United States of America* **91:**6245.

283. Ward, T., P. A. Pipkin, N. A. Clarkson, D. M. Stone, P. D. Minor, and J. W. Almond. 1994. Decay-accelerating factor CD55 is identified as the receptor for echovirus 7 using CELICS, a rapid immuno-focal cloning method. *EMBO Journal* **13:**5070.

284. Karnauchow, T. M., D. L. Tolson, B. A. Harrison, E. Altman, D. M. Lublin, and K. Dimock. 1996. The HeLa cell receptor for enterovirus 70 is decay-accelerating factor (CD55). *Journal of Virology* **70:**5143.

285. Bergelson, J. M., J. G. Mohanty, R. L. Crowell, N. F. St. John, D. M. Lublin, and R. W. Finberg. 1995. Coxsackievirus B3 adapted to growth in RD cells binds to decay-accelerating factor (CD55). *Journal of Virology* **69:**1903.

286. Shafren, D. R., R. C. Bates, M. V. Agrez, R. L. Herd, G. F. Burns, and R. D. Barry. 1995. Coxsackieviruses B1, B3, and B5 use decay accelerating factor as a receptor for cell attachment. *Journal of Virology* **69:**3873.

287. Bergelson, J. M., J. F. Modlin, W. Wieland-Alter, J. A. Cunningham, R. L. Crowell, and R. W. Finberg. 1997. Clinical coxsackievirus B isolates differ from laboratory strains in their interaction with two cell surface receptors. *Journal of Infectious Diseases* **175:**697.

288. Shafren, D. R., D. J. Dorahy, R. A. Ingham, G. F. Burns, and R. D. Barry. 1997. Coxsackievirus A21 binds to decay-accelerating factor but requires intercellular adhesion molecule 1 for cell entry. *Journal of Virology* **71:**4736.

289. Naniche, D., G. Varior-Krishnan, F. Cervoni, T. F. Wild, B. Rossi, C. Rabourdin-Combe, and D. Gerlier. 1993. Human membrane cofactor protein (CD46) acts as a cellular receptor for measles virus. *Journal of Virology* **67:**6025.

290. Dorig, R. E., A. Marcil, A. Chopra, and C. D. Richardson. 1993. The human CD46 molecule is a receptor for measles virus (Edmonston strain). *Cell* **75:**295.

291. Dorig, R. E., A. Marcil, and C. D. Richardson. 1994. CD46, a primate-specific receptor for measles virus. *Trends in Microbiology* **2:**312.

292. Pinter, C., A. G. Siccardi, L. Lopalco, R. Longhi, and A. Clivio. 1995. HIV glycoprotein 41 and complement factor H interact with each other and share functional as well as antigenic homology. *AIDS Research and Human Retroviruses* **11:**971.

293. Pinter, C., A. G. Siccardi, R. Longhi, and A. Clivio. 1995. Direct interaction of complement factor H with the C1 domain of HIV type 1 glycoprotein 120. *AIDS Research and Human Retroviruses* **11:**577.

294. Stoiber, H., R. Schneider, J. Janatova, and M. P. Dierich. 1995. Human complement proteins C3b, C4b, factor H and properdin react with specific sites in gp120 and gp41, the envelope proteins of HIV-1. *Immunobiology* **193:**98.

295. Stoiber, H., C. Pinter, A. G. Siccardi, A. Clivio, and M. P. Dierich. 1996. Efficient destruction of human immunodeficiency virus in human serum by inhibiting the protective action of complement factor H and decay accelerating factor (DAF, CD55). *Journal of Experimental Medicine* **183:**307.

296. Marschang, P., J. Sodroski, R. Wurzner, and M. P. Dierich. 1995. Decay-accelerating factor (CD55) protects human immunodeficiency virus type 1 from inactivation by human complement. *European Journal of Immunology* **25:**285.

297. Saifuddin, M., T. Hedayati, J. P. Atkinson, M. H. Holguin, C. J. Parker, and G. T. Spear. 1997. Human immunodeficiency virus type 1 incorporates both glycosyl phosphatidylinositol-anchored CD55 and CD59 and integral membrane CD46 at levels that protect from complement-mediated destruction. *Journal of General Virology* **78:**1907.

298. Sölder, B. M., E. C. Reisinger, D. Koefler, G. Bitterlich, H. Wachter, and M. P. Dierich. 1989. Complement receptors: another port of entry for HIV. *Lancet* **22:**271.

299. Delibrias, C. C., A. Mouhoub, E. Fischer, and M. D. Kazatchkine. 1994. CR1(CD35) and CR2(CD21) complement C3 receptors are expressed on normal human thymocytes and mediate infection of thymocytes with opsonized human immunodeficiency virus. *European Journal of Immunology* **24:**2784.

300. Horstmann, R. D., H. J. Sievertsen, J. Knobloch, and V. A. Fischetti. 1988. Antiphagocytic activity of streptococcal M protein: selective binding of complement control protein factor H. *Proceedings of the National Academy of Sciences of the United States of America* **85:**1657.

301. Thern, A., L. Stenberg, B. Dahlback, and G. Lindahl. 1995. Ig-binding surface proteins of Streptococcus pyogenes also bind human C4b-binding protein (C4BP), a regulatory component of the complement system. *Journal of Immunology* **154:**375.

302. Okada, N., M. K. Liszewski, J. P. Atkinson, and M. Caparon. 1995. Membrane cofactor protein (CD46) is a keratinocyte receptor for the M protein of the group A streptococcus. *Proceedings of the National Academy of Sciences of the United States of America* **92:**2489.

303. Carroll, M. C., R. D. Campbell, D. R. Bentley, and R. R. Porter. 1984. A molecular map of the human major histocompatibility complex class III region linking complement genes C4, C2 and factor B. *Nature* **307:**237.

304. Campbell, R. D., D. R. Bentley, and B. J. Morley. 1984. The factor B and C2 genes. *Philosophical Transactions of the Royal Society of London, Series B: Biological Sciences* **306:**367.

305. Porter, R. R. 1985. The complement components coded in the major

histocompatibility complexes and their biological activities. *Immunological Reviews* **87**:7.

306. Hobart, M. J., V. Joysey, and P. J. Lachmann. 1978. Inherited structural variation and linkage relationships of C7. *Journal of Immunogenetics* **5**:157.

307. Jeremiah, S. J., C. M. Abbott, Z. Murad, S. Povey, H. J. Thomas, E. Solomon, R. G. DiScipio, and G. H. Fey. 1990. The assignment of the genes coding for human complement components C6 and C7 to chromosome 5. *Annals of Human Genetics* **54**:141.

308. Tosi, M., C. Duponchel, T. Meo, and C. Julier. 1987. Complete cDNA sequence of human complement C1s and close physical linkage of the homologous genes C1s and C1r. *Biochemistry* **26**:8516.

309. DiScipio, R. G., and T. E. Hugli. 1989. The molecular architecture of human complement component C6. *Journal of Biological Chemistry* **264**:16197.

310. Bentley, D. R. 1986. Primary structure of human complement component C2. Homology to two unrelated protein families. *Biochemical Journal* **239**:339.

311. Rodriguez de Cordoba, S., D. M. Lublin, P. Rubinstein, and J. P. Atkinson. 1985. Human genes for three complement components that regulate the activation of C3 are tightly linked. *Journal of Experimental Medicine* **161**:1189.

312. Wong, W. W., L. B. Klickstein, J. A. Smith, J. H. Weis, and D. T. Fearon. 1985. Identification of a partial cDNA clone for the human receptor for complement fragments C3b/C4b. *Proceedings of the National Academy of Sciences of the United States of America* **82**:7711.

313. Weis, J. J., D. T. Fearon, L. B. Klickstein, W. W. Wong, S. A. Richards, A. de Bruyn Kops, J. A. Smith, and J. H. Weis. 1986. Identification of a partial cDNA clone for the C3d/Epstein–Barr virus receptor of human B lymphocytes: homology with the receptor for fragments C3b and C4b of the third and fourth components of complement. *Proceedings of the National Academy of Sciences of the United States of America* **83**:5639.

314. Weis, J. H., C. C. Morton, G. A. Bruns, J. J. Weis, L. B. Klickstein, W. W. Wong, and D. T. Fearon. 1987. A complement receptor locus: genes encoding C3b/C4b receptor and C3d/Epstein–Barr virus receptor map to 1q32. *Journal of Immunology* **138**:312.

315. Lublin, D. M., R. S. Lemons, M. M. Le Beau, V. M. Holers, M. L. Tykocinski, M. E. Medof, and J. P. Atkinson. 1987. The gene encoding decay-accelerating factor (DAF) is located in the complement-regulatory locus on the long arm of chromosome 1. *Journal of Experimental Medicine* **165**:1731.

316. Rey-Campos, J., P. Rubinstein, and S. Rodriguez de Cordoba. 1987. Decay-accelerating factor. Genetic polymorphism and linkage to the RCA (regulator of complement activation) gene cluster in humans. *Journal of Experimental Medicine* **166**:246.

317. Bora, N. S., D. M. Lublin, B. V. Kumar, R. D. Hockett, V. M. Holers, and J. P. Atkinson. 1989. Structural gene for human membrane cofactor protein (MCP) of complement maps to within 100 kb of the 3′ end of the C3b/C4b receptor gene. *Journal of Experimental Medicine* **169**:597.

318. Pardo-Manuel, F., J. Rey-Campos, A. Hillarp, B. Dahlback, and S. Rodriguez de Cordoba. 1990. Human genes for the alpha and beta chains of complement C4b-binding protein are closely linked in a head-to-tail arrangement. *Proceedings of the National Academy of Sciences of the United States of America* **87**:4529.

319. Rey-Campos, J., P. Rubinstein, and S. Rodriguez de Cordoba. 1988. A physical map of the human regulator of complement activation gene cluster linking the complement genes CR1, CR2, DAF, and C4BP. *Journal of Experimental Medicine* **167**:664.

320. Carroll, M. C., E. M. Alicot, P. J. Katzman, L. B. Klickstein, J. A. Smith, and D. T. Fearon. 1988. Organization of the genes encoding complement receptors type 1 and 2, decay-accelerating factor, and C4-binding protein in the RCA locus on human chromosome 1. *Journal of Experimental Medicine* **167**:1271.

321. Hourcade, D., A. D. Garcia, T. W. Post, P. Taillon-Miller, V. M. Holers, L. M. Wagner, N. S. Bora, and J. P. Atkinson. 1992. Analysis of the human regulators of complement activation (RCA) gene cluster with yeast artificial chromosomes (YACs). *Genomics* **12**:289.

322. Sanchez-Corral, P., F. Pardo-Manuel de Villena, J. Rey-Campos, and S. Rodriguez de Cordoba. 1993. C4BPAL1, a member of the human regulator of complement activation (RCA) gene cluster that resulted from the duplication of the gene coding for the alpha-chain of C4b-binding protein. *Genomics* **17**:185.

323. Rodriguez de Cordoba, S., and P. Rubinstein. 1987. New alleles of C4-binding protein and factor H and further linkage data in the regulator of complement activation (RCA) gene cluster in man. *Immunogenetics* **25**:267.

324. Rey-Campos, J., D. Baeza-Sanz, and S. Rodriguez de Cordoba. 1990. Physical linkage of the human genes coding for complement factor H and coagulation factor XIII B subunit. *Genomics* **7**:644.

325. Skerka, C., J. M. Moulds, P. Taillon-Miller, D. Hourcade, and P. F. Zipfel. 1995. The human factor H-related gene 2 (FHR2): structure and linkage to the coagulation factor XIIIb gene. *Immunogenetics* **42**:268.

326. Barlow, P. N., M. Baron, D. G. Norman, A. J. Day, A. C. Willis, R. B. Sim, and I. D. Campbell. 1991. Secondary structure of a complement control protein module by two-dimensional 1H NMR. *Biochemistry* **30**:997.

327. Barlow, P. N., D. G. Norman, A. Steinkasserer, T. J. Horne, J. Pearce, P. C. Driscoll, R. B. Sim, and I. D. Campbell. 1992. Solution structure of the fifth repeat of factor H: a second example of the complement control protein module. *Biochemistry* **31**:3626.

328. Barlow, P. N., A. Steinkasserer, D. G. Norman, B. Kieffer, A. P. Wiles, R. B. Sim, and I. D. Campbell. 1993. Solution structure of a pair of complement modules by nuclear magnetic resonance. *Journal of Molecular Biology* **232**:268.

329. Rodriguez de Cordoba, S., P. Sanchez-Corral, and J. Rey-Campos. 1991. Structure of the gene coding for the alpha polypeptide chain of the human complement component C4b-binding protein. *Journal of Experimental Medicine* **173**:1073.

330. Hillarp, A., and B. Dahlback. 1990. Cloning of cDNA coding for the beta chain of human complement component C4b-binding protein: sequence homology with the alpha chain. *Proceedings of the National Academy of Sciences of the United States of America* **87**:1183.

331. Sanchez-Corral, P., O. Criado Garcia, and S. Rodriguez de Cordoba. 1995. Isoforms of human C4b-binding protein. I. Molecular basis for the C4BP isoform pattern and its variations in human plasma. *Journal of Immunology* **155**:4030.

332. Dahlback, B., C. A. Smith, and H. J. Müller-Eberhard. 1983. Visualization of human C4b-binding protein and its complexes with vitamin K-dependent protein S and complement protein C4b. *Proceedings of the National Academy of Sciences of the United States of America* **80**:3461.

333. Perkins, S. J., L. P. Chung, and K. B. Reid. 1986. Unusual ultrastructure of complement-component-C4b-binding protein of human complement by synchrotron X-ray scattering and hydrodynamic analysis. *Biochemical Journal* **233**:799.

334. Morley, B. J., and R. D. Campbell. 1984. Internal homologies of the Ba fragment from human complement component Factor B, a class III MHC antigen. *EMBO Journal* **3**:153.

335. Mole, J. E., J. K. Anderson, E. A. Davison, and D. E. Woods. 1984. Complete primary structure for the zymogen of human complement factor B. *Journal of Biological Chemistry* **259**:3407.

336. DiScipio, R. G., D. N. Chakravarti, H. J. Müller-Eberhard, and G. H. Fey. 1988. The structure of human complement component C7 and the C5b-7 complex. *Journal of Biological Chemistry* **263**:549.

337. Haefliger, J. A., J. Tschopp, N. Vial, and D. E. Jenne. 1989. Complete primary structure and functional characterization of the sixth component of the human complement system. Identification of the C5b-binding domain in complement C6. *Journal of Biological Chemistry* **264**:18041.

338. Journet, A., and M. Tosi. 1986. Cloning and sequencing of full-length cDNA encoding the precursor of human complement component C1r. *Biochemical Journal* **240**:783.

339. Goldberger, G., G. A. Bruns, M. Rits, M. D. Edge, and D. J. Kwiatkowski. 1987. Human complement factor I: analysis of cDNA-derived primary structure and assignment of its gene to chromosome 4. *Journal of Biological Chemistry* **262**:10065.

340. Ichinose, A., B. A. McMullen, K. Fujiwaka, and E. W. Davie. 1986. The amino acid sequence of the β subunit of human factor XIII: a protein composed of ten repetitive segments. *Biochemistry* **25**:4633.

341. Kurosky, A., D. R. Barnett, T. H. Lee, B. Touchstone, R. E. Hay, M. S. Arnott, B. H. Bowman, and W. M. Fitch. 1980. Covalent structure of human haptoglobin: a serine protease homolog. *Proceedings of the National Academy of Sciences of the United States of America* **77**:3388.

342. Lozier, J., N. Takahashi, and F. W. Putman. 1984. Complete amino acid sequence of human plasma β2-glycoprotein I. *Proceedings of the National Academy of Sciences of the USA* **81**:3640.

343. Greene, W. C., and W. J. Leonard. 1986. The human interleukin-2 receptor. *Annual Review of Immunology* **4**:69.

344. Klickstein, L. B., W. W. Wong, J. A. Smith, J. H. Weis, J. G. Wilson, and D. T. Fearon. 1987. Human C3b/C4b receptor (CR1). Demonstration of long homologous repeating domains that are composed of the short consensus repeats characteriztics of C3/C4 binding proteins. *Journal of Experimental Medicine* **165**:1095.

345. Hourcade, D., D. R. Miesner, J. P. Atkinson, and V. M. Holers. 1988. Identification of an alternative polyadenylation site in the human C3b/C4b receptor (complement receptor type 1) transcriptional unit and prediction of a secreted form of complement receptor type 1. *Journal of Experimental Medicine* **168**:1255.

346. Klickstein, L. B., T. J. Bartow, V. Miletic, L. D. Rabson, J. A. Smith, and D. T. Fearon. 1988. Identification of distinct C3b and C4b recognition sites in the human C3b/C4b receptor (CR1, CD35) by deletion mutagenesis. *Journal of Experimental Medicine* **168**:1699.

347. Hourcade, D., D. R. Miesner, C. Bee, W. Zeldes, and J. P. Atkinson. 1990. Duplication and divergence of the amino-terminal coding region of the complement receptor 1 (CR1) gene. An example of concerted (horizontal) evolution within a gene. *Journal of Biological Chemistry* **265**:974.

348. Smith, G. P. 1976. Evolution of repeated DNA sequences by unequal crossover. *Science* **191**:528.

349. Szostak, J. W., and R. Wu. 1980. Unequal crossing over in the ribosomal DNA of Saccharomyces cerevisiae. *Nature* **284**:426.

350. Baltimore, D. 1981. Gene conversion: some implications for immunoglobulin genes. *Cell* **24**:592.

351. Weis, J. J., L. E. Toothaker, J. A. Smith, J. H. Weis, and D. T. Fearon. 1988. Structure of the human B lymphocyte receptor for C3d and the Epstein–Barr virus and relatedness to other members of the family of C3/C4 binding proteins [published erratum appears in *J Exp Med* 1988 Nov 1;168(5):1953–4]. *Journal of Experimental Medicine* **167**:1047.

352. Fujisaku, A., J. B. Harley, M. B. Frank, B. A. Gruner, B. Frazier, and V. M. Holers. 1989. Genomic organization and polymorphisms of the human C3d/Epstein–Barr virus receptor. *Journal of Biological Chemistry* **264**:2118.

353. Moore, M. D., N. R. Cooper, B. F. Tack, and G. R. Nemerow. 1987. Molecular cloning of the cDNA encoding the Epstein–Barr virus/C3d receptor (complement receptor type 2) of human B lymphocytes. *Proceedings of the National Academy of Sciences of the United States of America* **84**:9194.

354. Wong, W. W., J. M. Cahill, M. D. Rosen, C. A. Kennedy, E. T. Bonaccio, M. J. Morris, J. G. Wilson, L. B. Klickstein, and D. T. Fearon. 1989. Structure of the human CR1 gene. Molecular basis of the structural and quantitative polymorphisms and identification of a new CR1-like allele. *Journal of Experimental Medicine* **169**:847.

355. Vik, D. P., and W. W. Wong. 1993. Structure of the gene for the F allele of complement receptor type 1 and sequence of the coding region unique to the S allele. *Journal of Immunology* **151**:6214.

356. Hillarp, A., F. Pardo-Manuel, R. R. Ruiz, S. Rodriguez de Cordoba, and B. Dahlback. 1993. The human C4b-binding protein beta-chain gene. *Journal of Biological Chemistry* **268**:15017.

357. Post, T. W., M. A. Arce, M. K. Liszewski, E. S. Thompson, J. P. Atkinson, and D. M. Lublin. 1990. Structure of the gene for human complement protein decay accelerating factor. *Journal of Immunology* **144**:740.

358. Post, T. W., M. K. Liszewski, E. M. Adams, I. Tedja, E. A. Miller, and J. P. Atkinson. 1991. Membrane cofactor protein of the complement system: alternative splicing of serine/threonine/proline-rich exons and cytoplasmic tails produces multiple isoforms that correlate with protein phenotype. *Journal of Experimental Medicine* **174**:93.

359. Vik, D. P., J. B. Keeney, P. Munoz-Canoves, D. D. Chaplin, and B. F. Tack. 1988. Structure of the murine complement factor H gene. *Journal of Biological Chemistry* **263**:16720.

360. Nilsson, V. R., and H. J. Müller-Eberhard. 1965. Isolation of β1H globulin from human serum and its characterization as the fifth component of complement. *Journal of Experimental Medicine* **122**:277.

361. Pangburn, M. K., R. D. Schreiber, and H. J. Müller-Eberhard. 1977. Human complement C3b inactivator: isolation, characterization, and demonstration of an absolute requirement for the serum protein beta1H for cleavage of C3b and C4b in solution. *Journal of Experimental Medicine* **146**:257.

362. Whaley, K., and S. Ruddy. 1976. Modulation of C3b hemolytic activity by a plasma protein distinct from C3b inactivator. *Science* **193**:1011.

363. Conrad, D. H., J. R. Carlo, and S. Ruddy. 1978. Interaction of beta1H globulin with cell-bound C3b: quantitative analysis of binding and influence of alternative pathway components on binding. *Journal of Experimental Medicine* **147**:1792.

364. Pangburn, M. K. 1986. Differences between the binding sites of the complement regulatory proteins DAF, CR1, and factor H on C3 convertases. *Journal of Immunology* **136**:2216.

365. Harrison, R. A., and P. J. Lachmann. 1980. The physiological breakdown of the third component of human complement. *Molecular Immunology* **17**:9.

366. Ross, G. D., J. D. Lambris, J. A. Cain, and S. L. Newman. 1982. Generation of three different fragments of bound C3 with purified factor I or serum. I. Requirements for factor H vs CR1 cofactor activity. *Journal of Immunology* **129**:2051.

367. Fujita, T., I. Gigli, and V. Nussenzweig. 1978. Human C4-binding protein. II. Role in proteolysis of C4b by C3b-inactivator. *Journal of Experimental Medicine* **148**:1044.

368. Gigli, I., T. Fujita, and V. Nussenzweig. 1979. Modulation of the classical pathway C3 convertase by plasma proteins C4 binding protein and C3b inactivator. *Proceedings of the National Academy of Sciences of the United States of America* **76**:6596.

369. Seya, T., K. Nakamura, T. Masaki, C. Ichihara-Itoh, M. Matsumoto, and S. Nagasawa. 1995. Human factor H and C4b-binding protein serve as factor I-cofactors both encompassing inactivation of C3b and C4b. *Molecular Immunology* **32**:355.

370. Nicol, P. A., and P. J. Lachmann. 1973. The alternate pathway of complement activation. The role of C3 and its inactivator (KAF). *Immunology* **24**:259.

371. Fearon, D. T., and K. F. Austen. 1977. Activation of the alternative complement pathway due to resistance of zymosan-bound. *Proceedings of the National Academy of Sciences of the United States of America* **74**:1683.

372. Fearon, D. T., and K. F. Austen. 1977. Activation of the alternative complement pathway with rabbit erythrocytes by circumvention of the regulatory action of endogenous control proteins. *Journal of Experimental Medicine* **146**:22.

373. Schreiber, R. D., M. K. Pangburn, P. H. Lesavre, and H. J. Muller-Eberhard. 1978. Initiation of the alternative pathway of complement: recognition of activators by bound C3b and assembly of the entire pathway from six isolated proteins. *Proceedings of the National Academy of Sciences of the United States of America* **75**:3948.

374. Brai, M., G. Misiano, S. Maringhini, I. Cutaja, and G. Hauptmann. 1988. Combined homozygous factor H and heterozygous C2 deficiency in an Italian family. *Journal of Clinical Immunology* **8**:50.

375. Nielsen, H. E., K. C. Christensen, C. Koch, B. S. Thomsen, N. H. Heegaard, and J. Tranum-Jensen. 1989. Hereditary, complete deficiency of complement factor H associated with recurrent meningococcal disease. *Scandinavian Journal of Immunology* **30**:711.

376. Pichette, V., S. Querin, W. Schurch, G. Brun, G. Lehner-Netsch, and J. M. Delage. 1994. Familial hemolytic-uremic syndrome and homozygous factor H deficiency. *American Journal of Kidney Diseases* **24**:936.

377. Fijen, C. A., E. J. Kuijper, M. Te Bulte, M. M. van de Heuvel, A. C. Holdrinet, R. B. Sim, M. R. Daha, and J. Dankert. 1996. Heterozygous and homozygous factor H deficiency states in a Dutch family. *Clinical and Experimental Immunology* **105**:511.

378. Thompson, R. A., and M. H. Winterborn. 1981. Hypocomplementaemia due to a genetic deficiency of β1H globulin. *Clinical and Experimental Immunology* **46**:110.

379. Levy, M., L. Halbwachs-Mecarelli, M. C. Gubler, G. Kohout, A. Bensenouci, P. Niaudet, G. Hauptmann, and P. Lesavre. 1986. H deficiency in two brothers with atypical dense intramembranous deposit disease. *Kidney International* **30**:949.

380. Kazatchkine, M. D., D. T. Fearon, J. E. Silbert, and K. F. Austen. 1979. Surface-associated heparin inhibits zymosan-induced activation of the human alternative complement pathway by augmenting the regulatory action of the control proteins on particle-bound C3b. *Journal of Experimental Medicine* **150**:1202.

381. Fearon, D. T. 1978. Regulation by membrane sialic acid of beta1H-dependent decay-dissociation of amplification C3 convertase of the alternative complement pathway. *Proceedings of the National Academy of Sciences of the United States of America* **75**:1971.

382. Nydegger, U. E., D. T. Fearon, and K. F. Austen. 1978. Autosomal locus regulates inverse relationship between sialic acid content and capacity of

mouse erythrocytes to activate human alternative complement pathway. *Proceedings of the National Academy of Sciences of the United States of America* **75**:6078.

383. Pangburn, M. K., and H. J. Müller-Eberhard. 1978. Complement C3 convertase: cell surface restriction of beta1H control and generation of restriction on neuraminidase-treated cells. *Proceedings of the National Academy of Sciences of the United States of America* **75**:2416.

384. Carreno, M. P., D. Labarre, F. Maillet, M. Jozefowicz, and M. D. Kazatchkine. 1989. Regulation of the human alternative complement pathway: formation of a ternary complex between factor H, surface-bound C3b and chemical groups on nonactivating surfaces. *European Journal of Immunology* **19**:2145.

385. Meri, S., and M. K. Pangburn. 1990. Discrimination between activators and nonactivators of the alternative pathway of complement: regulation via a sialic acid/polyanion binding site on factor H. *Proceedings of the National Academy of Sciences of the United States of America* **87**:3982.

386. Pangburn, M. K., M. A. Atkinson, and S. Meri. 1991. Localization of the heparin-binding site on complement factor H. *Journal of Biological Chemistry* **266**:16847.

387. Pangburn, M. K. 1989. Analysis of recognition in the alternative pathway of complement. Effect of polysaccharide size. *Journal of Immunology* **142**:2766.

388. Pangburn, M. K. 1989. Analysis of the mechanism of recognition in the complement alternative pathway using C3b-bound low molecular weight polysaccharides. *Journal of Immunology* **142**:2759.

389. Fries, L. F., T. A. Gaither, C. H. Hammer, and M. M. Frank. 1984. C3b covalently bound to IgG demonstrates a reduced rate of inactivation by factors H and I. *Journal of Experimental Medicine* **160**:1640.

390. Ripoche, J., A. J. Day, T. J. Harris, and R. B. Sim. 1988. The complete amino acid sequence of human complement factor H. *Biochemical Journal* **249**:593.

391. Schwaeble, W., J. Zwirner, T. F. Schulz, R. P. Linke, M. P. Dierich, and E. H. Weiss. 1987. Human complement factor H: expression of an additional truncated gene product of 43 kDa in human liver. *European Journal of Immunology* **17**:1485.

392. Katz, Y., and R. C. Strunk. 1988. Synthesis and regulation of complement protein factor H in human skin fibroblasts. *Journal of Immunology* **141**:559.

393. Ripoche, J., J. A. Mitchell, A. Erdei, C. Madin, B. Moffatt, T. Mokoena, S. Gordon, and R. B. Sim. 1988. Interferon gamma induces synthesis of complement alternative pathway proteins by human endothelial cells in culture. *Journal of Experimental Medicine* **168**:1917.

394. Brooimans, R. A., A. A. van der Ark, W. A. Buurman, L. A. van Es, and M. R. Daha. 1990. Differential regulation of complement factor H and C3 production in human umbilical vein endothelial cells by IFN-gamma and IL-1. *Journal of Immunology* **144**:3835.

395. Whaley, K. 1980. Biosynthesis of the complement components and the regulatory proteins of the alternative complement pathway by human peripheral blood monocytes. *Journal of Experimental Medicine* **151**:501.

396. Lappin, D. F., and K. Whaley. 1990. Interferon-induced transcriptional and post-transcriptional modulation of factor H and C4 binding-protein synthesis in human monocytes. *Biochemical Journal* **271**:767.

397. Devine, D. V., and W. F. Rosse. 1987. Regulation of the activity of platelet-bound C3 convertase of the alternative pathway of complement by platelet factor H. *Proceedings of the National Academy of Sciences of the United States of America* **84**:5873.

398. Legoedec, J., P. Gasque, J. F. Jeanne, and M. Fontaine. 1995. Expression of the complement alternative pathway by human myoblasts in vitro: biosynthesis of C3, factor B, factor H and factor I. *European Journal of Immunology* **25**:3460.

399. Gasque, P., N. Julen, A. M. Ischenko, C. Picot, C. Mauger, C. Chauzy, J. Ripoche, and M. Fontaine. 1992. Expression of complement components of the alternative pathway by glioma cell lines. *Journal of Immunology* **149**:1381.

400. Schulz, T. F., O. Scheiner, J. Alsenz, J. D. Lambris, and M. P. Dierich. 1984. Use of monoclonal antibodies against factor H to investigate the role of a membrane-associated protein antigenically related to H in C3b-receptor function. *Journal of Immunology* **132**:392.

401. Malhotra, V., and R. B. Sim. 1985. Expression of complement factor H on the cell surface of the human monocytic cell line U937. *European Journal of Immunology* **15**:935.

402. Demares, M. J. 1989. Membrane-associated complement factor H on lymphoblastoid cell lines Raji expresses a co-factor activity for the factor I-mediated cleavage of C3b. *Immunology* **67**:553.

403. Erdei, A., N. Julen, P. Marschang, E. Feifel, K. Kerekes, and M. P. Dierich. 1994. A novel, complement factor H-related regulatory protein expressed on the surface of human B cell lines. *European Journal of Immunology* **24**:867.

404. Sim, R. B., and R. G. DiScipio. 1982. Purification and structural studies on the complement-system control protein beta 1H (Factor H). *Biochemical Journal* **205**:285.

405. Jouvin, M. H., M. D. Kazatchkine, A. Cahour, and N. Bernard. 1984. Lysine residues, but not carbohydrates, are required for the regulatory function of H on the amplification C3 convertase of complement. *Journal of Immunology* **133**:3250.

406. Kristensen, T., and B. F. Tack. 1986. Murine protein H is comprised of 20 repeating units, 61 amino acids in length. *Proceedings of the National Academy of Sciences of the United States of America* **83**:3963.

407. Schulz, T. F., W. Schwable, K. K. Stanley, E. Weiss, and M. P. Dierich. 1986. Human complement factor H: isolation of cDNA clones and partial cDNA sequence of the 38-kDa tryptic fragment containing the binding site for C3b. *European Journal of Immunology* **16**:1351.

408. Perkins, S. J., P. I. Haris, R. B. Sim, and D. Chapman. 1988. A study of the structure of human complement component factor H by Fourier transform infrared spectroscopy and secondary structure averaging methods. *Biochemistry* **27**:4004.

409. DiScipio, R. G. 1992. Ultrastructures and interactions of complement factors H and I. *Journal of Immunology* **149**:2592.

410. Perkins, S. J., A. S. Nealis, and R. B. Sim. 1991. Oligomeric domain structure of human complement factor H by X-ray and neutron solution scattering. *Biochemistry* **30**:2847.

411. Rodriguez de Cordoba, S., and P. Rubinstein. 1984. Genetic polymorphism of human factor H (beta 1H). *Journal of Immunology* **132**:1906.

412. Day, A. J., A. C. Willis, J. Ripoche, and R. B. Sim. 1988. Sequence polymorphism of human complement factor H. *Immunogenetics* **27**:211.

413. Tsunenari, S., T. Higashi, K. Kibayashi, H. Pang, M. Ding, and J. Jia. 1992. Genetic polymorphism of human FH (HF, β1H globulin) in Chinese Han population in northeast China. *Japanese Journal of Human Genetics* **37**:145.

414. Nakamura, S., O. Ohue, and A. Sawaguchi. 1990. Genetic polymorphism of human factor H (beta 1H globulin). *Human Heredity* **40**:121.

415. Alsenz, J., J. D. Lambris, T. F. Schulz, and M. P. Dierich. 1984. Localization of the complement-component-C3b-binding site and the cofactor activity for factor I in the 38kDa tryptic fragment of factor H. *Biochemical Journal* **224**:389.

416. Alsenz, J., T. F. Schulz, J. D. Lambris, R. B. Sim, and M. P. Dierich. 1985. Structural and functional analysis of the complement component factor H with the use of different enzymes and monoclonal antibodies to factor H. *Biochemical Journal* **232**:841.

417. Koistinen, V. 1992. Limited tryptic cleavage of complement factor H abrogates recognition of sialic acid-containing surfaces by the alternative pathway of complement. *Biochemical Journal* **283**:317.

418. Hong, K., T. Kinoshita, Y. Dohi, and K. Inoue. 1982. Effect of trypsinization on the activity of human factor H. *Journal of Immunology* **129**:647.

419. Gordon, D. L., R. M. Kaufman, T. K. Blackmore, J. Kwong, and D. M. Lublin. 1995. Identification of complement regulatory domains in human factor H. *Journal of Immunology* **155**:348.

420. Kuhn, S., C. Skerka, and P. F. Zipfel. 1995. Mapping of the complement regulatory domains in the human factor H-like protein 1 and in factor H1. *Journal of Immunology* **155**:5663.

421. Kuhn, S., and P. F. Zipfel. 1996. Mapping of the domains required for decay acceleration activity of the human factor H-like protein 1 and factor H. *European Journal of Immunology* **26**:2383.

422. Soames, C. J., and R. B. Sim. 1995. An investigation of the interaction between human complement factor H and C3b. *Biochemical Society Transactions* **23**:53S.

423. Sharma, A. K., and M. K. Pangburn. 1996. Identification of three physically and functionally distinct binding sites for C3b in human complement factor H by deletion mutagenesis. *Proceedings of the National Academy of Sciences of the United States of America* **93**:10996.

424. Blackmore, T. K., T. A. Sadlon, H. M. Ward, D. M. Lublin, and D. L. Gordon. 1996. Identification of a heparin binding domain in the seventh short consensus repeat of complement factor H. *Journal of Immunology* **157**:5422.

425. Blackmore, T. K., J. Hellwage, T. A. Sadlon, N. Higgs, P. F. Zipfel, H. M.

Ward, and D. L. Gordon. 1998. Identification of the second heparin-binding domain in human complement factor H1. *Journal of Immunology* **160:**3342.

426. Stoiber, H., A. Clivio, and M. P. Dierich. 1997. Role of complement in HIV infection. *Annual Review of Immunology* **15:**649.

427. Sharma, A. K., and M. K. Pangburn. 1997. Localization by site-directed mutagenesis of the site in human complement factor H that binds to Streptococcus pyogenes M protein. *Infection and Immunity* **65:**484.

428. Kotarsky, H., J. Hellwage, E. Johnsson, C. Skerka, H. G. Svensson, G. Lindahl, U. Sjobring, and P. F. Zipfel. 1998. Identification of a domain in human factor H and factor H-like protein-1 required for the interaction with streptococcal M proteins1. *Journal of Immunology* **160:**3349.

429. Blackmore, T. K., V. A. Fischetti, T. A. Sadlon, H. M. Ward, and D. L. Gordon. 1998. M protein of the group A Streptococcus binds to the seventh short consensus repeat of human complement factor H. *Infection and Immunity* **66:**1427.

430. Zipfel, P. F., and C. Skerka. 1994. Complement factor H and related proteins: an expanding family of complement-regulatory proteins? *Immunology Today* **15:**121.

431. Ripoche, J., A. J. Day, B. Moffatt, and R. B. Sim. 1987. mRNA coding for a truncated form of complement FH. *Biochemical Society Transactions:*651.

432. Skerka, C., R. D. Horstmann, and P. F. Zipfel. 1991. Molecular cloning of a human serum protein structurally related to complement factor H. *Journal of Biological Chemistry* **266:**12015.

433. Fontaine, M., M. J. Demares, V. Koistinen, A. J. Day, C. Davrinche, R. B. Sim, and J. Ripoche. 1989. Truncated forms of human complement factor H. *Biochemical Journal* **258:**927.

434. Estaller, C., W. Schwaeble, M. Dierich, and E. H. Weiss. 1991. Human complement factor H: two factor H proteins are derived from alternatively spliced transcripts. *European Journal of Immunology* **21:**799.

435. Misasi, R., H. P. Huemer, W. Schwaeble, E. Solder, C. Larcher, and M. P. Dierich. 1989. Human complement factor H: an additional gene product of 43 kDa isolated from human plasma shows cofactor activity for the cleavage of the third component of complement. *European Journal of Immunology* **19:**1765.

436. Hellwage, J., S. Kuhn, and P. F. Zipfel. 1997. The human complement regulatory factor-H-like protein 1, which represents a truncated form of factor H, displays cell-attachment activity. *Biochemical Journal* **326:**321.

437. DiScipio, R. G., P. J. Daffern, I. U. Schraufstatter, and P. Sriramarao. 1998. Human polymorphonuclear leukocytes adhere to complement factor H through an interaction that involves alpha(M)beta(2) (CD11b/CD18). *Journal of Immunology* **160:**4057.

438. Estaller, C., V. Koistinen, W. Schwaeble, M. P. Dierich, and E. H. Weiss. 1991. Cloning of the 1.4-kb mRNA species of human complement factor H reveals a novel member of the short consensus repeat family related to the carboxy terminal of the classical 150-kDa molecule. *Journal of Immunology* **146:**3190.

439. Skerka, C., C. Timmann, R. D. Horstmann, and P. F. Zipfel. 1992. Two additional human serum proteins structurally related to complement factor H. Evidence for a family of factor H-related genes. *Journal of Immunology* **148**:3313.

440. Timmann, C., M. Leippe, and R. D. Horstmann. 1991. Two major serum components antigenically related to complement factor H are different glycosylation forms of a single protein with no factor H-like complement regulatory functions. *Journal of Immunology* **146**:1265.

441. Skerka, C., S. Kuhn, K. Gunther, K. Lingelbach, and P. F. Zipfel. 1993. A novel short consensus repeat-containing molecule is related to human complement factor H. *Journal of Biological Chemistry* **268**:2904.

442. Skerka, C., J. Hellwage, W. Weber, A. Tilkorn, F. Buck, T. Marti, E. Kampen, U. Beisiegel, and P. F. Zipfel. 1997. The human factor H-related protein 4 (FHR-4). A novel short consensus repeat-containing protein is associated with human triglyceride-rich lipoproteins. *Journal of Biological Chemistry* **272**:5627.

443. Hellwage, J., C. Skerka, and P. F. Zipfel. 1997. Biochemical and functional characterization of the factor-H-related protein 4 (FHR-4). *Immunopharmacology* **38**:149.

444. Park, C. T., and S. D. Wright. 1996. Plasma lipopolysaccharide-binding protein is found associated with a particle containing apolipoprotein A-I, phospholipid, and factor H-related proteins. *Journal of Biological Chemistry* **271**:18054.

445. Prodinger, W. M., J. Hellwage, M. Spruth, M. P. Dierich, and P. F. Zipfel. 1998. The C-terminus of factor H: monoclonal antibodies inhibit heparin binding and identify epitopes common to factor H and factor H-related proteins. *Biochemical Journal* **331**:41.

446. Lambris, J. D., and G. D. Ross. 1982. Characterization of the lymphocyte membrane receptor for factor H (beta 1H-globulin) with an antibody to anti-factor H idiotype. *Journal of Experimental Medicine* **155**:1400.

447. Erdei, A., and R. B. Sim. 1987. Complement factor H-binding protein of Raji cells and tonsil B lymphocytes. *Biochemical Journal* **246**:149.

448. Avery, V. M., and D. L. Gordon. 1993. Characterization of factor H binding to human polymorphonuclear leukocytes. *Journal of Immunology* **151**:5545.

449. Lambris, J. D., N. J. Dobson, and G. D. Ross. 1980. Release of endogenous C3b inactivator from lymphocytes in response to triggering membrane receptors for $\beta$1H globulin. *Journal of Experimental Medicine* **152**:1625.

450. Hartung, H. P., U. Hadding, D. Bitter-Suermann, and D. Gemsa. 1984. Release of prostaglandin E and thromboxane from macrophages by stimulation with factor H. *Clinical and Experimental Immunology* **56**:453.

451. Schopf, R. E., K. P. Hammann, O. Scheiner, E. M. Lemmel, and M. P. Dierich. 1982. Activation of human monocytes by both human beta 1H and C3b. *Immunology* **46**:307.

452. Tsokos, G. C., G. Inghirami, C. D. Tsoukas, J. E. Balow, and J. D. Lambris. 1985. Regulation of immunoglobulin secretion by factor H of human complement. *Immunology* **55**:419.

453. Fischetti, V. A., R. D. Horstmann, and V. Pancholi. 1995. Location of the complement factor H binding site on streptococcal M6 protein. *Infection and Immunity* **63**:149.

454. Scharfstein, J., A. Ferreira, I. Gigli, and V. Nussenzweig. 1978. Human C4-binding protein. I. Isolation and characterization. *Journal of Experimental Medicine* **148**:207.

455. Sata, T., R. J. Havel, L. Kotite, and J. P. Kane. 1976. New protein in human blood plasma, rich in proline, with lipid-binding properties. *Proceedings of the National Academy of Sciences of the United States of America* **73**:1063.

456. Funakoshi, M., J. Sasaki, and K. Arakawa. 1988. Proline-rich protein is a glycoprotein and an acute phase reactant. *Biochimica et Biophysica Acta* **963**:98.

457. Matsuguchi, T., S. Okamura, T. Aso, T. Sata, and Y. Niho. 1989. Molecular cloning of the cDNA coding for proline-rich protein (PRP): identity of PRP as C4b-binding protein. *Biochemical and Biophysical Research Communications* **165**:138.

458. Shiraishi, S., and R. M. Stroud. 1975. Cleavage products of C4b produced by enzymes in human serum. *Immunochemistry* **12**:935.

459. Nagasawa, S., and R. M. Stroud. 1977. Mechanism of action of the C3b inactivator: requirement for a high molecular weight cofactor (C3b-C4bINA cofactor) and production of a new C3b derivative (C3b'). *Immunochemistry* **14**:749.

460. Fujita, T., and V. Nussenzweig. 1979. The role of C4-binding protein and beta 1H in proteolysis of C4b and C3b. *Journal of Experimental Medicine* **150**:267.

461. Nagasawa, S., and R. M. Stroud. 1980. Purification and characterization of a macromolecular weight cofactor for C3b-inactivator, C4bC3bINA-cofactor, of human plasma. *Molecular Immunology* **17**:1365.

462. Ziccardi, R. J., B. Dahlback, and H. J. Müller-Eberhard. 1984. Characterization of the interaction of human C4b-binding protein with physiological ligands. *Journal of Biological Chemistry* **259**:13674.

463. Walker, F. J. 1980. Regulation of activated protein C by a new protein. A possible function for bovine protein S. *Journal of Biological Chemistry* **255**:5521.

464. Walker, F. J. 1981. Regulation of activated protein C by protein S. The role of phospholipid in factor Va inactivation. *Journal of Biological Chemistry* **256**:11128.

465. Comp, P. C., R. R. Nixon, M. R. Cooper, and C. T. Esmon. 1984. Familial protein S deficiency is associated with recurrent thrombosis. *Journal of Clinical Investigation* **74**:2082.

466. Dahlback, B. 1986. Inhibition of protein C cofactor function of human and bovine protein S by C4b-binding protein. *Journal of Biological Chemistry* **261**:12022.

467. Dahlback, B., and J. Stenflo. 1981. High molecular weight complex in human plasma between vitamin K-dependent protein S and complement component C4b-binding protein. *Proceedings of the National Academy of Sciences of the United States of America* **78**:2512.

468. Dahlback, B. 1983. Purification of human C4b-binding protein and formation of its complex with vitamin K-dependent protein S. *Biochemical Journal* **209**:847.

469. Dahlback, B., and B. Hildebrand. 1983. Degradation of human complement component C4b in the presence of the C4b-binding protein-protein S complex. *Biochemical Journal* **209**:857.

470. Schwalbe, R., B. Dahlback, A. Hillarp, and G. Nelsestuen. 1990. Assembly of protein S and C4b-binding protein on membranes. *Journal of Biological Chemistry* **265**:16074.

471. Nelson, R. M., and G. L. Long. 1991. Solution-phase equilibrium binding interaction of human protein S with C4b-binding protein. *Biochemistry* **30**:2384.

472. Pepys, M. B., and M. L. Baltz. 1983. Acute phase proteins with special reference to C-reactive protein and related proteins (pentaxins) and serum amyloid A protein. *Advances in Immunology* **34**:141.

473. Schwalbe, R. A., B. Dahlback, and G. L. Nelsestuen. 1990. Independent association of serum amyloid P component, protein S, and complement C4b with complement C4b-binding protein and subsequent association of the complex with membranes. *Journal of Biological Chemistry* **265**:21749.

474. Garcia de Frutos, P., and B. Dahlback. 1994. Interaction between serum amyloid P component and C4b-binding protein associated with inhibition of factor I-mediated C4b degradation. *Journal of Immunology* **152**:2430.

475. Trapp, R. G., M. Fletcher, J. Forristal, and C. D. West. 1987. C4 binding protein deficiency in a patient with atypical Behcet's disease. *Journal of Rheumatology* **14**:135.

476. Comp, P. C., J. Forristall, C. D. West, and R. G. Trapp. 1990. Free protein S levels are elevated in familial C4b-binding protein deficiency. *Blood* **76**:2527.

477. Sim, R. B., K. Kolble, M. A. McAleer, O. Dominguez, and V. M. Dee. 1993. Genetics and deficiencies of the soluble regulatory proteins of the complement system. *International Reviews of Immunology* **10**:65.

478. Barnum, S. R., and B. Dahlback. 1990. C4b-binding protein, a regulatory component of the classical pathway of complement, is an acute-phase protein and is elevated in systemic lupus erythematosus. *Complement and Inflammation* **7**:71.

479. Stankiewicz, A. J., M. Steiner, E. V. Lally, and S. R. Kaplan. 1991. Abnormally high level of C4b binding protein and deficiency of free fraction of protein S in a patient with systemic lupus erythematosus and recurrent thromboses. *Journal of Rheumatology* **18**:82.

480. Daha, M. R., H. M. Hazevoet, J. Hermans, L. A. van Es, and A. Cats. 1983. Relative importance of C4 binding protein in the modulation of the classical pathway C3 convertase in patients with systemic lupus erythematosus. *Clinical and Experimental Immunology* **54**:248.

481. Matsuda, J., K. Gohchi, M. Gotoh, M. Tsukamoto, and N. Saitoh. 1994. Plasma concentrations of total/free and functional protein S are not decreased in systemic lupus erythematosus patients with lupus anticoagulant and/or antiphospholipid antibodies. *Annals of Hematology* **69**:311.

482. Saeki, T., S. Hirose, M. Nukatsuka, Y. Kusunoki, and S. Nagasawa. 1989. Evidence that C4b-binding protein is an acute phase protein. *Biochemical and Biophysical Research Communications* **164**:1446.

483. Garcia de Frutos, P., R. I. Alim, Y. Hardig, B. Zoller, and B. Dahlback. 1994. Differential regulation of alpha and beta chains of C4b-binding protein during acute-phase response resulting in stable plasma levels of free anti-coagulant protein S. *Blood* **84**:815.

484. Criado Garcia, O., P. Sanchez-Corral, and S. Rodriguez de Cordoba. 1995. Isoforms of human C4b-binding protein. II. Differential modulation of the C4BPA and C4BPB genes by acute phase cytokines. *Journal of Immunology* **155**:4037.

485. Chung, L. P., and K. B. Reid. 1985. Structural and functional studies on C4b-binding protein, a regulatory component of the human complement system. *Bioscience Reports* **5**:855.

486. Chung, L. P., J. Gagnon, and K. B. Reid. 1985. Amino acid sequence studies of human C4b-binding protein: N-terminal sequence analysis and alignment of the fragments produced by limited proteolysis with chymotrypsin and the peptides produced by cyanogen bromide treatment. *Molecular Immunology* **22**:427.

487. Chung, L. P., D. R. Bentley, and K. B. Reid. 1985. Molecular cloning and characterization of the cDNA coding for C4b-binding protein, a regulatory protein of the classical pathway of the human complement system. *Biochemical Journal* **230**:133.

488. Lintin, S. J., A. R. Lewin, and K. B. M. Reid. 1988. Derivation of the sequence of the signal peptide in human C4b-binding protein and interspecies cross-hybridisation of the C4bp cDNA sequence. *FEBS Letters* **232**:328.

489. Hillarp, A., and B. Dahlback. 1988. Novel subunit in C4b-binding protein required for protein S binding. *Journal of Biological Chemistry* **263**:12759.

490. Hillarp, A., M. Hessing, and B. Dahlback. 1989. Protein S binding in relation to the subunit composition of human C4b-binding protein. *FEBS Letters* **259**:53.

491. Rodriguez de Cordoba, S., A. Ferreira, V. Nussenzweig, and P. Rubinstein. 1983. Genetic polymorphism of human C4-binding protein. *Journal of Immunology* **131**:1565.

492. Aso, T., S. Okamura, T. Matsuguchi, N. Sakamoto, T. Sata, and Y. Niho. 1991. Genomic organization of the alpha chain of the human C4b-binding protein gene. *Biochemical and Biophysical Research Communications* **174**:222.

493. Rodriguez de Cordoba, S., M. Perez-Blas, R. Ramos-Ruiz, P. Sanchez-Corral, F. Pardo-Manuel de Villena, and J. Rey-Campos. 1994. The gene coding for the beta-chain of C4b-binding protein (C4BPB) has become a pseudogene in the mouse. *Genomics* **21**:501.

494. Hillarp, A., H. Wiklund, A. Thern, and B. Dahlback. 1997. Molecular cloning of rat C4b binding protein α and β chains. *Journal of Immunology* **158**:1315.

495. Moffat, G. J., and B. F. Tack. 1992. Regulation of C4b-binding protein gene expression by the acute-phase mediators tumor necrosis factor-alpha, interleukin-6, and interleukin-1. *Biochemistry* **31**:12376.

496. Nagasawa, S., K. Mizuguchi, C. Ichihara, and J. Koyama. 1982. Limited chymotryptic cleavage of human C4-binding protein: isolation of a carbohydrate-containing core domain and an active fragment. *Journal of Biochemistry* **92**:1329.

497. Nagasawa, S., H. Unno, C. Ichihara, J. Koyama, and T. Koide. 1983. Human C4b-binding protein, C4bp. Chymotryptic cleavage and location of the 48 kDa active fragment within C4bp. *FEBS Letters* **164**:135.

498. Garcia de Frutos, P., Y. Hardig, and B. Dahlback. 1995. Serum amyloid P component binding to C4b-binding protein. *Journal of Biological Chemistry* **270**:26950.

499. Dahlback, B., and H. J. Muller-Eberhard. 1984. Ultrastructure of C4b-binding protein fragments formed by limited proteolysis using chymotrypsin. *Journal of Biological Chemistry* **259**:11631.

500. Hessing, M., D. Kanters, H. Takeya, C. van 't Veer, T. M. Hackeng, S. Iwanaga, and B. N. Bouma. 1993. The region Ser333-Arg356 of the alpha-chain of human C4b-binding protein is involved in the binding of complement C4b. *FEBS Letters* **317**:228.

501. Hardig, Y., A. Hillarp, and B. Dahlback. 1997. The amino-terminal module of the C4b-binding protein alpha-chain is crucial for C4b binding and factor I-cofactor function. *Biochemical Journal* **323**:469.

502. Kristensen, T., R. T. Ogata, L. P. Chung, K. B. Reid, and B. F. Tack. 1987. cDNA structure of murine C4b-binding protein, a regulatory component of the serum complement system. *Biochemistry* **26**:4668.

503. Kai, S., T. Fujita, I. Gigli, and V. Nussenzweig. 1980. Mouse C3b/C4b inactivator: purification and properties. *Journal of Immunology* **125**:2409.

504. Fernandez, J. A., and J. H. Griffin. 1994. A protein S binding site on C4b-binding protein involves beta chain residues 31-45. *Journal of Biological Chemistry* **269**:2535.

505. Hardig, Y., and B. Dahlback. 1996. The amino-terminal module of the C4b-binding protein beta-chain contains the protein S-binding site. *Journal of Biological Chemistry* **271**:20861.

506. Hillarp, A., A. Thern, and B. Dahlback. 1994. Bovine C4b binding protein. Molecular cloning of the alpha- and beta-chains provides structural background for lack of complex formation with protein S. *Journal of Immunology* **153**:4190.

507. Johnsson, E., A. Thern, B. Dahlback, L. O. Heden, M. Wikstrom, and G. Lindahl. 1996. A highly variable region in members of the streptococcal M protein family binds the human complement regulator C4BP. *Journal of Immunology* **157**:3021.

508. Cleary, P., and D. Retnoningrum. 1994. Group A streptococcal immunoglobulin-binding proteins: adhesins, molecular mimicry or sensory proteins? *Trends in Microbiology* **2**:131.

509. Accardo, P., P. Sanchez-Corral, O. Criado, E. Garcia, and S. Rodriguez de Cordoba. 1996. Binding of human complement component C4b-binding protein (C4BP) to Streptococcus pyogenes involves the C4b-binding site. *Journal of Immunology* **157**:4935.

510. Nelson, R. A. 1953. The immune adherence phenomenon: an immuno-

logically specific reaction between micro-organisms and erythrocytes leading to enhanced phagocytosis. *Science* **118**:733.

511. Nelson, D. S. 1963. Immune adherence. *Advances in Immunology* **3**:131.

512. Duke, H. L., and J. M. Wallace. 1930. 'Red-cell adhesion' in trypanosomiasis of man and animals. *Parasitology* **22**:414.

513. Wallace, J. M., and A. Wormall. 1931. 'Red-cell adhesion' in trypanosomiasis of man and animals; some experiments on mechanism of reaction. *Parasitology* **23**:346.

514. Brown, H. C., and J. C. Broom. 1938. Studies in trypanosomiasis. II. Observations on the red cell adhesion test. *Transactions of the Royal Society of Tropical Medicine and Hygiene* **32**:209.

515. Lay, W. H., and V. Nussenzweig. 1968. Receptors for complement of leukocytes. *Journal of Experimental Medicine* **128**:991.

516. Fearon, D. T. 1979. Regulation of the amplification C3 convertase of human complement by an inhibitory protein isolated from human erythrocyte membrane. *Proceedings of the National Academy of Sciences of the United States of America* **76**:5867.

517. Fearon, D. T. 1980. Identification of the membrane glycoprotein that is the C3b receptor of the human erythrocyte, polymorphonuclear leukocyte, B lymphocyte, and monocyte. *Journal of Experimental Medicine* **152**:20.

518. Medof, M. E., K. Iida, C. Mold, and V. Nussenzweig. 1982. Unique role of the complement receptor CR1 in the degradation of C3b associated with immune complexes. *Journal of Experimental Medicine* **156**:1739.

519. Medicus, R. G., J. Melamed, and M. A. Arnaout. 1983. Role of human factor I and C3b receptor in the cleavage of surface-bound C3bi molecules. *European Journal of Immunology* **13**:465.

520. Cooper, N. R. 1969. Immune adherence by the fourth component of complement. *Science* **165**:396.

521. Ross, G. D., and M. J. Polley. 1975. Specificity of human lymphocyte complement receptors. *Journal of Experimental Medicine* **141**:1163.

522. Iida, K., and V. Nussenzweig. 1981. Complement receptor is an inhibitor of the complement cascade. *Journal of Experimental Medicine* **153**:1138.

523. Iida, K., and V. Nussenzweig. 1983. Functional properties of membrane-associated complement receptor CR1. *Journal of Immunology* **130**:1876.

524. Medof, M. E., and V. Nussenzweig. 1984. Control of the function of substrate-bound C4b-C3b by the complement receptor Cr1. *Journal of Experimental Medicine* **159**:1669.

525. Kinoshita, T., M. E. Medof, K. Hong, and V. Nussenzweig. 1986. Membrane-bound C4b interacts endogenously with complement receptor CR1 of human red cells. *Journal of Experimental Medicine* **164**:1377.

526. Ross, G. D., S. L. Newman, J. D. Lambris, J. E. Devery-Pocius, J. A. Cain, and P. J. Lachmann. 1983. Generation of three different fragments of bound C3 with purified factor I or serum. II. Location of binding sites in the C3 fragments for factors B and H, complement receptors, and bovine conglutinin. *Journal of Experimental Medicine* **158**:334.

527. Medof, M. E., G. M. Prince, and J. J. Oger. 1982. Kinetics of interaction of immune complexes with complement receptors on human blood cells:

modification of complexes during interaction with red cells. *Clinical and Experimental Immunology* **48**:715.

528. Medof, M. E., G. M. Prince, and C. Mold. 1982. Release of soluble immune complexes from immune adherence receptors on human erythrocytes is mediated by C3b inactivator independently of Beta 1H and is accompanied by generation of C3c. *Proceedings of the National Academy of Sciences of the United States of America* **79**:5047.

529. Arnaout, M. A., J. Melamed, B. F. Tack, and H. R. Colten. 1981. Characterization of the human complement (C3b) receptor with a fluid phase C3b dimer. *Journal of Immunology* **127**:1348.

530. Arnaout, M. A., N. Dana, J. Melamed, R. Medicus, and H. R. Colten. 1983. Low ionic strength or chemical cross-linking of monomeric C3b increases its binding affinity to the human complement C3b receptor. *Immunology* **48**:229.

531. Dierich, M. P., and R. A. Reisfeld. 1975. C3-mediated cytoadherence. Formation of C3 receptor aggregates as prerequisite for cell attachment. *Journal of Experimental Medicine* **142**:242.

532. Paccaud, J. P., J. L. Carpentier, and J. A. Schifferli. 1988. Direct evidence for the clustered nature of complement receptors type 1 on the erythrocyte membrane. *Journal of Immunology* **141**:3889.

533. Chevalier, J., and M. D. Kazatchkine. 1989. Distribution in clusters of complement receptor type one (CR1) on human erythrocytes. *Journal of Immunology* **142**:2031.

534. Klickstein, L. B., S. F. Barbashov, T. Liu, R. M. Jack, and A. Nicholson-Weller. 1997. Complement receptor type 1 (CR1, CD35) is a receptor for C1q. *Immunity* **7**:345.

535. Ehlenberger, A. G., and V. Nussenzweig. 1977. The role of membrane receptors for C3b and C3d in phagocytosis. *Journal of Experimental Medicine* **145**:357.

536. Ross, G. D., M. J. Walport, and N. Hogg. 1989. Receptors for IgG Fc and fixed C3. In *Human Monocytes*. M. Zembala, and G. L. Asherson, eds. Academic Press, London.

537. Tedder, T. F., D. T. Fearon, G. L. Gartland, and M. D. Cooper. 1983. Expression of C3b receptors on human B cells and myelomonocytic cells but not natural killer cells. *Journal of Immunology* **130**:1668.

538. Wilson, J. G., T. F. Tedder, and D. T. Fearon. 1983. Characterization of human T lymphocytes that express the C3b receptor. *Journal of Immunology* **131**:684.

539. Fischer, E., M. Capron, L. Prin, J. P. Kusnierz, and M. D. Kazatchkine. 1986. Human eosinophils express CR1 and CR3 complement receptors for cleavage fragments of C3. *Cellular Immunology* **97**:297.

540. Fearon, D. T., and L. A. Collins. 1983. Increased expression of C3b receptors on polymorphonuclear leukocytes induced by chemotactic factors and by purification procedures. *Journal of Immunology* **130**:370.

541. O'Shea, J. J., E. J. Brown, B. E. Seligmann, J. A. Metcalf, M. M. Frank, and J. I. Gallin. 1985. Evidence for distinct intracellular pools of receptors for C3b and C3bi in human neutrophils. *Journal of Immunology* **134**:2580.

542. Berger, M., E. M. Wetzler, E. Welter, J. R. Turner, and A. M. Tartakoff. 1991. Intracellular sites for storage and recycling of C3b receptors in human neutrophils. *Proceedings of the National Academy of Sciences of the United States of America* **88**:3019.

543. Sengelov, H., L. Kjeldsen, W. Kroeze, M. Berger, and N. Borregaard. 1994. Secretory vesicles are the intracellular reservoir of complement receptor 1 in human neutrophils. *Journal of Immunology* **153**:804.

544. Kumar, A., E. Wetzler, and M. Berger. 1997. Isolation and characterization of complement receptor type 1 (CR1) storage vesicles from human neutrophils using antibodies to the cytoplasmic tail of CR1. *Blood* **89**:4555.

545. Gelfand, M. C., M. M. Frank, and I. Green. 1975. A receptor for the third component of complement in the human renal glomerulus. *Journal of Experimental Medicine* **142**:1029.

546. Appay, M. D., M. D. Kazatchkine, M. Levi-Strauss, N. Hinglais, and J. Bariety. 1990. Expression of CR1 (CD35) mRNA in podocytes from adult and fetal human kidneys. *Kidney International* **38**:289.

547. Reynes, M., J. P. Aubert, J. H. Cohen, J. Audouin, V. Tricottet, J. Diebold, and M. D. Kazatchkine. 1985. Human follicular dendritic cells express CR1, CR2, and CR3 complement receptor antigens. *Journal of Immunology* **135**:2687.

548. Gasque, P., P. Chan, C. Mauger, M. T. Schouft, S. Singhrao, M. P. Dierich, B. P. Morgan, and M. Fontaine. 1996. Identification and characterization of complement C3 receptors on human astrocytes. *Journal of Immunology* **156**:2247.

549. Yoon, S. H., and D. T. Fearon. 1985. Characterization of a soluble form of the C3b/C4b receptor (CR1) in human plasma. *Journal of Immunology* **134**:3332.

550. Pascual, M., M. A. Duchosal, G. Steiger, E. Giostra, A. Pechere, J. P. Paccaud, C. Danielsson, and J. A. Schifferli. 1993. Circulating soluble CR1 (CD35). Serum levels in diseases and evidence for its release by human leukocytes. *Journal of Immunology* **151**:1702.

551. Hamer, I., J. P. Paccaud, D. Belin, C. Maeder, and J. L. Carpentier. 1998. Soluble form of complement C3b/C4b receptor (CR1) results from a proteolytic cleavage in the C-terminal region of CR1 transmembrane domain. *Biochemical Journal* **329**:183.

552. Danielsson, C., M. Pascual, L. French, G. Steiger, and J. A. Schifferli. 1994. Soluble complement receptor type 1 (CD35) is released from leukocytes by surface cleavage. *European Journal of Immunology* **24**:2725.

553. Sadallah, S., E. Lach, H. U. Lutz, S. Schwarz, P. A. Guerne, and J. A. Schifferli. 1997. CR1, CD35 in synovial fluid from patients with inflammatory joint diseases. *Arthritis and Rheumatism* **40**:520.

554. Dykman, T. R., J. L. Cole, K. Iida, and J. P. Atkinson. 1983. Polymorphism of human erythrocyte C3b/C4b receptor. *Proceedings of the National Academy of Sciences of the United States of America* **80**:1698.

555. Wong, W. W., J. G. Wilson, and D. T. Fearon. 1983. Genetic regulation of a structural polymorphism of human C3b receptor. *Journal of Clinical Investigation* **72**:685.

556. Seya, T., V. M. Holers, and J. P. Atkinson. 1985. Purification and functional analysis of the polymorphic variants of the C3b/C4b receptor (CR1) and comparison with H, C4b-binding protein (C4bp), and decay accelerating factor (DAF). *Journal of Immunology* **135**:2661.

557. Dykman, T. R., J. A. Hatch, and J. P. Atkinson. 1984. Polymorphism of the human C3b/C4b receptor. Identification of a third allele and analysis of receptor phenotypes in families and patients with systemic lupus erythematosus. *Journal of Experimental Medicine* **159**:691.

558. Dykman, T. R., J. A. Hatch, M. S. Aqua, and J. P. Atkinson. 1985. Polymorphism of the C3b/C4b receptor (CR1): characterization of a fourth allele. *Journal of Immunology* **134**:1787.

559. Wong, W. W., C. A. Kennedy, E. T. Bonaccio, J. G. Wilson, L. B. Klickstein, J. H. Weis, and D. T. Fearon. 1986. Analysis of multiple restriction fragment length polymorphisms of the gene for the human complement receptor type I. Duplication of genomic sequences occurs in association with a high molecular mass receptor allotype. *Journal of Experimental Medicine* **164**:1531.

560. Holers, V. M., D. D. Chaplin, J. F. Leykam, B. A. Gruner, V. Kumar, and J. P. Atkinson. 1987. Human complement C3b/C4b receptor (CR1) mRNA polymorphism that correlates with the CR1 allelic molecular weight polymorphism. *Proceedings of the National Academy of Sciences of the United States of America* **84**:2459.

561. Lublin, D. M., R. C. Griffith, and J. P. Atkinson. 1986. Influence of glycosylation on allelic and cell-specific Mr variation, receptor processing, and ligand binding of the human complement C3b/C4b receptor. *Journal of Biological Chemistry* **261**:5736.

562. Dykman, T. R., J. L. Cole, K. Iida, and J. P. Atkinson. 1983. Structural heterogeneity of the C3b/C4b receptor (CR 1) on human peripheral blood cells. *Journal of Experimental Medicine* **157**:2160.

563. Iida, K., R. Mornaghi, and V. Nussenzweig. 1982. Complement receptor (CR1) deficiency in erythrocytes from patients with systemic lupus erythematosus. *Journal of Experimental Medicine* **155**:1427.

564. Wilson, J. G., W. W. Wong, P. H. Schur, and D. T. Fearon. 1982. Mode of inheritance of decreased C3b receptors on erythrocytes of patients with systemic lupus erythematosus. *New England Journal of Medicine* **307**:981.

565. Walport, M. J., G. D. Ross, C. Mackworth-Young, J. V. Watson, N. Hogg, and P. J. Lachmann. 1985. Family studies of erythrocyte complement receptor type 1 levels: reduced levels in patients with SLE are acquired, not inherited. *Clinical and Experimental Immunology* **59**:547.

566. Wilson, J. G., E. E. Murphy, W. W. Wong, L. B. Klickstein, J. H. Weis, and D. T. Fearon. 1986. Identification of a restriction fragment length polymorphism by a CR1 cDNA that correlates with the number of CR1 on erythrocytes. *Journal of Experimental Medicine* **164**:50.

567. Rodriguez de Cordoba, S., and P. Rubinstein. 1986. Quantitative variations of the C3b/C4b receptor (CR1) in human erythrocytes are controlled by genes within the regulator of complement activation (RCA) gene cluster. *Journal of Experimental Medicine* **164**:1274.

568. Moldenhauer, F., J. David, A. H. Fielder, P. J. Lachmann, and M. J. Walport. 1987. Inherited deficiency of erythrocyte complement receptor type 1 does not cause susceptibility to systemic lupus erythematosus. *Arthritis and Rheumatism* **30**:961.

569. Miyakawa, Y., A. Yamada, K. Kosaka, F. Tsuda, E. Kosugi, and M. Mayumi. 1981. Defective immune-adherence (C3b) receptor on erythrocytes from patients with systemic lupus erythematosus. *Lancet* **2**:493.

570. Walport, M. J., and K. A. Davies. 1996. Complement and immune complexes. *Research in Immunology* **147**:103.

571. Inada, Y., M. Kamiyama, T. Kanemitsu, C. L. Hyman, and W. S. Clark. 1982. Studies on immune adherence (C3b) receptor activity of human erythrocytes: relationship between receptor activity and presence of immune complexes in serum. *Clinical and Experimental Immunology* **50**:189.

572. Ross, G. D., W. J. Yount, M. J. Walport, J. B. Winfield, C. J. Parker, C. R. Fuller, R. P. Taylor, B. L. Myones, and P. J. Lachmann. 1985. Disease-associated loss of erythrocyte complement receptors (CR1, C3b receptors) in patients with systemic lupus erythematosus and other diseases involving autoantibodies and/or complement activation. *Journal of Immunology* **135**:2005.

573. Holme, E., A. Fyfe, A. Zoma, J. Veitch, J. Hunter, and K. Whaley. 1986. Decreased C3b receptors (CR1) on erythrocytes from patients with systemic lupus erythematosus. *Clinical and Experimental Immunology* **63**:41.

574. Walport, M., Y. C. Ng, and P. J. Lachmann. 1987. Erythrocytes transfused into patients with SLE and haemolytic anaemia lose complement receptor type 1 from their cell surface. *Clinical and Experimental Immunology* **69**:501.

575. Yen, J. H., H. W. Liu, S. F. Lin, J. R. Chen, and T. P. Chen. 1989. Erythrocyte complement receptor type 1 in patients with systemic lupus erythematosus. *Journal of Rheumatology* **16**:1320.

576. Moldenhauer, F., M. Botto, and M. J. Walport. 1988. The rate of loss of CR1 from ageing erythrocytes in vivo in normal subjects and SLE patients: no correlation with structural or numerical polymorphisms. *Clinical and Experimental Immunology* **72**:74.

577. Davies, K. A., V. Hird, S. Stewart, G. B. Sivolapenko, P. Jose, A. A. Epenetos, and M. J. Walport. 1990. A study of in vivo immune complex formation and clearance in man. *Journal of Immunology* **144**:4613.

578. Ripoche, J., and R. B. Sim. 1986. Loss of complement receptor type 1 (CR1) on ageing of erythrocytes. Studies of proteolytic release of the receptor. *Biochemical Journal* **235**:815.

579. Cohen, J. H., H. U. Lutz, J. L. Pennaforte, A. Bouchard, and M. D. Kazatchkine. 1992. Peripheral catabolism of CR1 (the C3b receptor, CD35) on erythrocytes from healthy individuals and patients with systemic lupus erythematosus (SLE). *Clinical and Experimental Immunology* **87**:422.

580. Krych, M., D. Hourcade, and J. P. Atkinson. 1991. Sites within the complement C3b/C4b receptor important for the specificity of ligand binding. *Proceedings of the National Academy of Sciences of the United States of America* **88**:4353.

581. Krych, M., L. Clemenza, D. Howdeshell, R. Hauhart, D. Hourcade, and J. P. Atkinson. 1994. Analysis of the functional domains of complement receptor type 1 (C3b/C4b receptor; CD35) by substitution mutagenesis. *Journal of Biological Chemistry* **269**:13273.

582. Subramanian, V. B., L. Clemenza, M. Krych, and J. P. Atkinson. 1996. Substitution of two amino acids confers C3b binding to the C4b binding site of CR1 (CD35). Analysis based on ligand binding by chimpanzee erythrocyte complement receptor. *Journal of Immunology* **157**:1242.

583. Krych, M., R. Hauhart, and J. P. Atkinson. 1998. Structure-function analysis of the active sites of complement receptor type 1. *Journal of Biological Chemistry* **273**:8623.

584. Kalli, K. R., P. H. Hsu, T. J. Bartow, J. M. Ahearn, A. K. Matsumoto, L. B. Klickstein, and D. T. Fearon. 1991. Mapping of the C3b-binding site of CR1 and construction of a (CR1)2-F(ab')2 chimeric complement inhibitor. *Journal of Experimental Medicine* **174**:1451.

585. Delibrias, C. C., M. D. Kazatchkine, and E. Fischer. 1993. Evidence for the role of CR1 (CD35), in addition to CR2 (CD21), in facilitating infection of human T cells with opsonized HIV. *Scandinavian Journal of Immunology* **38**:183.

586. Thieblemont, N., N. Haeffner-Cavaillon, A. Ledur, L. A.-S. J, H. W. Ziegler-Heitbrock, and M. D. Kazatchkine. 1993. CR1 (CD35) and CR3 (CD11b/CD18) mediate infection of human monocytes and monocytic cell lines with complement-opsonized HIV independently of CD4. *Clinical and Experimental Immunology* **92**:106.

587. Embretson, J., M. Zupancic, J. L. Ribas, A. Burke, P. Racz, K. Tenner-Racz, and A. T. Haase. 1993. Massive covert infection of helper T lymphocytes and macrophages by HIV during the incubation period of AIDS. *Nature* **362**:359.

588. Heath, S. L., J. G. Tew, J. G. Tew, A. K. Szakal, and G. F. Burton. 1995. Follicular dendritic cells and human immunodeficiency virus infectivity. *Nature* **377**:740.

589. Cameron, P., M. Pope, A. Granelli-Piperno, and R. M. Steinman. 1996. Dendritic cells and the replication of HIV-1. *Journal of Leukocyte Biology* **59**:158.

590. Schorey, J. S., M. C. Carroll, and E. J. Brown. 1997. A macrophage invasion mechanism of pathogenic mycobacteria. *Science* **277**:1091.

591. Zaffran, Y., and J. J. Ellner. 1997. A coat of many complements. *Nature Medicine* **3**:1078.

592. Rowe, J. A., J. M. Moulds, C. I. Newbold, and L. H. Miller. 1997. P. falciparum rosetting mediated by a parasite-variant erythrocyte membrane protein and complement-receptor 1. *Nature* **388**:292.

593. Ross, G. D., M. J. Polley, E. M. Rabellino, and H. M. Grey. 1973. Two different complement receptors on human lymphocytes. *Journal of Experimental Medicine* **138**:798.

594. Eden, A., G. W. Miller, and V. Nussenzweig. 1973. Human lymphocytes bear membrane receptors for C3b and C3d. *Journal of Clinical Investigation* **52**:3239.

595. Okada, H., and K. Nishioka. 1973. Complement receptors on cell membranes. 1. Evidence for two complement receptors. *Journal of Immunology* **111**:1444.

596. Lambris, J. D., N. J. Dobson, and G. D. Ross. 1981. Isolation of lymphocyte membrane complement receptor type two (the C3d receptor) and preparation of receptor-specific antibody. *Proceedings of the National Academy of Sciences of the United States of America* **78**:1828.

597. Barel, M., C. Charriaut, and R. Frade. 1981. Isolation and characterization of a C3b receptor-like molecule from membranes of a human B lymphoblastoid cell line (Raji). *FEBS Letters* **136**:111.

598. Nadler, L. M., P. Stashenko, R. Hardy, A. van Agthoven, C. Terhorst, and S. F. Schlossman. 1981. Characterization of a human B cell-specific antigen (B2) distinct from B1. *Journal of Immunology* **126**:1941.

599. Iida, K., L. Nadler, and V. Nussenzweig. 1983. Identification of the membrane receptor for the complement fragment C3d by means of a monoclonal antibody. *Journal of Experimental Medicine* **158**:1021.

600. Weis, J. J., T. F. Tedder, and D. T. Fearon. 1984. Identification of a 145,000 Mr membrane protein as the C3d receptor (CR2) of human B lymphocytes. *Proceedings of the National Academy of Sciences of the United States of America* **81**:881.

601. Micklem, K. J., R. B. Sim, and E. Sim. 1984. Analysis of C3-receptor activity on human B-lymphocytes and isolation of the complement receptor type 2 (CR2). *Biochemical Journal* **224**:75.

602. Frade, R., B. L. Myones, M. Barel, L. Krikorian, C. Charriaut, and G. D. Ross. 1985. gp140, a C3b-binding membrane component of lymphocytes, is the B cell C3dg/C3d receptor (Cr2) and is distinct from the neutrophil C3dg receptor (Cr4). *European Journal of Immunology* **15**:1192.

603. Myones, B. L., and G. D. Ross. 1987. Identification of a spontaneously shed fragment of B cell complement receptor type two (CR2) containing the C3d-binding site. *Complement* **4**:87.

604. Matsumoto, A. K., J. Kopicky-Burd, R. H. Carter, D. A. Tuveson, T. F. Tedder, and D. T. Fearon. 1991. Intersection of the complement and immune systems: a signal transduction complex of the B lymphocyte-containing complement receptor type 2 and CD19. *Journal of Experimental Medicine* **173**:55.

605. Bradbury, L. E., G. S. Kansas, S. Levy, R. L. Evans, and T. F. Tedder. 1992. The CD19/CD21 signal transducing complex of human B lymphocytes includes the target of antiproliferative antibody-1 and Leu-13 molecules. *Journal of Immunology* **149**:2841.

606. Matsumoto, A. K., D. R. Martin, R. H. Carter, L. B. Klickstein, J. M. Ahearn, and D. T. Fearon. 1993. Functional dissection of the CD21/CD19/TAPA-1/Leu-13 complex of B lymphocytes. *Journal of Experimental Medicine* **178**:1407.

607. Tedder, T. F., M. Inaoki, and S. Sato. 1997. The CD19-CD21 complex regulates signal transduction thresholds governing humoral immunity and autoimmunity. *Immunity* **6**:107.

608. Carter, R. H., M. O. Spycher, Y. C. Ng, R. Hoffman, and D. T. Fearon.

1988. Synergistic interaction between complement receptor type 2 and membrane IgM on B lymphocytes. *Journal of Immunology* **141**:457.

609. Carter, R. H., D. A. Tuveson, D. J. Park, S. G. Rhee, and D. T. Fearon. 1991. The CD19 complex of B lymphocytes. Activation of phospholipase C by a protein tyrosine kinase-dependent pathway that can be enhanced by the membrane IgM complex. *Journal of Immunology* **147**:3663.

610. Aubry, J. P., S. Pochon, P. Graber, K. U. Jansen, and J. Y. Bonnefoy. 1992. CD21 is a ligand for CD23 and regulates IgE production. *Nature* **358**:505.

611. Pochon, S., P. Graber, M. Yeager, K. Jansen, A. R. Bernard, J. P. Aubry, and J. Y. Bonnefoy. 1992. Demonstration of a second ligand for the low affinity receptor for immunoglobulin E (CD23) using recombinant CD23 reconstituted into fluorescent liposomes. *Journal of Experimental Medicine* **176**:389.

612. Aubry, J. P., S. Pochon, J. F. Gauchat, A. Nueda-Marin, V. M. Holers, P. Graber, C. Siegfried, and J. Y. Bonnefoy. 1994. CD23 interacts with a new functional extracytoplasmic domain involving N-linked oligosaccharides on CD21. *Journal of Immunology* **152**:5806.

613. Prodinger, W. M., C. Larcher, M. Schwendinger, and M. P. Dierich. 1996. Ligation of the functional domain of complement receptor type 2 (CR2, CD21) is relevant for complex formation in T cell lines. *Journal of Immunology* **156**:2580.

614. Jackson, C. G., H. D. Ochs, and R. J. Wedgwood. 1979. Immune response of a patient with deficiency of the fourth component of complement and systemic lupus erythematosus. *New England Journal of Medicine* **300**:1124.

615. Bottger, E. C., T. Hoffmann, U. Hadding, and D. Bitter-Suermann. 1985. Influence of genetically inherited complement deficiencies on humoral immune response in guinea pigs. *Journal of Immunology* **135**:4100.

616. Pepys, M. B. 1974. Role of complement in induction of antibody production in vivo. Effect of cobra factor and other C3-reactive agents on thymus-dependent and thymus-independent antibody responses. *Journal of Experimental Medicine* **140**:126.

617. Heyman, B., E. J. Wiersma, and T. Kinoshita. 1990. In vivo inhibition of the antibody response by a complement receptor-specific monoclonal antibody. *Journal of Experimental Medicine* **172**:665.

618. Hebell, T., J. M. Ahearn, and D. T. Fearon. 1991. Suppression of the immune response by a soluble complement receptor of B lymphocytes. *Science* **254**:102.

619. Tedder, T. F., L. T. Clement, and M. D. Cooper. 1984. Expression of C3d receptors during human B cell differentiation: immunofluorescence analysis with the HB-5 monoclonal antibody. *Journal of Immunology* **133**:678.

620. Sixbey, J. W., D. S. Davis, L. S. Young, L. Hutt-Fletcher, T. F. Tedder, and A. B. Rickinson. 1987. Human epithelial cell expression of an Epstein–Barr virus receptor. *Journal of General Virology* **68**:805.

621. Billaud, M., P. Busson, D. Huang, N. Mueller-Lantzch, G. Rousselet, O. Pavlish, H. Wakasugi, J. M. Seigneurin, T. Tursz, and G. M. Lenoir. 1989. Epstein–Barr virus (EBV)-containing nasopharyngeal carcinoma cells express the B-cell activation antigen blast2/CD23 and low levels of the EBV receptor CR2. *Journal of Virology* **63**:4121.

622. Birkenbach, M., X. Tong, L. E. Bradbury, T. F. Tedder, and E. Kieff. 1992. Characterization of an Epstein–Barr virus receptor on human epithelial cells. *Journal of Experimental Medicine* **176**:1405.

623. Liu, Y. J., J. Xu, O. de Bouteiller, C. L. Parham, G. Grouard, O. Djossou, B. de Saint-Vis, S. Lebecque, J. Banchereau, and K. W. Moore. 1997. Follicular dendritic cells specifically express the long CR2/CD21 isoform. *Journal of Experimental Medicine* **185**:165.

624. Tsoukas, C. D., and J. D. Lambris. 1988. Expression of CR2/EBV receptors on human thymocytes detected by monoclonal antibodies. *European Journal of Immunology* **18**:1299.

625. Watry, D., J. A. Hedrick, S. Siervo, G. Rhodes, J. J. Lamberti, J. D. Lambris, and C. D. Tsoukas. 1991. Infection of human thymocytes by Epstein–Barr virus. *Journal of Experimental Medicine* **173**:971.

626. Fischer, E., C. Delibrias, and M. D. Kazatchkine. 1991. Expression of CR2 (the C3dg/EBV receptor, CD21) on normal human peripheral blood T lymphocytes. *Journal of Immunology* **146**:865.

627. Tsoukas, C. D., and J. D. Lambris. 1993. Expression of EBV/C3d receptors on T cells: biological significance. *Immunology Today* **14**:56.

628. Delibrias, C. C., E. Fischer, G. Bismuth, and M. D. Kazatchkine. 1992. Expression, molecular association, and functions of C3 complement receptors CR1 (CD35) and CR2 (CD21) on the human T cell line HPB-ALL. *Journal of Immunology* **149**:768.

629. Ling, N. R., and B. Brown. 1992. Properties of soluble CR2 in human serum. *Immunobiology* **185**:403.

630. Fremeaux-Bacchi, V., I. Bernard, F. Maillet, J. C. Mani, M. Fontaine, J. Y. Bonnefoy, M. D. Kazatchkine, and E. Fischer. 1996. Human lymphocytes shed a soluble form of CD21 (the C3dg/Epstein–Barr virus receptor, CR2) that binds iC3b and CD23. *European Journal of Immunology* **26**:1497.

631. Larcher, C., F. Fend, M. Mitterer, N. Prang, F. Schwarzmann, and H. P. Huemer. 1995. Role of Epstein–Barr virus and soluble CD21 in persistent polyclonal B-cell lymphocytosis. *British Journal of Haematology* **90**:532.

632. Weis, J. J., and D. T. Fearon. 1985. The identification of N-linked oligosaccharides on the human CR2/Epstein–Barr virus receptor and their function in receptor metabolism, plasma membrane expression, and ligand binding. *Journal of Biological Chemistry* **260**:13824.

633. Toothaker, L. E., A. J. Henjes, and J. J. Weis. 1989. Variability of CR2 gene products is due to alternative exon usage and different CR2 alleles. *Journal of Immunology* **142**:3668.

634. Lowell, C. A., L. B. Klickstein, R. H. Carter, J. A. Mitchell, D. T. Fearon, and J. M. Ahearn. 1989. Mapping of the Epstein–Barr virus and C3dg binding sites to a common domain on complement receptor type 2. *Journal of Experimental Medicine* **170**:1931.

635. Carel, J. C., B. L. Myones, B. Frazier, and V. M. Holers. 1990. Structural requirements for C3d,g/Epstein–Barr virus receptor (CR2/CD21) ligand binding, internalization, and viral infection. *Journal of Biological Chemistry* **265**:12293.

636. Molina, H., S. J. Perkins, J. Guthridge, J. Gorka, T. Kinoshita, and V. M. Holers. 1995. Characterization of a complement receptor 2 (CR2, CD21) ligand binding site for C3. An initial model of ligand interaction with two linked short consensus repeat modules. *Journal of Immunology* **154**:5426.

637. Nemerow, G. R., C. Mold, V. K. Schwend, V. Tollefson, and N. R. Cooper. 1987. Identification of gp350 as the viral glycoprotein mediating attachment of Epstein–Barr virus (EBV) to the EBV/C3d receptor of B cells: sequence homology of gp350 and C3 complement fragment C3d. *Journal of Virology* **61**:1416.

638. Rao, P. E., S. D. Wright, E. F. Westberg, and G. Goldstein. 1985. OKB7, a monoclonal antibody that reacts at or near the C3d binding site of human CR2. *Cellular Immunology* **93**:549.

639. Martin, D. R., A. Yuryev, K. R. Kalli, D. T. Fearon, and J. M. Ahearn. 1991. Determination of the structural basis for selective binding of Epstein–Barr virus to human complement receptor type 2. *Journal of Experimental Medicine* **174**:1299.

640. Frade, R., M. Barel, B. Ehlin-Henriksson, and G. Klein. 1985. gp140, the C3d receptor of human B lymphocytes, is also the Epstein–Barr virus receptor. *Proceedings of the National Academy of Sciences of the United States of America* **82**:1490.

641. Nemerow, G. R., R. Wolfert, M. E. McNaughton, and N. R. Cooper. 1985. Identification and characterization of the Epstein–Barr virus receptor on human B lymphocytes and its relationship to the C3d complement receptor (CR2). *Journal of Virology* **55**:347.

642. Hoffmann, E. M. 1969. Inhibition of complement by a substance isolated from human erythrocytes. I. Extraction from human erythrocyte stromata. *Immunochemistry* **6**:391.

643. Hoffmann, E. M. 1969. Inhibition of complement by a substance isolated from human erythrocytes. II. Studies on the site and mechanism of action. *Immunochemistry* **6**:405.

644. Hoffmann, E. M., and H. M. Etlinger. 1973. Extraction of complement inhibitory factors from the erythrocytes of non-human species. *Journal of Immunology* **111**:946.

645. Nicholson-Weller, A., J. Burge, and K. F. Austen. 1981. Purification from guinea pig erythrocyte stroma of a decay-accelerating factor for the classical c3 convertase, C4b,2a. *Journal of Immunology* **127**:2035.

646. Nicholson-Weller, A., J. Burge, D. T. Fearon, P. F. Weller, and K. F. Austen. 1982. Isolation of a human erythrocyte membrane glycoprotein with decay-accelerating activity for C3 convertases of the complement system. *Journal of Immunology* **129**:184.

647. Medof, M. E., T. Kinoshita, and V. Nussenzweig. 1984. Inhibition of complement activation on the surface of cells after incorporation of decay-accelerating factor (DAF) into their membranes. *Journal of Experimental Medicine* **160**:1558.

648. Medof, M. E., E. I. Walter, W. L. Roberts, R. Haas, and T. L. Rosenberry. 1986. Decay accelerating factor of complement is anchored to cells by a C-terminal glycolipid. *Biochemistry* **25**:6740.

649. Davitz, M. A., M. G. Low, and V. Nussenzweig. 1986. Release of decay-accelerating factor (DAF) from the cell membrane by phosphatidylinositol-specific phospholipase C (PIPLC). Selective modification of a complement regulatory protein. *Journal of Experimental Medicine* **163**:1150.

650. Fujita, T., T. Inoue, K. Ogawa, K. Iida, and N. Tamura. 1987. The mechanism of action of decay-accelerating factor (DAF). DAF inhibits the assembly of C3 convertases by dissociating C2a and Bb. *Journal of Experimental Medicine* **166**:1221.

651. Kinoshita, T., M. E. Medof, and V. Nussenzweig. 1986. Endogenous association of decay-accelerating factor (DAF) with C4b and C3b on cell membranes. *Journal of Immunology* **136**:3390.

652. Nicholson-Weller, A., J. P. March, C. E. Rosen, D. B. Spicer, and K. F. Austen. 1985. Surface membrane expression by human blood leukocytes and platelets of decay-accelerating factor, a regulatory protein of the complement system. *Blood* **65**:1237.

653. Kinoshita, T., M. E. Medof, R. Silber, and V. Nussenzweig. 1985. Distribution of decay-accelerating factor in the peripheral blood of normal individuals and patients with paroxysmal nocturnal hemoglobinuria. *Journal of Experimental Medicine* **162**:75.

654. Nicholson-Weller, A., D. A. Russian, and K. F. Austen. 1986. Natural killer cells are deficient in the surface expression of the complement regulatory protein, decay accelerating factor (DAF). *Journal of Immunology* **137**:1275.

655. Tomita, A., N. Okada, and H. Okada. 1991. Comparative studies of decay-accelerating factor and HLA-DR within the CD8-brightly positive population. *European Journal of Immunology* **21**:1843.

656. Finberg, R. W., W. White, and A. Nicholson-Weller. 1992. Decay-accelerating factor expression on either effector or target cells inhibits cytotoxicity by human natural killer cells. *Journal of Immunology* **149**:2055.

657. Medof, M. E., E. I. Walter, J. L. Rutgers, D. M. Knowles, and V. Nussenzweig. 1987. Identification of the complement decay-accelerating factor (DAF) on epithelium and glandular cells and in body fluids. *Journal of Experimental Medicine* **165**:848.

658. Cosio, F. G., D. D. Sedmak, J. D. Mahan, and N. S. Nahman, Jr. 1989. Localization of decay accelerating factor in normal and diseased kidneys. *Kidney International* **36**:100.

659. Ichida, S., Y. Yuzawa, H. Okada, K. Yoshioka, and S. Matsuo. 1994. Localization of the complement regulatory proteins in the normal human kidney. *Kidney International* **46**:89.

660. Asch, A. S., T. Kinoshita, E. A. Jaffe, and V. Nussenzweig. 1986. Decay-accelerating factor is present on cultured human umbilical vein endothelial cells. *Journal of Experimental Medicine* **163**:221.

661. Holmes, C. H., K. L. Simpson, S. D. Wainwright, C. G. Tate, J. M. Houlihan, I. H. Sawyer, I. P. Rogers, F. A. Spring, D. J. Anstee, and M. J. Tanner. 1990. Preferential expression of the complement regulatory protein decay accelerating factor at the fetomaternal interface during human pregnancy. *Journal of Immunology* **144**:3099.

662. Rooney, I. A., A. Davies, and B. P. Morgan. 1992. Membrane attack complex (MAC)-mediated damage to spermatozoa: protection of the cells by the presence on their membranes of MAC inhibitory proteins. *Immunology* **75**:499.

663. Cervoni, F., T. J. Oglesby, P. Fenichel, G. Dohr, B. Rossi, J. P. Atkinson, and B. L. Hsi. 1993. Expression of decay-accelerating factor (CD55) of the complement system on human spermatozoa. *Journal of Immunology* **151**:939.

664. Sayama, K., S. Shiraishi, Y. Shirakata, Y. Kobayashi, and Y. Miki. 1991. Characterization of decay-accelerating factor (DAF) in human skin. *Journal of Investigative Dermatology* **96**:61.

665. Sayama, K., S. Shiraishi, and Y. Miki. 1992. Distribution of complement regulators (CD46, CD55 and CD59) in skin appendages, and in benign and malignant skin neoplasms. *British Journal of Dermatology* **127**:1.

666. Lass, J. H., E. I. Walter, T. E. Burris, H. E. Grossniklaus, M. I. Roat, D. L. Skelnik, L. Needham, M. Singer, and M. E. Medof. 1990. Expression of two molecular forms of the complement decay-accelerating factor in the eye and lacrimal gland. *Investigative Ophthalmology and Visual Science* **31**:1136.

667. Hindmarsh, E. J., and R. M. Marks. 1998. Decay-accelerating factor is a component of subendothelial extracellular matrix in vitro, and is augmented by activation of endothelial protein kinase C. *European Journal of Immunology* **28**:1052.

668. Medof, M. E., E. I. Walter, J. L. Rutgers, D. M. Knowles, and V. Nussenzweig. 1987. Identification of the complement decay-accelerating factor (DAF) on epithelium and glandular cells and in body fluids. *Journal of Experimental Medicine* **165**:848.

669. Pangburn, M. K., R. D. Schreiber, and H. J. Muller-Eberhard. 1983. Deficiency of an erythrocyte membrane protein with complement regulatory activity in paroxysmal nocturnal hemoglobinuria. *Proceedings of the National Academy of Sciences of the United States of America* **80**:5430.

670. Nicholson-Weller, A., J. P. March, S. I. Rosenfeld, and K. F. Austen. 1983. Affected erythrocytes of patients with paroxysmal nocturnal hemoglobinuria are deficient in the complement regulatory protein, decay accelerating factor. *Proceedings of the National Academy of Sciences of the United States of America* **80**:5066.

671. Rosse, W. F. 1990. Paroxysmal nocturnal hemoglobinuria and decay-accelerating factor. *Annual Review of Medicine* **41**:431.

672. Rosse, W. F. 1992. Paroxysmal nocturnal hemoglobinuria. *Current Topics in Microbiology and Immunology* 178:163.

673. Nicholson-Weller, A., D. B. Spicer, and K. F. Austen. 1985. Deficiency of the complement regulatory protein, 'decay-accelerating factor,' on membranes of granulocytes, monocytes, and platelets in paroxysmal nocturnal hemoglobinuria. *New England Journal of Medicine* **312**:1091.

674. Okuda, K., A. Kanamaru, E. Ueda, T. Kitani, and K. Nagai. 1990. Membrane expression of decay-accelerating factor on neutrophils from normal individuals and patients with paroxysmal nocturnal hemoglobinuria. *Blood* **75**:1186.

675. Telen, M. J., S. E. Hall, A. M. Green, J. J. Moulds, and W. F. Rosse. 1988. Identification of human erythrocyte blood group antigens on decay-accelerating factor (DAF) and an erythrocyte phenotype negative for DAF. *Journal of Experimental Medicine* **167**:1993.

676. Petty, A. C., G. L. Daniels, D. J. Anstee, and P. Tippett. 1993. Use of the MAIEA technique to confirm the relationship between the Cromer antigens and decay-accelerating factor and to assign provisionally antigens to the short-consensus repeats. *Vox Sanguinis* **65**:309.

677. Telen, M. J., N. Rao, M. Udani, E. S. Thompson, R. M. Kaufman, and D. M. Lublin. 1994. Molecular mapping of the Cromer blood group Cra and Tca epitopes of decay accelerating factor: toward the use of recombinant antigens in immunohematology. *Blood* **84**:3205.

678. Caras, I. W., M. A. Davitz, L. Rhee, G. Weddell, D. W. Martin, Jr., and V. Nussenzweig. 1987. Cloning of decay-accelerating factor suggests novel use of splicing to generate two proteins. *Nature* **325**:545.

679. Medof, M. E., D. M. Lublin, V. M. Holers, D. J. Ayers, R. R. Getty, J. F. Leykam, J. P. Atkinson, and M. L. Tykocinski. 1987. Cloning and characterization of cDNAs encoding the complete sequence of decay-accelerating factor of human complement. *Proceedings of the National Academy of Sciences of the United States of America* **84**:2007.

680. Udenfriend, S., and K. Kodukula. 1995. How glycosylphosphatidylinositol-anchored membrane proteins are made. *Annual Review of Biochemistry* **64**:563.

681. Moran, P., and I. W. Caras. 1991. A nonfunctional sequence converted to a signal for glycophosphatidylinositol membrane anchor attachment. *Journal of Cell Biology* **115**:329.

682. Coyne, K. E., A. Crisci, and D. M. Lublin. 1993. Construction of synthetic signals for glycosyl-phosphatidylinositol anchor attachment. Analysis of amino acid sequence required for anchoring. *Journal of Biological Chemistry* 268.

683. Amthauer, R., K. Kodukula, L. Gerber, and S. Udenfriend. 1993. Evidence that the putative COOH-terminal signal transamidase involved in glycosylphosphatidylinositol protein synthesis is present in the endoplasmic reticulum. *Proceedings of the National Academy of Sciences of the USA* **90**:3973.

684. Walter, E. I., W. L. Roberts, T. L. Rosenberry, W. D. Ratnoff, and M. E. Medof. 1990. Structural basis for variations in the sensitivity of human decay accelerating factor to phosphatidylinositol-specific phospholipase C cleavage. *Journal of Immunology* **144**:1030.

685. Lublin, D. M., J. Krsek-Staples, M. K. Pangburn, and J. P. Atkinson. 1986. Biosynthesis and glycosylation of the human complement regulatory protein decay-accelerating factor. *Journal of Immunology* **137**:1629.

686. Seya, T., T. Farries, M. Nickells, and J. P. Atkinson. 1987. Additional forms of human decay-accelerating factor (DAF). *Journal of Immunology* **139**:1260.

687. Nakano, Y., Y. Sugita, Y. Ishikawa, N. H. Choi, T. Tobe, and M. Tomita. 1991. Isolation of two forms of decay-accelerating factor (DAF) from human urine. *Biochimica et Biophysica Acta* **1074**:326.

688. Kinoshita, T., S. I. Rosenfeld, and V. Nussenzweig. 1987. A high m.w. form of decay-accelerating factor (DAF-2) exhibits size abnormalities in paroxysmal nocturnal hemoglobinuria erythrocytes. *Journal of Immunology* **138**:2994.

689. Nickells, M. W., J. I. Alvarez, D. M. Lublin, and J. P. Atkinson. 1994. Characterization of DAF-2, a high molecular weight form of decay-accelerating factor (DAF; CD55), as a covalently cross-linked dimer of DAF-1. *Journal of Immunology* **152**:676.

690. Coyne, K. E., S. E. Hall, S. Thompson, M. A. Arce, T. Kinoshita, T. Fujita, D. J. Anstee, W. Rosse, and D. M. Lublin. 1992. Mapping of epitopes, glycosylation sites, and complement regulatory domains in human decay accelerating factor. *Journal of Immunology* **149**:2906.

691. Brodbeck, W. G., D. Liu, J. Sperry, C. Mold, and M. E. Medof. 1996. Localization of classical and alternative pathway regulatory activity within the decay-accelerating factor. *Journal of Immunology* **156**:2528.

692. Kuttner-Kondo, L., M. E. Medof, W. Brodbeck, and M. Shoham. 1996. Molecular modeling and mechanism of action of human decay-accelerating factor. *Protein Engineering* **9**:1143.

693. Hamann, J., B. Vogel, G. M. van Schijndel, and R. A. van Lier. 1996. The seven-span transmembrane receptor CD97 has a cellular ligand (CD55, DAF). *Journal of Experimental Medicine* **184**:1185.

694. Eichler, W., G. Aust, and D. Hamann. 1994. Characterization of an early activation-dependent antigen on lymphocytes defined by the monoclonal antibody BL-Ac(F2). *Scandanavian Journal of Immunology* **39**:111.

695. Hamann, J., W. Eichler, D. Hamann, H. M. J. Kerstens, P. J. Poddighe, J. M. N. Hoovers, E. Hartmann, M. Strauss, and R. A. W. van Lier. 1995. Expression cloning and chromosomal mapping of the leucocyte activation antigen CD97, a new seven-span transmembrane molecule of the secretin receptor superfamily with an unusual extracellular domain. *Journal of Immunology* **155**:1942.

696. Lublin, D. M., and K. E. Coyne. 1991. Phospholipid-anchored and transmembrane versions of either decay-accelerating factor or membrane cofactor protein show equal efficiency in protection from complement-mediated cell damage. *Journal of Experimental Medicine* **174**:35.

697. Parolini, I., M. Sargiacomo, M. P. Lisanti, and C. Peschle. 1996. Signal transduction and glycophosphatidylinositol-linked proteins (lyn, lck, CD4, CD45, G proteins, and CD55) selectively localize in Triton-insoluble plasma membrane domains of human leukemic cell lines and normal granulocytes. *Blood* **87**:3783.

698. Kuraya, M., and T. Fujita. 1998. Signal transduction via a protein associated with a glycosylphosphatidylinositol-anchored protein, decay-accelerating factor (DAF/CD55). *International Immunology* **10**:473.

699. Morgan, B. P., C. W. van den Berg, E. V. Davies, M. B. Hallett, and V. Horejsi. 1993. Cross-linking of CD59 and of other glycosyl phosphatidylinositol-anchored molecules on neutrophils triggers cell activation via tyrosine kinase. *European Journal of Immunology* **23**:2841.

700. van den Berg, C. W., T. Cinek, M. B. Hallett, V. Horejsi, and B. P. Morgan.

1995. Exogenous CD59 incorporated into U937 cells through its glycosyl phosphatidylinositol anchor becomes associated with signalling molecules in a time dependent manner. *Biochemical Society Transactions* 23:269S.

701.   Nowicki, B., A. Hart, K. E. Coyne, D. M. Lublin, and S. Nowicki. 1993. Short consensus repeat-3 domain of recombinant decay-accelerating factor is recognized by Escherichia coli recombinant Dr adhesin in a model of a cell–cell interaction. *Journal of Experimental Medicine* 178:2115.

702.   Kaul, A., B. J. Nowicki, M. G. Martens, P. Goluszko, A. Hart, M. Nagamani, D. Kumar, T. Q. Pham, and S. Nowicki. 1994. Decay-accelerating factor is expressed in the human endometrium and may serve as the attachment ligand for Dr pili of Escherichia coli. *American Journal of Reproductive Immunology* 32:194.

703.   Pham, T., A. Kaul, A. Hart, P. Goluszko, J. Moulds, S. Nowicki, D. M. Lublin, and B. J. Nowicki. 1995. dra-related X adhesins of gestational pyelonephritis-associated Escherichia coli recognize SCR-3 and SCR-4 domains of recombinant decay-accelerating factor. *Infection and Immunity* 63:1663.

704.   Cole, J. L., G. A. Housley, Jr., T. R. Dykman, R. P. MacDermott, and J. P. Atkinson. 1985. Identification of an additional class of C3-binding membrane proteins of human peripheral blood leukocytes and cell lines. *Proceedings of the National Academy of Sciences of the United States of America* 82:859.

705.   Liszewski, M. K., T. W. Post, and J. P. Atkinson. 1991. Membrane cofactor protein (MCP or CD46): newest member of the regulators of complement activation gene cluster. *Annual Review of Immunology* 9:431.

706.   Liszewski, M. K., and J. P. Atkinson. 1992. Membrane cofactor protein. *Current Topics in Microbiology and Immunology* 178:45.

707.   Seya, T., J. R. Turner, and J. P. Atkinson. 1986. Purification and characterization of a membrane protein (gp45-70) that is a cofactor for cleavage of C3b and C4b. *Journal of Experimental Medicine* 163:837.

708.   Yu, G. H., V. M. Holers, T. Seya, L. Ballard, and J. P. Atkinson. 1986. Identification of a third component of complement-binding glycoprotein of human platelets. *Journal of Clinical Investigation* 78:494.

709.   Johnson, P. M., H. M. Cheng, C. M. Molloy, C. M. M. Stern, and M. B. Slade. 1981. Human trophoblast-specific surface antigens identified using monoclonal antibodies. *American Journal of Reproductive Immunology* 1:246.

710.   Stern, P. L., N. Beresford, S. Thompson, P. M. Johnson, P. D. Webb, and N. Hole. 1986. Characterization of the human trophoblast-leukocyte antigenic molecules defined by a monoclonal antibody. *Journal of Immunology* 137:1604.

711.   Anderson, D. J., J. S. Michaelson, and P. M. Johnson. 1989. Trophoblast/leukocyte-common antigen is expressed by human testicular germ cells and appears on the surface of acrosome-reacted sperm. *Biology of Reproduction* 41:285.

712.   Sparrow, R. L., and I. F. McKenzie. 1983. Hu Ly-m5: a unique antigen physically associated with HLA molecules. *Human Immunology* 7:1.

713. Andrews, P. W., B. B. Knowles, M. Parkar, B. Pym, K. Stanley, and P. N. Goodfellow. 1985. A human cell-surface antigen defined by a monoclonal antibody and controlled by a gene on human chromosome 1. *Annals of Human Genetics* **49**:31.

714. Hsi, B. L., C. J. Yeh, P. Fenichel, M. Samson, and C. Grivaux. 1988. Monoclonal antibody GB24 recognizes a trophoblast-lymphocyte cross-reactive antigen. *American Journal of Reproductive Immunology and Microbiology* **18**:21.

715. Fenichel, P., B. L. Hsi, D. Farahifar, M. Donzeau, D. Barrier-Delpech, and C. J. Yehy. 1989. Evaluation of the human sperm acrosome reaction using a monoclonal antibody, GB24, and fluorescence-activated cell sorter. *Journal of Reproduction and Fertility* **87**:699.

716. Fenichel, P., G. Dohr, C. Grivaux, F. Cervoni, M. Donzeau, and B. L. Hsi. 1990. Localization and characterization of the acrosomal antigen recognized by GB24 on human spermatozoa. *Molecular Reproduction and Development* **27**:173.

717. Purcell, D. F., N. J. Deacon, S. M. Andrew, and I. F. McKenzie. 1990. Human non-lineage antigen, CD46 (HuLy-m5): purification and partial sequencing demonstrates structural homology with complement-regulating glycoproteins. *Immunogenetics* **31**:21.

718. Purcell, D. F., I. F. McKenzie, D. M. Lublin, P. M. Johnson, J. P. Atkinson, T. J. Oglesby, and N. J. Deacon. 1990. The human cell-surface glycoproteins HuLy-m5, membrane co-factor protein (MCP) of the complement system, and trophoblast leucocyte-common (TLX) antigen, are CD46. *Immunology* **70**:155.

719. Cho, S. W., T. J. Oglesby, B. L. Hsi, E. M. Adams, and J. P. Atkinson. 1991. Characterization of three monoclonal antibodies to membrane co-factor protein (MCP) of the complement system and quantification of MCP by radioassay. *Clinical and Experimental Immunology* **83**:257.

720. Seya, T., M. Okada, M. Matsumoto, K. S. Hong, T. Kinoshita, and J. P. Atkinson. 1991. Preferential inactivation of the C5 convertase of the alternative complement pathway by factor I and membrane cofactor protein (MCP). *Molecular Immunology* **28**:1137.

721. Oglesby, T. J., C. J. Allen, M. K. Liszewski, D. J. White, and J. P. Atkinson. 1992. Membrane cofactor protein (CD46) protects cells from complement-mediated attack by an intrinsic mechanism. *Journal of Experimental Medicine* **175**:1547.

722. Kojima, A., K. Iwata, T. Seya, M. Matsumoto, H. Ariga, J. P. Atkinson, and S. Nagasawa. 1993. Membrane cofactor protein (CD46) protects cells predominantly from alternative complement pathway-mediated C3-fragment deposition and cytolysis. *Journal of Immunology* **151**:1519.

723. Seya, T., L. L. Ballard, N. S. Bora, V. Kumar, W. Cui, and J. P. Atkinson. 1988. Distribution of membrane cofactor protein of complement on human peripheral blood cells. An altered form is found on granulocytes. *European Journal of Immunology* **18**:1289.

724. Seya, T., T. Hara, M. Matsumoto, and H. Akedo. 1990. Quantitative analysis of membrane cofactor protein (MCP) of complement. High

expression of MCP on human leukemia cell lines, which is down-regulated during cell differentiation. *Journal of Immunology* **145**:238.

725. Johnstone, R. W., B. E. Loveland, and I. F. McKenzie. 1993. Identification and quantification of complement regulator CD46 on normal human tissues. *Immunology* **79**:341.

726. McNearney, T., L. Ballard, T. Seya, and J. P. Atkinson. 1989. Membrane cofactor protein of complement is present on human fibroblast, epithelial, and endothelial cells. *Journal of Clinical Investigation* **84**:538.

727. Nakanishi, I., A. Moutabarrik, T. Hara, M. Hatanaka, T. Hayashi, T. Syouji, N. Okada, E. Kitamura, Y. Tsubakihara, M. Matsumoto, *et al.* 1994. Identification and characterization of membrane cofactor protein (CD46) in the human kidneys. *European Journal of Immunology* **24**:1529.

728. Sayama, K., S. Shiraishi, Y. Shirakata, Y. Kobayashi, T. Seya, and Y. Miki. 1991. Expression and characterization of membrane co-factor protein (MCP) in human skin. *Journal of Investigative Dermatology* **97**:722.

729. Varsano, S., I. Frolkis, and D. Ophir. 1995. Expression and distribution of cell-membrane complement regulatory glycoproteins along the human respiratory tract. *American Journal of Respiratory and Critical Care Medicine* **152**:1087.

730. Bora, N. S., C. L. Gobleman, J. P. Atkinson, J. S. Pepose, and H. J. Kaplan. 1993. Differential expression of the complement regulatory proteins in the human eye. *Investigative Ophthalmology and Visual Science* **34**:3579.

731. Gordon, D. L., T. A. Sadlon, S. L. Wesselingh, S. M. Russell, R. W. Johnstone, and D. F. Purcell. 1992. Human astrocytes express membrane cofactor protein (CD46), a regulator of complement activation. *Journal of Neuroimmunology* **36**:199.

732. Buchholz, C. J., D. Gerlier, A. Hu, T. Cathomen, M. K. Liszewski, J. P. Atkinson, and R. Cattaneo. 1996. Selective expression of a subset of measles virus receptor-competent CD46 isoforms in human brain. *Virology* **217**:349.

733. Hsi, B. L., J. S. Hunt, and J. P. Atkinson. 1991. Differential expression of complement regulatory proteins on subpopulations of human trophoblast cells. *Journal of Reproductive Immunology* **19**:209.

734. Hunt, J. S., and B. L. Hsi. 1990. Evasive strategies of trophoblast cells: selective expression of membrane antigens. *American Journal of Reproductive Immunology* **23**:57.

735. Ballard, L., T. Seya, J. Teckman, D. M. Lublin, and J. P. Atkinson. 1987. A polymorphism of the complement regulatory protein MCP (membrane cofactor protein or gp45-70). *Journal of Immunology* **138**:3850.

736. Simpson, K. L., and C. H. Holmes. 1994. Presence of the complement-regulatory protein membrane cofactor protein (MCP, CD46) as a membrane-associated product in seminal plasma. *Journal of Reproduction and Fertility* **102**:419.

737. Kitamura, M., M. Namiki, K. Matsumiya, K. Tanaka, M. Matsumoto, T. Hara, H. Kiyohara, M. Okabe, A. Okuyama, and T. Seya. 1995. Membrane cofactor protein (CD46) in seminal plasma is a prostasome-bound form

with complement regulatory activity and measles virus neutralizing activity. *Immunology* **84**:626.

738. Hara, T., S. Kuriyama, H. Kiyohara, Y. Nagase, M. Matsumoto, and T. Seya. 1992. Soluble forms of membrane cofactor protein (CD46, MCP) are present in plasma, tears, and seminal fluid in normal subjects. *Clinical and Experimental Immunology* **89**:490.

739. Seya, T., T. Hara, K. Iwata, S. Kuriyama, T. Hasegawa, Y. Nagase, S. Miyagawa, M. Matsumoto, M. Hatanaka, J. P. Atkinson, et al. 1995. Purification and functional properties of soluble forms of membrane cofactor protein (CD46) of complement: identification of forms increased in cancer patients' sera. *International Immunology* **7**:727.

740. McLaughlin, P. J., S. J. Holland, C. T. Taylor, K. S. Olah, D. I. Lewis-Jones, T. Hara, T. Seya, and P. M. Johnson. 1996. Soluble CD46 (membrane cofactor protein, MCP) in human reproductive tract fluids. *Journal of Reproductive Immunology* **31**:209.

741. Purcell, D. F., S. M. Russell, N. J. Deacon, M. A. Brown, D. J. Hooker, and I. F. McKenzie. 1991. Alternatively spliced RNAs encode several isoforms of CD46 (MCP), a regulator of complement activation. *Immunogenetics* **33**:335.

742. Russell, S. M., R. L. Sparrow, I. F. McKenzie, and D. F. Purcell. 1992. Tissue-specific and allelic expression of the complement regulator CD46 is controlled by alternative splicing. *European Journal of Immunology* **22**:1513.

743. Ballard, L. L., N. S. Bora, G. H. Yu, and J. P. Atkinson. 1988. Biochemical characterization of membrane cofactor protein of the complement system. *Journal of Immunology* **141**:3923.

744. Hara, T., Y. Suzuki, T. Semba, M. Hatanaka, M. Matsumoto, and T. Seya. 1995. High expression of membrane cofactor protein of complement (CD46) in human leukaemia cell lines: implication of an alternatively spliced form containing the STA domain in CD46 up-regulation. *Scandinavian Journal of Immunology* **42**:581.

745. Xing, P. X., S. Russell, J. Prenzoska, and J. F. McKenzie. 1994. Discrimination between alternatively spliced STP-A and -B isoforms of CD46. *Immunology* **83**:122.

746. Matsumoto, M., T. Seya, and S. Nagasawa. 1992. Polymorphism and proteolytic fragments of granulocyte membrane cofactor protein (MCP, CD46) of complement. *Biochemical Journal* **281**:493.

747. Cervoni, F., T. J. Oglesby, E. M. Adams, C. Milesifluet, M. Nickells, P. Fenichel, J. P. Atkinson, and B. L. Hsi. 1992. Identification and characterization of membrane cofactor protein of human spermatozoa. *Journal of Immunology* **148**:1431.

748. Cervoni, F., P. Fenichel, C. Akhoundi, B. L. Hsi, and B. Rossi. 1993. Characterization of a cDNA clone coding for human testis membrane cofactor protein (MCP, CD46). *Molecular Reproduction and Development* **34**:107.

749. Hara, T., Y. Suzuki, T. Nakazawa, H. Nishimura, S. Nagasawa, M. Nishiguchi, M. Matsumoto, M. Hatanaka, M. Kitamura, and T. Seya. 1998. Post-translational modification and intracellular localization of a splice product of CD46 cloned from human testis: role of the intracellular domains in O-glycosylation. *Immunology* **93**:546.

750. Wong, T. C., S. Yant, B. J. Harder, J. KorteSarfaty, and A. Hirano. 1997. The cytoplasmic domains of complement regulatory protein CD46 interact with multiple kinases in macrophages. *Journal of Leukocyte Biology* **62**:892.

751. Liszewski, M. K., I. Tedja, and J. P. Atkinson. 1994. Membrane cofactor protein (CD46) of complement. Processing differences related to alternatively spliced cytoplasmic domains. *Journal of Biological Chemistry* **269**:10776.

752. Gorelick, A., T. Oglesby, W. Rashbaum, J. Atkinson, and J. P. Buyon. 1995. Ontogeny of membrane cofactor protein: phenotypic divergence in the fetal heart. *Lupus* **4**:293.

753. Johnstone, R. W., S. M. Russell, B. E. Loveland, and I. F. McKenzie. 1993. Polymorphic expression of CD46 protein isoforms due to tissue-specific RNA splicing. *Molecular Immunology* **30**:1231.

754. Adams, E. M., M. C. Brown, M. Nunge, M. Krych, and J. P. Atkinson. 1991. Contribution of the repeating domains of membrane cofactor protein (CD46) of the complement system to ligand binding and cofactor activity. *Journal of Immunology* **147**:3005.

755. Hsu, E. C., R. E. Dorig, F. Sarangi, A. Marcil, C. Iorio, and C. D. Richardson. 1997. Artificial mutations and natural variations in the CD46 molecules from human and monkey cells define regions important for measles virus binding. *Journal of Virology* **71**:6144.

756. Murakami, Y., T. Seya, M. Kurita, A. Fukui, S. Ueda, and S. Nagasawa. 1998. Molecular cloning of membrane cofactor protein (MCP; CD46) on B95a cell, an Epstein–Barr virus-transformed marmoset B cell line: B95a-MCP is susceptible to infection by the CAM, but not the Nagahata strain of the measles virus. *Biochemical Journal* **330**:1351.

757. Iwata, K., T. Seya, S. Ueda, H. Ariga, and S. Nagasawa. 1994. Modulation of complement regulatory function and measles virus receptor function by the serine-threonine-rich domains of membrane cofactor protein (CD46). *Biochemical Journal* **304**:169.

758. Liszewski, M. K., and J. P. Atkinson. 1996. Membrane cofactor protein (MCP; CD46). Isoforms differ in protection against the classical pathway of complement. *Journal of Immunology* **156**:4415.

759. Rooney, I. A., J. P. Atkinson, E. S. Krul, G. Schonfeld, K. Polakoski, J. E. Saffitz, and B. P. Morgan. 1993. Physiologic relevance of the membrane attack complex inhibitory protein CD59 in human seminal plasma: CD59 is present on extracellular organelles (prostasomes), binds cell membranes, and inhibits complement-mediated lysis. *Journal of Experimental Medicine* **177**:1409.

760. Simpson, K. L., and C. H. Holmes. 1994. Differential expression of complement regulatory proteins decay-accelerating factor (CD55), membrane cofactor protein (CD46) and CD59 during human spermatogenesis. *Immunology* **81**:452.

761. Holmes, C. H., K. L. Simpson, H. Okada, N. Okada, S. D. Wainwright, D. F. Purcell, and J. M. Houlihan. 1992. Complement regulatory proteins at the feto-maternal interface during human placental development: distribu-

tion of CD59 by comparison with membrane cofactor protein (CD46) and decay accelerating factor (CD55). *European Journal of Immunology* **22**:1579.

762. Kitamura, M., K. Matsumiya, M. Yamanaka, S. Takahara, T. Hara, M. Matsumoto, M. Namiki, A. Okuyama, and T. Seya. 1997. Possible association of infertility with sperm-specific abnormality of CD46. *Journal of Reproductive Immunology* **33**:83.

763. Manchester, M., A. Valsamakis, R. Kaufman, M. K. Liszewski, J. Alvarez, J. P. Atkinson, D. M. Lublin, and M. B. Oldstone. 1995. Measles virus and C3 binding sites are distinct on membrane cofactor protein (CD46). *Proceedings of the National Academy of Sciences of the United States of America* **92**:2303.

764. Iwata, K., T. Seya, Y. Yanagi, J. M. Pesando, P. M. Johnson, M. Okabe, S. Ueda, H. Ariga, and S. Nagasawa. 1995. Diversity of sites for measles virus binding and for inactivation of complement C3b and C4b on membrane cofactor protein CD46. *Journal of Biological Chemistry* **270**:15148.

765. Manchester, M., J. E. Gairin, J. B. Patterson, J. Alvarez, M. K. Liszewski, D. S. Eto, J. P. Atkinson, and M. B. Oldstone. 1997. Measles virus recognizes its receptor, CD46, via two distinct binding domains within SCR1-2. *Virology* **233**:174.

766. Maisner, A., J. Schneider-Schaulies, M. K. Liszewski, J. P. Atkinson, and G. Herrler. 1994. Binding of measles virus to membrane cofactor protein (CD46): importance of disulfide bonds and N-glycans for the receptor function. *Journal of Virology* **68**:6299.

767. Maisner, A., and G. Herrler. 1995. Membrane cofactor protein with different types of N-glycans can serve as measles virus receptor. *Virology* **210**:479.

768. Maisner, A., J. Alvarez, M. K. Liszewski, D. J. Atkinson, J. P. Atkinson, and G. Herrler. 1996. The N-glycan of the SCR 2 region is essential for membrane cofactor protein (CD46) to function as a measles virus receptor. *Journal of Virology* **70**:4973.

769. Nelson, R. A., J. Jensen, I. Gigli, and N. Tamura. 1966. Methods for the separation, purification and measurement of nine components of hemolytic complement in guinea pig serum. *Immunochemistry* **3**:111.

770. Tamura, N., and R. A. Nelson. 1967. Three naturally occurring inhibitors of components of complement in guinea pig and rabbit serum. *Journal of Immunology* **99**:582.

771. Lachmann, P. J., and H. J. Müller-Eberhard. 1968. The demonstration in human serum of 'conglutinogen-activating factor' and its effect on the third component of complement. *Journal of Immunology* **100**:691.

772. Ruddy, S., and K. F. Austen. 1969. C3 inactivator of man. I. Hemolytic measurement by the inactivation of cell-bound C3. *Journal of Immunology* **102**:533.

773. Ruddy, S., L. G. Hunsicker, and K. F. Austen. 1972. C3b inactivator of man. 3. Further purification and production of antibody to C3b INA. *Journal of Immunology* **108**:657.

774. Hirani, S., J. D. Lambris, and H. J. Müller-Eberhard. 1985. Localization of

the conglutinin binding site on the third component of human complement. *Journal of Immunology* **134**:1105.

775. Alper, C. A., N. Abramson, R. B. Johnston, Jr., J. H. Jandl, and F. S. Rosen. 1970. Increased susceptibility to infection associated with abnormalities of complement-mediated functions and of the third component of complement (C3). *New England Journal of Medicine* **282**:350.

776. Alper, C. A., N. Abramson, R. B. Johnston, Jr., J. H. Jandl, and F. S. Rosen. 1970. Studies in vivo and in vitro on an abnormality in the metabolism of C3 in a patient with increased susceptibility to infection. *Journal of Clinical Investigation* **49**:1975.

777. Gitlin, J. D., F. S. Rosen, and P. J. Lachmann. 1975. The mechanism of action of the C3b inactivator (conglutinogen-activating factor) on its naturally occurring substrate, the major fragment of the third component of complement (C3b). *Journal of Experimental Medicine* **141**:1221.

778. Ruddy, S., and K. F. Austen. 1971. C3b inactivator of man. II. Fragments produced by C3b inactivator cleavage of cell-bound or fluid phase C3b. *Journal of Immunology* **107**:742.

779. Nagasawa, S., C. Ichihara, and R. M. Stroud. 1980. Cleavage of C4b by C3b inactivator: production of a nicked form of C4b, C4b', as an intermediate cleavage product of C4b by C3b inactivator. *Journal of Immunology* **125**:578.

780. Medicus, R. G., O. Götze, and H. J. Müller-Eberhard. 1976. Alternative pathway of complement: recruitment of precursor properdin by the labile C3/C5 convertase and the potentiation of the pathway. *Journal of Experimental Medicine* **144**:1076.

781. Farries, T. C., P. J. Lachmann, and R. A. Harrison. 1988. Analysis of the interactions between properdin, the third component of complement (C3), and its physiological activation products. *Biochemical Journal* **252**:47.

782. Soames, C. J., and R. B. Sim. 1997. Interactions between human complement components factor H, factor I and C3b. *Biochemical Journal* **326**:553.

783. Farries, T. C., T. Seya, R. A. Harrison, and J. P. Atkinson. 1990. Competition for binding sites on C3b by CR1, CR2, MCP, factor B and Factor H. *Complement and Inflammation* **7**:30.

784. Davis, A. E., 3rd. 1981. The C3b inactivator of the human complement system: homology with serine proteases. *FEBS Letters* **134**:147.

785. Yuan, J. M., L. M. Hsiung, and J. Gagnon. 1986. CNBr cleavage of the light chain of human complement factor I and alignment of the fragments. *Biochemical Journal* **233**:339.

786. Catterall, C. F., A. Lyons, R. B. Sim, A. J. Day, and T. J. Harris. 1987. Characterization of primary amino acid sequence of human complement control protein factor I from an analysis of cDNA clones. *Biochemical Journal* **242**:849.

787. Goldberger, G., M. A. Arnaout, D. Aden, R. Kay, M. Rits, and H. R. Colten. 1984. Biosynthesis and postsynthetic processing of human C3b/C4b inactivator (factor I) in three hepatoma cell lines. *Journal of Biological Chemistry* **259**:6492.

788. Jones, N. H., M. L. Clabby, D. P. Dialynas, H. J. Huang, L. A. Herzenberg, and J. L. Strominger. 1986. Isolation of complementary DNA clones encoding the human lymphocyte glycoprotein T1/Leu-1. *Nature* **323**:346.

789. Aruffo, A., M. B. Melnick, P. S. Linsley, and B. Seed. 1991. The lymphocyte glycoprotein CD6 contains a repeated domain structure characteristic of a new family of cell surface and secreted proteins. *Journal of Experimental Medicine* **174**:949.

790. Perkins, S. J., K. F. Smith, and R. B. Sim. 1993. Molecular modelling of the domain structure of factor I of human complement by X-ray and neutron solution scattering. *Biochemical Journal* **295**:101.

791. Shiang, R., J. C. Murray, C. C. Morton, K. H. Buetow, J. J. Wasmuth, A. H. Olney, W. G. Sanger, and G. Goldberger. 1989. Mapping of the human complement factor I gene to 4q25. *Genomics* **4**:82.

792. Vyse, T. J., G. P. Bates, M. J. Walport, and B. J. Morley. 1994. The organization of the human complement factor I gene (IF): a member of the serine protease gene family. *Genomics* **24**:90.

793. Pillemer, L., L. Blum, I. H. Lepow, O. A. Ross, E. W. Todd, and A. C. Wardlaw. 1954. The properdin system and immunity: demonstration and isolation of a new serum protein, properdin, and its role in immune phenomena. *Science* **120**:279.

794. Pensky, J., L. Wurz, L. Pillemer, and I. H. Lepow. 1959. The properdin system and immunity. XII. Assay, properties and partial purification of hydrazine-sensitive serum factor (factor A) in the properdin system. *Zeitschrift für Immunitatsforsch* **118**:329.

795. Blum, L., L. Pillemer, and I. H. Lepow. 1959. The properdin system and immunity. XIII. Assay and properties of a heat labile serum factor (factor B) in the properdin system. *Zeitschrift für Immunitatsforsch* **118**:349.

796. Pensky, J., C. F. Hinz, E. W. Todd, R. J. Wedgwood, J. T. Boyer, and I. H. Lepow. 1968. Properties of highly purified human properdin. *Journal of Immunology* **100**:142.

797. Gewurz, H., H. S. Shin, and S. E. Mergenhagen. 1968. Interactions of the complement system with endotoxic lipopolysaccharide: consumption of each of the six terminal complement components. *Journal of Experimental Medicine* **128**:1049.

798. Sandberg, A. L., A. G. Osler, H. S. Shin, and B. Oliveira. 1970. The biologic activities of guinea pig antibodies. II. Modes of complement interaction with gamma 1 and gamma 2-immunoglobulins. *Journal of Immunology* **104**:329.

799. Reid, K. B. 1971. Complement fixation by the F(ab')2-fragment of pepsin-treated rabbit antibody. *Immunology* **20**:649.

800. Marcus, R. L., H. S. Shin, and M. M. Mayer. 1971. An alternate complement pathway: C-3 cleaving activity, not due to C4,2a, on endotoxic lipopolysaccharide after treatment with guinea pig serum; relation to properdin. *Proceedings of the National Academy of Sciences of the United States of America* **68**:1351.

801. Müller-Eberhard, H. J., and O. Götze. 1972. C3 proactivator convertase and its mode of action. *Journal of Experimental Medicine* **135**:1003.

802. Götze, O., and H. J. Müller-Eberhard. 1971. The C3-activator system: an alternate pathway of complement activation. *Journal of Experimental Medicine* **134** Suppl:90s.

803. Alper, C. A., I. Goodkofsky, and I. H. Lepow. 1973. The relationship of glycine-rich glycoprotein to factor B in the properdin system and to the cobra factor-binding protein of human serum. *Journal of Experimental Medicine* **137**:424.

804. Goodkofsky, I., and I. H. Lepow. 1971. Functional relationship of factor B in the properdin system to C3 proactivator of human serum. *Journal of Immunology* **107**:1200.

805. Fearon, D. T., K. F. Austen, and S. Ruddy. 1973. Formation of a hemolytically active cellular intermediate by the interaction between properdin factors B and D and the activated third component of complement. *Journal of Experimental Medicine* **138**:1305.

806. Götze, O., and H. J. Müller-Eberhard. 1974. The role of properdin in the alternate pathway of complement activation. *Journal of Experimental Medicine* **139**:44.

807. Fearon, D. T., K. F. Austen, and S. Ruddy. 1974. Properdin factor D. II. Activation to D by properdin. *Journal of Experimental Medicine* **140**:426.

808. Fearon, D. T., and K. F. Austen. 1975. Properdin: binding to C3b and stabilization of the C3b-dependent C3 convertase. *Journal of Experimental Medicine* **142**:856.

809. Fearon, D. T., and K. F. Austen. 1975. Properdin: initiation of alternative complement pathway. *Proceedings of the National Academy of Sciences of the United States of America* **72**:3220.

810. Schreiber, R. D., R. G. Medicus, O. Götze, and H. J. Müller-Eberhard. 1975. Properdin- and nephritic factor-dependent C3 convertases: requirement of native C3 for enzyme formation and the function of bound C3b as properdin receptor. *Journal of Experimental Medicine* **142**:760.

811. Schreiber, R. D., O. Götze, and H. J. Müller-Eberhard. 1976. Alternative pathway of complement: demonstration and characterization of initiating factor and its properdin-independent function. *Journal of Experimental Medicine* **144**:1062.

812. Pangburn, M. K. 1989. Analysis of the natural polymeric forms of human properdin and their functions in complement activation. *Journal of Immunology* **142**:202.

813. Smith, K. F., K. F. Nolan, K. B. M. Reid, and S. J. Perkins. 1991. Neutron and X-ray scattering studies on the human complement protein properdin provide an analysis of the thrombospondin repeat. *Biochemistry* **30**:8000.

814. Schwaeble, W., H. P. Huemer, J. Most, M. P. Dierich, M. Strobel, C. Claus, K. B. Reid, and H. W. Ziegler-Heitbrock. 1994. Expression of properdin in human monocytes. *European Journal of Biochemistry* **219**:759.

815. Schwaeble, W., W. G. Dippold, M. K. Schafer, H. Pohla, D. Jonas, B. Luttig, E. Weihe, H. P. Huemer, M. P. Dierich, and K. B. Reid. 1993. Properdin, a positive regulator of complement activation, is expressed in human T cell lines and peripheral blood T cells. *Journal of Immunology* **151**:2521.

816. Maves, K. K., and J. M. Weiler. 1994. Human liver-derived HEP G2 cells

produce functional properdin. *Journal of Laboratory and Clinical Medicine* **124**:837.

817. Wirthmueller, U., B. Dewald, M. Thelen, M. K. Schafer, C. Stover, K. Whaley, J. North, P. Eggleton, K. B. Reid, and W. J. Schwaeble. 1997. Properdin, a positive regulator of complement activation, is released from secondary granules of stimulated peripheral blood neutrophils. *Journal of Immunology* **158**:4444.

818. Goundis, D., S. M. Holt, Y. Boyd, and K. B. Reid. 1989. Localization of the properdin structural locus to Xp11.23-Xp21.1. *Genomics* **5**:56.

819. Sjoholm, A. G., C. Soderstrom, and L. A. Nilsson. 1988. A second variant of properdin deficiency: the detection of properdin at low concentrations in affected males. *Complement* **5**:130.

820. Sjoholm, A. G. 1990. Inherited complement deficiency states: implications for immunity and immunological disease. *Apmis* **98**:861.

821. Farries, T. C., and J. P. Atkinson. 1989. Biosynthesis of properdin. *Journal of Immunology* **142**:842.

822. Medicus, R. G., A. F. Esser, H. N. Fernandez, and H. J. Müller-Eberhard. 1980. Native and activated properdin: interconvertibility and identity of amino- and carboxy-terminal sequences. *Journal of Immunology* **124**:602.

823. Minta, J. O., and I. H. Lepow. 1974. Studies on the sub-unit structure of human properdin. *Immunochemistry* **11**:361.

824. Minta, J. O. 1976. Purification of native properdin by reversed affinity chromatography and its activation by proteolytic enzymes. *Journal of Immunology* **117**:405.

825. DiScipio, R. G. 1982. Properdin is a trimer. *Molecular Immunology* **19**:631.

826. Chapatis, J., and I. H. Lepow. 1976. Multiple sedimenting species of properdin in human serum and interaction of purified properdin with the third component of complement. *Journal of Experimental Medicine* **143**:241.

827. Smith, C. A., M. K. Pangburn, C. W. Vogel, and H. J. Müller-Eberhard. 1984. Molecular architecture of human properdin, a positive regulator of the alternative pathway of complement. *Journal of Biological Chemistry* **259**:4582.

828. Smith, K. F., K. F. Nolan, K. B. Reid, and S. J. Perkins. 1991. Neutron and X-ray scattering studies on the human complement protein properdin provide an analysis of the thrombospondin repeat. *Biochemistry* **30**:8000.

829. Minta, J. O. 1975. Changes in the immunochemical properties of highly purified properdin in human serum. *Journal of Immunology* **114**:1415.

830. Götze, O., R. G. Medicus, and H. J. Müller-Eberhard. 1977. Alternative pathway of complement: nonenzymatic, reversible transition of precursor to active properdin. *Journal of Immunology* **118**:525.

831. Farries, T. C., J. T. Finch, P. J. Lachmann, and R. A. Harrison. 1987. Resolution and analysis of 'native' and 'activated' properdin. *Biochemical Journal* **243**:507.

832. Whiteman, L. Y., D. B. Purkall, and S. Ruddy. 1991. Association of activated properdin with complexes of properdin with C3. *Journal of Immunology* **147**:1344.

833. Whiteman, L. Y., D. B. Purkall, and S. Ruddy. 1995. Covalent linkage of

C3 to properdin during complement activation. *European Journal of Immunology* **25**:1481.

834. Reid, K. B., and J. Gagnon. 1981. Amino acid sequence studies of human properdin–N-terminal sequence analysis and alignment of the fragments produced by limited proteolysis with trypsin and the peptides produced by cyanogen bromide treatment. *Molecular Immunology* **18**:949.

835. Nolan, K. F., W. Schwaeble, S. Kaluz, M. P. Dierich, and K. B. Reid. 1991. Molecular cloning of the cDNA coding for properdin, a positive regulator of the alternative pathway of human complement. *European Journal of Immunology* **21**:771.

836. Perkins, S. J., A. S. Nealis, P. I. Haris, D. Chapman, D. Goundis, and K. B. Reid. 1989. Secondary structure in properdin of the complement cascade and related proteins: a study by Fourier transform infrared spectroscopy. *Biochemistry* **28**:7176.

837. Lawler, J., and R. O. Hynes. 1986. The structure of human thrombospondin, an adhesive glycoprotein with multiple calcium-binding sites and homologies with several different proteins. *Journal of Cell Biology* **103**:1635.

838. Goundis, D., and K. B. Reid. 1988. Properdin, the terminal complement components, thrombospondin and the circumsporozoite protein of malaria parasites contain similar sequence motifs. *Nature* **335**:82.

839. Higgins, J. M., H. Wiedemann, R. Timpl, and K. B. Reid. 1995. Characterization of mutant forms of recombinant human properdin lacking single thrombospondin type I repeats. Identification of modules important for function. *Journal of Immunology* **155**:5777.

840. Nolan, K. F., S. Kaluz, J. M. Higgins, D. Goundis, and K. B. Reid. 1992. Characterization of the human properdin gene. *Biochemical Journal* **287**:291.

841. DiScipio, R. G. 1981. The binding of human complement proteins C5, factor B, beta 1H and properdin to complement fragment C3b on zymosan. *Biochemical Journal* **199**:485.

842. Farries, T. C., P. J. Lachmann, and R. A. Harrison. 1988. Analysis of the interaction between properdin and factor B, components of the alternative-pathway C3 convertase of complement. *Biochemical Journal* **253**:667.

843. Rich, K. A., F. W. T. George, J. L. Law, and W. J. Martin. 1990. Cell-adhesive motif in region II of malarial circumsporozoite protein. *Science* **249**:1574.

844. Prater, C. A., J. Plotkin, D. Jaye, and W. A. Frazier. 1991. The properdin-like type I repeats of human thrombospondin contain a cell attachment site. *Journal of Cell Biology* **112**:1031.

845. Hugli, T. E. 1984. Structure and function of the anaphylatoxins. *Springer Seminars in Immunopathology* **7**:193.

846. Hugli, T. E. 1981. The structural basis for anaphylatoxin and chemotactic functions of C3a, C4a, and C5a. *Critical Reviews in Immunology* **1**:321.

847. Hugli, T. E. 1975. Human anaphylatoxin (C3a) from the third component of complement. Primary structure. *Journal of Biological Chemistry* **250**:8293.

848. Moon, K. E., J. P. Gorski, and T. E. Hugli. 1981. Complete primary structure of human C4a anaphylatoxin. *Journal of Biological Chemistry* **256**:8685.

849. Fernandez, H. N., and T. E. Hugli. 1978. Primary structural analysis of the polypeptide portion of human C5a anaphylatoxin. Polypeptide sequence determination and assignment of the oligosaccharide attachment site in C5a. *Journal of Biological Chemistry* **253**:6955.

850. Gerard, C., and T. E. Hugli. 1981. Identification of classical anaphylatoxin as the des-Arg form of the C5a molecule: evidence of a modulator role for the oligosaccharide unit in human des-Arg74-C5a. *Proceedings of the National Academy of Sciences of the United States of America* **78**:1833.

851. Swerlick, R. A., K. B. Yancey, and T. J. Lawley. 1988. A direct in vivo comparison of the inflammatory properties of human C5a and C5a des Arg in human skin. *Journal of Immunology* **140**:2376.

852. Burgi, B., T. Brunner, and C. A. Dahinden. 1994. The degradation product of the C5a anaphylatoxin C5a(desarg) retains basophil-activating properties. *European Journal of Immunology* **24**:1583.

853. Huey, R., C. M. Bloor, M. S. Kawahara, and T. E. Hugli. 1983. Potentiation of the anaphylatoxins in vivo using an inhibitor of serum carboxypeptidase N (SCPN). I. Lethality and pathologic effects on pulmonary tissue. *American Journal of Pathology* **112**:48.

854. Erdös, E. G., and E. M. Sloane. 1962. An enzyme in human blood plasma that inactivates bradykinin and kallidin. *Biochemical Pharmacology* **11**:585.

855. Bokisch, V. A., H. J. Müller-Eberhard, and C. G. Cochrane. 1969. Isolation of a fragment (C3a) of the third component of human complement containing anaphylatoxin and chemotactic activity and description of an anaphylatoxin inactivator of human serum. *Journal of Experimental Medicine* **129**:1109.

856. Bokisch, V. A., and H. J. Müller-Eberhard. 1970. Anaphylatoxin inactivator of human plasma: its isolation and characterization as a carboxypeptidase. *Journal of Clinical Investigation* **49**:2427.

857. Vallota, E. H., and H. J. Müller-Eberhard. 1973. Formation of C3a and C5a anaphylatoxins in whole human serum after inhibition of the anaphylatoxin inactivator. *Journal of Experimental Medicine* **137**:1109.

858. Oshima, G., J. Kato, and E. G. Erdös. 1974. Subunits of human plasma carboxypeptidase N (kininase I; anaphylatoxin inactivator). *Biochimica et Biophysica Acta* **365**:344.

859. Oshima, G., J. Kato, and E. G. Erdös. 1975. Plasma carboxypeptidase N, subunits and characteristics. *Archives of Biochemistry and Biophysics* **170**:132.

860. Plummer, T. H., Jr., and M. Y. Hurwitz. 1978. Human plasma carboxypeptidase N. Isolation and characterization. *Journal of Biological Chemistry* **253**:3907.

861. Levin, Y., R. A. Skidgel, and E. G. Erdos. 1982. Isolation and characterization of the subunits of human plasma carboxypeptidase N (kininase i). *Proceedings of the National Academy of Sciences of the United States of America* **79**:4618.

862. Skidgel, R. A., C. D. Bennett, J. W. Schilling, F. L. Tan, D. K. Weerasinghe, and E. G. Erdos. 1988. Amino acid sequence of the N-terminus and

selected tryptic peptides of the active subunit of human plasma carboxy-peptidase N: comparison with other carboxypeptidases. *Biochemical and Biophysical Research Communications* **154**:1323.

863. Gebhard, W., M. Schube, and M. Eulitz. 1989. cDna cloning and complete primary structure of the small, active subunit of human carboxypeptidase N (kininase 1). *European Journal of Biochemistry* **178**:603.

864. Tan, F., D. K. Weerasinghe, R. A. Skidgel, H. Tamei, R. K. Kaul, I. B. Roninson, J. W. Schilling, and E. G. Erdos. 1990. The deduced protein sequence of the human carboxypeptidase N high molecular weight subunit reveals the presence of leucine-rich tandem repeats. *Journal of Biological Chemistry* **265**:13.

865. Takahashi, N., Y. Takahashi, and F. W. Putnam. 1985. Periodicity of leucine and tandem repetition of a 24-amino acid segment in the primary structure of leucine-rich alpha 2-glycoprotein of human serum. *Proceedings of the National Academy of Sciences of the United States of America* **82**:1906.

866. Neame, P. J., H. U. Choi, and L. C. Rosenberg. 1989. The primary structure of the core protein of the small, leucine-rich proteoglycan (PG I) from bovine articular cartilage. *Journal of Biological Chemistry* **264**:8653.

867. Fisher, L. W., J. D. Termine, and M. F. Young. 1989. Deduced protein sequence of bone small proteoglycan I (biglycan) shows homology with proteoglycan II (decorin) and several nonconnective tissue proteins in a variety of species. *Journal of Biological Chemistry* **264**:4571.

868. Lee, F. S., E. A. Fox, H. M. Zhou, D. J. Strydom, and B. L. Vallee. 1988. Primary structure of human placental ribonuclease inhibitor. *Biochemistry* **27**:8545.

869. Lopez-Trascasa, M., D. H. Bing, M. Rivard, and A. Nicholson-Weller. 1989. Factor J: isolation and characterization of a new polypeptide inhibitor of complement C1. *Journal of Biological Chemistry* **264**:16214.

870. Jimenez-Clavero, M. A., C. Gonzalez-Rubio, G. Fontan, and M. Lopez-Trascasa. 1994. Factor J, an inhibitor of the complement classical pathway: the quantitation by an ELISA inhibition assay in normal human serum. *Clinical Biochemistry* **27**:169.

871. Jimenez-Clavero, M. A., C. Gonzalez-Rubio, G. Fontan, and M. Lopez-Trascasa. 1994. Factor J, a human inhibitor of complement C1, is a cationic, highly glycosylated protein. *Immunology Letters* **42**:185.

872. Gonzalez-Rubio, C., M. A. Jimenez-Clavero, G. Fontan, and M. Lopez-Trascasa. 1994. The inhibitory effect of factor J on the alternative complement pathway. *Journal of Biological Chemistry* **269**:26017.

873. Gonzalez-Rubio, C., R. Gonzalez-Muniz, M. A. Jimenez-Clavero, G. Fontan, and M. Lopez-Trascasa. 1996. Factor J, an inhibitor of the classical and alternative complement pathway, does not inhibit esterolysis by factor D. *Biochimica et Biophysica Acta* **1295**:174.

874. Jimenez-Clavero, M. A., C. Gonzalez-Rubio, S. Larrucea, C. Gamallo, G. Fontan, and M. Lopez-Trascasa. 1995. Cell surface molecules related to factor J in human lymphoid cells and cell lines. *Journal of Immunology* **155**:2143.

875. Lintin, S. J., A. R. Lewin, and K. B. Reid. 1988. Derivation of the sequence of the signal peptide in human C4b-binding protein and interspecies cross-hybridisation of the C4bp cDNA sequence. *FEBS Letters* **232**:328.

876. Esser, A. F. 1989. The membrane attack pathway of complement. *Year in Immunology* **6**:229.

877. Matsuo, S., S. Ichida, H. Takizawa, N. Okada, L. Baranyi, A. Iguchi, B. P. Morgan, and H. Okada. 1994. In vivo effects of monoclonal antibodies that functionally inhibit complement regulatory proteins in rats. *Journal of Experimental Medicine* **180**:1619.

878. Nishikawa, K., S. Matsuo, N. Okada, B. P. Morgan, and H. Okada. 1996. Local inflammation caused by a monoclonal antibody that blocks the function of the rat membrane inhibitor of C3 convertase. *Journal of Immunology* **156**:1182.

879. Morgan, B. P. 1989. Complement membrane attack on nucleated cells: resistance, recovery and non-lethal effects. *Biochemical Journal* **264**:1.

880. Holmes, R. 1967. Preparation from human serum of an alpha-one protein which induces the immediate growth of unadapted cells in vitro. *Journal of Cell Biology* **32**:297.

881. Barnes, D. W., J. Silnutzer, C. See, and M. Shaffer. 1983. Characterization of human serum spreading factor with monoclonal antibody. *Proceedings of the National Academy of Sciences of the United States of America* **80**:1362.

882. Barnes, D. W., and J. Silnutzer. 1983. Isolation of human serum spreading factor. *Journal of Biological Chemistry* **258**:12548.

883. Hayman, E. G., M. D. Pierschbacher, Y. Ohgren, and E. Ruoslahti. 1983. Serum spreading factor (vitronectin) is present at the cell surface and in tissues. *Proceedings of the National Academy of Sciences of the United States of America* **80**:4003.

884. Suzuki, S., A. Oldberg, E. G. Hayman, M. D. Pierschbacher, and E. Ruoslahti. 1985. Complete amino acid sequence of human vitronectin deduced from cDNA. Similarity of cell attachment sites in vitronectin and fibronectin. *EMBO Journal* **4**:2519.

885. Suzuki, S., W. S. Argraves, R. Pytela, H. Arai, T. Krusius, M. D. Pierschbacher, and E. Ruoslahti. 1986. cDNA and amino acid sequences of the cell adhesion protein receptor recognizing vitronectin reveal a transmembrane domain and homologies with other adhesion protein receptors. *Proceedings of the National Academy of Sciences of the United States of America* **83**:8614.

886. Kolb, W. P., and H. J. Müller-Eberhard. 1975. The membrane attack complex of complement: Isolation and subunit composition of the C5b-9 complex. *Journal of Experimental Medicine* **141**:724.

887. Podack, E. R., and H. J. Müller-Eberhard. 1979. Isolation of human S-protein, an inhibitor of the membrane attack complex of complement. *Journal of Biological Chemistry* **254**:9808.

888. Podack, E. R., and H. J. Müller-Eberhard. 1978. Binding of desoxycholate, phosphatidylcholine vesicles, lipoprotein and of the S-protein to complexes of terminal complement components. *Journal of Immunology* **121**:1025.

889. Podack, E. R., B. Dahlback, and J. H. Griffin. 1986. Interaction of S-protein of complement with thrombin and antithrombin III during coagulation. Protection of thrombin by S-protein from antithrombin III inactivation. *Journal of Biological Chemistry* **261**:7387.

890. Dahlback, B., and E. R. Podack. 1985. Characterization of human S protein, an inhibitor of the membrane attack complex of complement. Demonstration of a free reactive thiol group. *Biochemistry* **24**:2368.

891. Tomasini, B. R., and D. F. Mosher. 1986. On the identity of vitronectin and S-protein: immunological crossreactivity and functional studies. *Blood* **68**:737.

892. Preissner, K. T., N. Heimburger, E. Anders, and G. Muller-Berghaus. 1986. Physicochemical, immunochemical and functional comparison of human S-protein and vitronectin. Evidence for the identity of both plasma proteins. *Biochemical and Biophysical Research Communications* **134**:951.

893. Preissner, K. P., E. R. Podack, and H. J. Müller-Eberhard. 1989. SC5b-7, SC5b-8 and SC5b-9 complexes of complement: ultrastructure and localization of the S-protein (vitronectin) within the macromolecules. *European Journal of Immunology* **19**:69.

894. Jenne, D., and K. K. Stanley. 1985. Molecular cloning of S-protein, a link between complement, coagulation and cell-substrate adhesion. *EMBO Journal* **4**:3153.

895. Shaffer, M. C., T. P. Foley, and D. W. Barnes. 1984. Quantitation of spreading factor in human biologic fluids. *Journal of Laboratory and Clinical Medicine* **103**:783.

896. Hogasen, K., T. E. Mollnes, J. Tschopp, and M. Harboe. 1993. Quantitation of vitronectin and clusterin. Pitfalls and solutions in enzyme immunoassays for adhesive proteins. *Journal of Immunological Methods* **160**:107.

897. Parker, C. J., O. L. Stone, V. F. White, and N. J. Bernshaw. 1989. Vitronectin (S protein) is associated with platelets. *British Journal of Haematology* **71**:245.

898. Pettersen, H. B., E. Johnson, T. E. Mollnes, P. Garred, G. Hetland, and S. S. Osen. 1990. Synthesis of complement by alveolar macrophages from patients with sarcoidosis. *Scandinavian Journal of Immunology* **31**:15.

899. Gladson, C. L., and D. A. Cheresh. 1991. Glioblastoma expression of vitronectin and the alpha v beta 3 integrin. Adhesion mechanism for transformed glial cells. *Journal of Clinical Investigation* **88**:1924.

900. Gasque, P., M. Fontaine, and B. P. Morgan. 1995. Complement expression in human brain. Biosynthesis of terminal pathway components and regulators in human glial cells and cell lines. *Journal of Immunology* **154**:4726.

901. Seiffert, D., G. M. Bordin, and D. J. Loskutoff. 1996. Evidence that extrahepatic cells express vitronectin mRNA at rates approaching those of hepatocytes. *Histochemistry and Cell Biology* **105**:195.

902. Seiffert, D., M. Geisterfer, J. Gauldie, E. Young, and T. J. Podor. 1995. IL-6 stimulates vitronectin gene expression in vivo. *Journal of Immunology* **155**:3180.

903. Peake, P. W., J. D. Greenstein, B. A. Pussell, and J. A. Charlesworth. 1996. The behaviour of human vitronectin in vivo: Effects of complement

activation, conformation and phosphorylation. *Clinical and Experimental Immunology* **106:**416.

904. Dahlback, K., H. Lofberg, and B. Dahlback. 1986. Localization of vitronectin (S-protein of complement) in normal human skin. *Acta Dermato-Venereologica* **66:**461.

905. Niculescu, F., H. G. Rus, and R. Vlaicu. 1987. Immunohistochemical localization of C5b-9, S-protein, C3d and apolipoprotein B in human arterial tissues with atherosclerosis. *Atherosclerosis* **65:**1.

906. Sato, R., Y. Komine, T. Imanaka, and T. Takano. 1990. Monoclonal antibody EMR1a/212D recognizing site of deposition of extracellular lipid in atherosclerosis. Isolation and characterization of a cDNA clone for the antigen. *Journal of Biological Chemistry* **265:**21232.

907. Falk, R. J., E. Podack, A. P. Dalmasso, and J. C. Jennette. 1987. Localization of S protein and its relationship to the membrane attack complex of complement in renal tissue. *American Journal of Pathology* **127:**182.

908. Murphy, B. F., D. J. Davies, W. Morrow, and A. J. d'Apice. 1989. Localization of terminal complement components S-protein and SP-40,40 in renal biopsies. *Pathology* **21:**275.

909. Dahlback, K., H. Lofberg, J. Alumets, and B. Dahlback. 1989. Immunohistochemical demonstration of age-related deposition of vitronectin (S-protein of complement) and terminal complement complex on dermal elastic fibers. *Journal of Investigative Dermatology* **92:**727.

910. Hayman, E. G., M. D. Pierschbacher, S. Suzuki, and E. Ruoslahti. 1985. Vitronectin – a major cell attachment-promoting protein in fetal bovine serum. *Experimental Cell Research* **160:**245.

911. Gebb, C., E. G. Hayman, E. Engvall, and E. Ruoslahti. 1986. Interaction of vitronectin with collagen. *Journal of Biological Chemistry* **261:**16698.

912. Bale, M. D., L. A. Wohlfahrt, D. F. Mosher, B. Tomasini, and R. C. Sutton. 1989. Identification of vitronectin as a major plasma protein adsorbed on polymer surfaces of different copolymer composition. *Blood* **74:**2698.

913. Pierschbacher, M. D., E. G. Hayman, and E. Ruoslahti. 1985. The cell attachment determinant in fibronectin. *Journal of Cellular Biochemistry* **28:**115.

914. Preissner, K. T., and G. Muller-Berghaus. 1987. Neutralization and binding of heparin by S protein/vitronectin in the inhibition of factor Xa by antithrombin III. Involvement of an inducible heparin-binding domain of S protein/vitronectin. *Journal of Biological Chemistry* **262:**12247.

915. Tomasini, B. R., and D. F. Mosher. 1988. Conformational states of vitronectin: preferential expression of an antigenic epitope when vitronectin is covalently and noncovalently complexed with thrombin-antithrombin III or treated with urea. *Blood* **72:**903.

916. Jenne, D., F. Hugo, and S. Bhakdi. 1985. Monoclonal antibodies to human plasma protein X alias complement S-protein. *Bioscience Reports* **5:**343.

917. Preissner, K. T., and D. Jenne. 1991. Vitronectin: a new molecular connection in haemostasis. *Thrombosis and Haemostasis* **66:**189.

918. Preissner, K. T. 1991. Structure and biological role of vitronectin. *Annual Review of Cell Biology* **7:**275.

919. Hildebrand, A., K. T. Preissner, G. Muller-Berghaus, and H. Teschemacher. 1989. A novel beta-endorphin binding protein. Complement S protein (= vitronectin) exhibits specific non-opioid binding sites for beta-endorphin upon interaction with heparin or surfaces. *Journal of Biological Chemistry* **264**:15429.

920. Hildebrand, A. 1989. Identification of the beta-endorphin-binding subunit of the SC5b-9 complement complex: S protein exhibits specific beta-endorphin-binding sites upon complex formation with complement proteins. *Biochemical and Biophysical Research Communications* **159**:799.

921. Fuquay, J. I., D. T. Loo, and D. W. Barnes. 1986. Binding of Staphylococcus aureus by human serum spreading factor in an in vitro assay. *Infection and Immunity* **52**:714.

922. Chhatwal, G. S., K. T. Preissner, G. Muller-Berghaus, and H. Blobel. 1987. Specific binding of the human S protein (vitronectin) to streptococci, Staphylococcus aureus, and Escherichia coli. *Infection and Immunity* **55**:1878.

923. Sheehan, M., C. A. Morris, B. A. Pussell, and J. A. Charlesworth. 1995. Complement inhibition by human vitronectin involves non-heparin binding domains. *Clinical and Experimental Immunology* **101**:136.

924. Ware, C. F., R. A. Wetsel, and W. P. Kolb. 1981. Physicochemical characterization of fluid phase (SC5b-9) and membrane derived (MC5b-9) attack complexes of human complement purified by immunoadsorbent affinity chromatography or selective detergent extraction. *Molecular Immunology* **18**:521.

925. Ware, C. F., and W. P. Kolb. 1981. Assembly of the functional membrane attack complex of human complement: formation of disulfide-linked C9 dimers. *Proceedings of the National Academy of Sciences of the United States of America* **78**:6426.

926. Falk, R. J., A. P. Dalmasso, Y. Kim, C. H. Tsai, J. I. Scheinman, H. Gewurz, and A. F. Michael. 1983. Neoantigen of the polymerized ninth component of complement. Characterization of a monoclonal antibody and immunohistochemical localization in renal disease. *Journal of Clinical Investigation* **72**:560.

927. Mollnes, T. E., T. Lea, and M. Harboe. 1984. Detection and quantification of the terminal C5b-9 complex of human complement by a sensitive enzyme-linked immunosorbent assay. *Scandinavian Journal of Immunology* **20**:157.

928. Johnson, E., V. Berge, and K. Hogasen. 1994. Formation of the terminal complement complex on agarose beads: further evidence that vitronectin (complement S-protein) inhibits C9 polymerization. *Scandinavian Journal of Immunology* **39**:281.

929. Milis, L., C. A. Morris, M. C. Sheehan, J. A. Charlesworth, and B. A. Pussell. 1993. Vitronectin-mediated inhibition of complement: evidence for different binding sites for C5b-7 and C9. *Clinical and Experimental Immunology* **92**:114.

930. Su, H. R. 1996. S-protein/vitronectin interaction with the C5b and the C8 of the complement membrane attack complex. *International Archives of Allergy and Immunology* **110**:314.

931. Bhakdi, S., R. Kaflein, T. S. Halstensen, F. Hugo, K. T. Preissner, and T. E. Mollnes. 1988. Complement S-protein (vitronectin) is associated with cytolytic membrane-bound C5b-9 complexes. *Clinical and Experimental Immunology* **74**:459.

932. Tschopp, J., D. Masson, S. Schafer, M. Peitsch, and K. T. Preissner. 1988. The heparin binding domain of S-protein/vitronectin binds to complement components C7, C8, and C9 and perforin from cytolytic T-cells and inhibits their lytic activities. *Biochemistry* **27**:4103.

933. Zheng, X., T. L. Saunders, S. A. Camper, L. C. Samuelson, and D. Ginsburg. 1995. Vitronectin is not essential for normal mammalian development and fertility. *Proceedings of the National Academy of Sciences of the United States of America* **92**:12426.

934. Heldin, C. H., A. Wasteson, L. Fryklund, and B. Westermark. 1981. Somatomedin B: mitogenic activity derived from contaminant epidermal growth factor. *Science* **213**:1122.

935. Buckley, M. F., K. A. Loveland, W. J. McKinstry, O. M. Garson, and J. W. Goding. 1990. Plasma cell membrane glycoprotein PC-1. cDNA cloning of the human molecule, amino acid sequence, and chromosomal location. *Journal of Biological Chemistry* **265**:17506.

936. Biesecker, G. 1990. The complement SC5b-9 complex mediates cell adhesion through a vitronectin receptor. *Journal of Immunology* **145**:209.

937. Jenne, D., A. Hille, K. K. Stanley, and W. B. Huttner. 1989. Sulfation of two tyrosine-residues in human complement S-protein (vitronectin). *European Journal of Biochemistry* **185**:391.

938. Skorstengaard, K., T. Halkier, P. Hojrup, and D. Mosher. 1990. Sequence location of a putative transglutaminase cross-linking site in human vitronectin. *FEBS Letters* **262**:269.

939. Murphy, G., and V. Knauper. 1997. Relating matrix metalloproteinase structure to function: why the 'hemopexin' domain? *Matrix Biology* **15**:511.

940. Hunt, L. T., W. C. Barker, and H. R. Chen. 1987. A domain structure common to hemopexin, vitronectin, interstitial collagenase, and a collagenase homolog. *Protein Sequences and Data Analysis* **1**:21.

941. Preissner, K. T., and D. Jenne. 1991. Structure of vitronectin and its biological role in haemostasis. *Thrombosis and Haemostasis* **66**:123.

942. Liang, O. D., S. Rosenblatt, G. S. Chatwal, and K. T. Preissner. 1997. Identification of novel heparin-binding domains of vitronectin. *FEBS Letters* **407**:169.

943. McGuire, E. A., M. E. Peacock, R. C. Inhorn, N. R. Siegel, and D. M. Tollefsen. 1988. Phosphorylation of vitronectin by a protein kinase in human plasma. Identification of a unique phosphorylation site in the heparin-binding domain. *Journal of Biological Chemistry* **263**:1942.

944. Shaltiel, S., I. Schvartz, B. Korc-Grodzicki, and T. Kreizman. 1993. Evidence for an extra-cellular function for protein kinase A. *Molecular and Cellular Biochemistry* **128**.

945. Barnes, D. W., and J. Reing. 1985. Human spreading factor: synthesis and response by HepG2 hepatoma cells in culture. *Journal of Cellular Physiology* **125**:207.

946.   Kubota, K., S. Katayama, M. Matsuda, and M. Hayashi. 1988. Three types of vitronectin in human blood. *Cell Structure and Function* **13**:123.

947.   Tollefsen, D. M., C. J. Weigel, and M. H. Kabeer. 1990. The presence of methionine or threonine at position 381 in vitronectin is correlated with proteolytic cleavage at arginine 379. *Journal of Biological Chemistry* **265**:9778.

948.   Kubota, K., M. Hayashi, N. Oishi, and Y. Sakaki. 1990. Polymorphism of the human vitronectin gene causes vitronectin blood type. *Biochemical and Biophysical Research Communications* **167**:1355.

949.   Conlan, M. G., B. R. Tomasini, R. L. Schultz, and D. F. Mosher. 1988. Plasma vitronectin polymorphism in normal subjects and patients with disseminated intravascular coagulation. *Blood* **72**:185.

950.   Fink, T. M., D. E. Jenne, and P. Lichter. 1992. The human vitronectin (complement S-protein) gene maps to the centromeric region of 17q. *Human Genetics* **88**:569.

951.   Jenne, D., and K. K. Stanley. 1987. Nucleotide sequence and organization of the human S-protein gene: repeating peptide motifs in the 'pexin' family and a model for their evolution. *Biochemistry* **26**:6735.

952.   Seiffert, D., J. Poenninger, and B. R. Binder. 1993. Organization of the gene encoding mouse vitronectin. *Gene* **134**:303.

953.   Blaschuk, O., K. Burdzy, and I. B. Fritz. 1983. Purification and characterization of a cell-aggregating factor (clusterin), the major glycoprotein in ram rete testis fluid. *Journal of Biological Chemistry* **258**:7714.

954.   Blaschuk, O. W., and I. B. Fritz. 1984. Isoelectric forms of clusterin isolated from ram rete testis fluid and from secretions of primary cultures of ram and rat Sertoli-cell-enriched preparations. *Canadian Journal of Biochemistry and Cell Biology* **62**:456.

955.   Murphy, B. F., L. Kirszbaum, I. D. Walker, and A. J. d'Apice. 1988. SP-40,40, a newly identified normal human serum protein found in the SC5b-9 complex of complement and in the immune deposits in glomerulonephritis. *Journal of Clinical Investigation* **81**:1858.

956.   Choi, N. H., Y. Nakano, T. Tobe, T. Mazda, and M. Tomita. 1990. Incorporation of SP-40,40 into the soluble membrane attack complex (SMAC, SC5b-9) of complement. *International Immunology* **2**:413.

957.   Choi, N. H., T. Mazda, and M. Tomita. 1989. A serum protein SP40,40 modulates the formation of membrane attack complex of complement on erythrocytes. *Molecular Immunology* **26**:835.

958.   Hogasen, K., C. Homann, T. E. Mollnes, N. Graudal, A. K. Hogasen, P. Hasselqvist, A. C. Thomsen, and P. Garred. 1996. Serum clusterin and vitronectin in alcoholic cirrhosis. *Liver* **16**:140.

959.   Silkensen, J. R., G. B. Schwochau, and M. E. Rosenberg. 1994. The role of clusterin in tissue injury. *Biochemistry and Cell Biology* **72**:483.

960.   Ahuja, H. S., M. Tenniswood, R. Lockshin, and Z. F. Zakeri. 1994. Expression of clusterin in cell differentiation and cell death. *Biochemistry and Cell Biology* **72**:523.

961.   Rosenberg, M. E., and J. Silkensen. 1995. Clusterin: physiologic and pathophysiologic considerations. *International Journal of Biochemistry and Cell Biology* **27**:633.

962. Tsuruta, J. K., K. Wong, I. B. Fritz, and M. D. Griswold. 1990. Structural analysis of sulphated glycoprotein 2 from amino acid sequence. Relationship to clusterin and serum protein 40,40. *Biochemical Journal* **268:**571.

963. Rosenberg, M. E., J. C. Manivel, F. A. Carone, and Y. S. Kanwar. 1995. Genesis of renal cysts is associated with clusterin expression in experimental cystic disease. *Journal of the American Society of Nephrology* **5:**1669.

964. Swertfeger, D. K., D. P. Witte, W. D. Stuart, H. A. Rockman, and J. A. Harmony. 1996. Apolipoprotein J/clusterin induction in myocarditis: a localized response gene to myocardial injury. *American Journal of Pathology* **148:**1971.

965. Koch-Brandt, C., and C. Morgans. 1996. Clusterin: a role in cell survival in the face of apoptosis? *Progress in Molecular and Subcellular Biology* **16:**130.

966. Little, S. A., and P. E. Mirkes. 1995. Clusterin expression during programmed and teratogen-induced cell death in the postimplantation rat embryo. *Teratology* **52:**41.

967. Tornqvist, E., L. Liu, H. Aldskogius, H. V. Holst, and M. Svensson. 1996. Complement and clusterin in the injured nervous system. *Neurobiology of Aging* **17:**695.

968. Choi-Miura, N. H., and T. Oda. 1996. Relationship between multifunctional protein 'clusterin' and Alzheimer disease. *Neurobiology of Aging* **17:**717.

969. Bellander, B. M., H. von Holst, P. Fredman, and M. Svensson. 1996. Activation of the complement cascade and increase of clusterin in the brain following a cortical contusion in the adult rat. *Journal of Neurosurgery* **85:**468.

970. Boggs, L. N., K. S. Fuson, M. Baez, L. Churgay, D. McClure, G. Becker, and P. C. May. 1996. Clusterin (Apo J) protects against in vitro amyloid-beta (1-40) neurotoxicity. *Journal of Neurochemistry* **67:**1324.

971. O'Bryan, M. K., H. W. Baker, J. R. Saunders, L. Kirszbaum, I. D. Walker, P. Hudson, D. Y. Liu, M. D. Glew, A. J. d'Apice, and B. F. Murphy. 1990. Human seminal clusterin (SP-40,40). Isolation and characterization. *Journal of Clinical Investigation* **85:**1477.

972. O'Bryan, M. K., B. F. Murphy, D. Y. Liu, G. N. Clarke, and H. W. Baker. 1994. The use of anticlusterin monoclonal antibodies for the combined assessment of human sperm morphology and acrosome integrity. *Human Reproduction* **9:**1490.

973. de Silva, H. V., J. A. Harmony, W. D. Stuart, C. M. Gil, and J. Robbins. 1990. Apolipoprotein J: structure and tissue distribution. *Biochemistry* **29:**5380.

974. de Silva, H. V., W. D. Stuart, C. R. Duvic, J. R. Wetterau, M. J. Ray, D. G. Ferguson, H. W. Albers, W. R. Smith, and J. A. Harmony. 1990. A 70-kDa apolipoprotein designated ApoJ is a marker for subclasses of human plasma high density lipoproteins. *Journal of Biological Chemistry* **265:**13240.

975. James, R. W., A. C. Hochstrasser, I. Borghini, B. Martin, D. Pometta, and D. Hochstrasser. 1991. Characterization of a human high density

lipoprotein-associated protein, NA1/NA2. Identity with SP-40,40, an inhibitor of complement-mediated cytolysis. *Arteriosclerosis and Thrombosis* **11**:645.

976. Jenne, D. E., B. Lowin, M. C. Peitsch, A. Bottcher, G. Schmitz, and J. Tschopp. 1991. Clusterin (complement lysis inhibitor) forms a high density lipoprotein complex with apolipoprotein A-I in human plasma. *Journal of Biological Chemistry* **266**:11030.

977. Silkensen, J. R., K. M. Skubitz, A. P. Skubitz, D. H. Chmielewski, J. C. Manivel, J. A. Dvergsten, and M. E. Rosenberg. 1995. Clusterin promotes the aggregation and adhesion of renal porcine epithelial cells. *Journal of Clinical Investigation* **96**:2646.

978. Partridge, S. R., M. S. Baker, M. J. Walker, and M. R. Wilson. 1996. Clusterin, a putative complement regulator, binds to the cell surface of Staphylococcus aureus clinical isolates. *Infection and Immunity* **64**:4324.

979. Murphy, B. F., J. R. Saunders, M. K. O'Bryan, L. Kirszbaum, I. D. Walker, and A. J. d'Apice. 1989. SP-40,40 is an inhibitor of C5b-6-initiated haemolysis. *International Immunology* **1**:551.

980. Tschopp, J., A. Chonn, S. Hertig, and L. E. French. 1993. Clusterin, the human apolipoprotein and complement inhibitor, binds to complement C7, C8 beta, and the b domain of C9. *Journal of Immunology* **151**:2159.

981. Saunders, J. R., A. Aminian, J. L. McRae, K. A. O'Farrell, W. R. Adam, and B. F. Murphy. 1994. Clusterin depletion enhances immune glomerular injury in the isolated perfused kidney. *Kidney International* **45**:817.

982. Kirszbaum, L., J. A. Sharpe, B. Murphy, A. J. d'Apice, B. Classon, P. Hudson, and I. D. Walker. 1989. Molecular cloning and characterization of the novel, human complement-associated protein, SP-40,40: a link between the complement and reproductive systems. *EMBO Journal* **8**:711.

983. Jenne, D. E., and J. Tschopp. 1989. Molecular structure and functional characterization of a human complement cytolysis inhibitor found in blood and seminal plasma: identity to sulfated glycoprotein 2, a constituent of rat testis fluid. *Proceedings of the National Academy of Sciences of the United States of America* **86**:7123.

984. Fink, T. M., M. Zimmer, J. Tschopp, J. Etienne, D. E. Jenne, and P. Lichter. 1993. Human clusterin (CLI) maps to 8p21 in proximity to the lipoprotein lipase (LPL) gene. *Genomics* **16**:526.

985. Wong, P., D. Taillefer, J. Lakins, J. Pineault, G. Chader, and M. Tenniswood. 1994. Molecular characterization of human TRPM-2/clusterin, a gene associated with sperm maturation, apoptosis and neurodegeneration. *European Journal of Biochemistry* **221**:917.

986. Tycko, B., L. Feng, L. Nguyen, A. Francis, A. Hays, W. Y. Chung, M. X. Tang, Y. Stern, A. Sahota, H. Hendrie, and R. Mayeux. 1996. Polymorphisms in the human apolipoprotein-J/clusterin gene: ethnic variation and distribution in Alzheimer's disease. *Human Genetics* **98**:430.

987. Rosenfeld, S. I., C. H. Packman, and J. P. Leddy. 1983. Inhibition of the lytic action of cell-bound terminal complement components by human high density lipoproteins and apoproteins. *Journal of Clinical Investigation* **71**:795.

988. Weiler, J. M., R. E. Edens, R. J. Linhardt, and D. P. Kapelanski. 1992.

Heparin and modified heparin inhibit complement activation in vivo. *Journal of Immunology* **148**:3210.

989. Baker, P. J., T. F. Lint, J. Siegel, M. W. Kies, and H. Gewurz. 1976. Potentiation of C56-initiated lysis by leucocyte cationic proteins, myelin basic proteins and lysine-rich histones. *Immunology* **30**:467.

990. Asghar, S. S. 1984. Pharmacological manipulation of complement system. *Pharmacological Reviews* **36**:223.

991. Thompson, R. A., and P. J. Lachmann. 1970. Reactive lysis: the complement-mediated lysis of unsensitized cells. I. The characterization of the indicator factor and its identification as C7. *Journal of Experimental Medicine* **131**:629.

992. Lachmann, P. J., and R. A. Thompson. 1970. Reactive lysis: the complement-mediated lysis of unsensitized cells. II. The characterization of activated reactor as C56 and the participation of C8 and C9. *Journal of Experimental Medicine* **131**:643.

993. Yamamoto, K. I. 1977. Lytic activity of C5-9 complexes for erythrocytes from the species other than sheep: C9 rather than C8-dependent variation in lytic activity. *Journal of Immunology* **119**:1482.

994. Blaas, P., B. Berger, S. Weber, H. H. Peter, and G. M. Hansch. 1988. Paroxysmal nocturnal hemoglobinuria. Enhanced stimulation of platelets by the terminal complement components is related to the lack of C8bp in the membrane. *Journal of Immunology* **140**:3045.

995. Tandon, N., B. P. Morgan, and A. P. Weetman. 1992. Expression and function of membrane attack complex inhibitory proteins on thyroid follicular cells. *Immunology* **75**:372.

996. Schieren, G., M. Schonermark, M. Braunger, and G. M. Hansch. 1994. Expression of the complement regulator factor C8 binding protein on human glomerular cells protects them from complement-mediated killing. *Experimental Nephrology* **2**:299.

997. Zalman, L. S., L. M. Wood, M. M. Frank, and H. J. Müller-Eberhard. 1987. Deficiency of the homologous restriction factor in paroxysmal nocturnal hemoglobinuria. *Journal of Experimental Medicine* **165**:572.

998. Hansch, G. M., S. Schonermark, and D. Roelcke. 1987. Paroxysmal nocturnal hemoglobinuria type III. Lack of an erythrocyte membrane protein restricting the lysis by C5b-9. *Journal of Clinical Investigation* **80**:7.

999. Zalman, L. S., M. A. Brothers, and H. J. Müller-Eberhard. 1989. Isolation of homologous restriction factor from human urine. Immunochemical properties and biologic activity. *Journal of Immunology* **143**:1943.

1000. Watts, M. J., J. R. Dankert, and B. P. Morgan. 1990. Isolation and characterization of a membrane-attack-complex-inhibiting protein present in human serum and other biological fluids. *Biochemical Journal* **265**:471.

1001. Hansch, G. M., P. F. Weller, and A. Nicholson-Weller. 1988. Release of C8 binding protein (C8bp) from the cell membrane by phosphatidylinositol-specific phospholipase C. *Blood* **72**:1089.

1002. Schonermark, S., S. Filsinger, B. Berger, and G. M. Hansch. 1988. The C8-binding protein of human erythrocytes: interaction with the components of the complement-attack phase. *Immunology* **63**:585.

1003. Young, J. D., S. Jiang, C. C. Liu, and C. S. Hasselkus-Light. 1990. Anti-idiotypic antibodies derived against C8, C9 and perforin bind homologous restriction factor. *Journal of Immunological Methods* **128**:133.

1004. Brickner, A., and J. M. Sodetz. 1984. Function of subunits within the eighth component of human complement: selective removal of the gamma chain reveals it has no direct role in cytolysis. *Biochemistry* **23**:832.

1005. Brickner, A., and J. M. Sodetz. 1985. Functional domains of the alpha sub-unit of the eighth component of human complement: identification and characterization of a distinct binding site for the gamma chain. *Biochemistry* **24**:4603.

1006. Zalman, L. S. 1992. Homologous restriction factor. *Current Topics in Micro-biology and Immunology* **178**:87.

1007. Rosenfeld, S. I., D. E. Jenkins, Jr., and J. P. Leddy. 1986. Enhanced reactive lysis of paroxysmal nocturnal hemoglobinuria erythrocytes. Studies on C9 binding and incorporation into high molecular weight complexes. *Journal of Experimental Medicine* **164**:981.

1008. Rosenfeld, S. I., D. E. Jenkins, Jr., and J. P. Leddy. 1985. Enhanced reactive lysis of paroxysmal nocturnal hemoglobinuria erythrocytes by C5b-9 does not involve increased C7 binding or cell-bound C3b. *Journal of Immunology* **134**:506.

1009. Takeda, J., and T. Kinoshita. 1995. GPI-anchor biosynthesis. *Trends in Biochemical Sciences* **20**:367.

1010. Kinoshita, T., N. Inoue, and J. Takeda. 1996. Role of phosphatidylinositol-linked proteins in paroxysmal nocturnal hemoglobinuria pathogenesis. *Annual Review of Medicine* **47**:1.

1011. Zalman, L. S., M. A. Brothers, and H. J. Müller-Eberhard. 1988. Self-protection of cytotoxic lymphocytes: a soluble form of homologous restriction factor in cytoplasmic granules. *Proceedings of the National Academy of Sciences of the United States of America* **85**:4827.

1012. Jiang, S., P. M. Persechini, B. Perussia, and J. D. Young. 1989. Resistance of cytolytic lymphocytes to perforin-mediated killing. Murine cytotoxic T lymphocytes and human natural killer cells do not contain functional soluble homologous restriction factor or other specific soluble protective factors. *Journal of Immunology* **143**:1453.

1013. Jiang, S. B., P. M. Persechini, A. Zychlinsky, C. C. Liu, B. Perussia, and J. D. Young. 1988. Resistance of cytolytic lymphocytes to perforin-mediated killing. Lack of correlation with complement-associated homologous species restriction. *Journal of Experimental Medicine* **168**:2207.

1014. Krahenbuhl, O. P., H. H. Peter, and J. Tschopp. 1989. Absence of homo-logous restriction factor does not affect CTL-mediated cytolysis. *European Journal of Immunology* **19**:217.

1015. Sugita, Y., Y. Nakano, and M. Tomita. 1988. Isolation from human erythro-cytes of a new membrane protein which inhibits the formation of comple-ment transmembrane channels. *Journal of Biochemistry* **104**:633.

1016. Holguin, M. H., L. R. Fredrick, N. J. Bernshaw, L. A. Wilcox, and C. J. Parker. 1989. Isolation and characterization of a membrane protein from normal human erythrocytes that inhibits reactive lysis of the erythrocytes of paroxysmal nocturnal hemoglobinuria. *Journal of Clinical Investigation* **84**:7.

1017. Okada, N., R. Harada, T. Fujita, and H. Okada. 1989. Monoclonal antibodies capable of causing hemolysis of neuraminidase-treated human erythrocytes by homologous complement. *Journal of Immunology* **143**:2262.

1018. Okada, H., Y. Nagami, K. Takahashi, N. Okada, T. Hideshima, H. Takizawa, and J. Kondo. 1989. 20 KDa homologous restriction factor of complement resembles T cell activating protein. *Biochemical and Biophysical Research Communications* **162**:1553.

1019. Davies, A., D. L. Simmons, G. Hale, R. A. Harrison, H. Tighe, P. J. Lachmann, and H. Waldmann. 1989. CD59, an LY-6-like protein expressed in human lymphoid cells, regulates the action of the complement membrane attack complex on homologous cells. *Journal of Experimental Medicine* **170**:637.

1020. Stefanova, I., I. Hilgert, H. Kristofova, R. Brown, M. G. Low, and V. Horejsi. 1989. Characterization of a broadly expressed human leucocyte surface antigen MEM-43 anchored in membrane through phosphatidylinositol. *Molecular Immunology* **26**:153.

1021. Nose, M., M. Katoh, N. Okada, M. Kyogoku, and H. Okada. 1990. Tissue distribution of HRF20, a novel factor preventing the membrane attack of homologous complement, and its predominant expression on endothelial cells in vivo. *Immunology* **70**:145.

1022. Meri, S., H. Waldmann, and P. J. Lachmann. 1991. Distribution of protectin (CD59), a complement membrane attack inhibitor, in normal human tissues. *Laboratory Investigation* **65**:532.

1023. Meri, S., B. P. Morgan, A. Davies, R. H. Daniels, M. G. Olavesen, H. Waldmann, and P. J. Lachmann. 1990. Human protectin (CD59), an 18,000–20,000 MW complement lysis restricting factor, inhibits C5b-8 catalysed insertion of C9 into lipid bilayers. *Immunology* **71**:1.

1024. Holguin, M. H., L. A. Wilcox, N. J. Bernshaw, W. F. Rosse, and C. J. Parker. 1989. Relationship between the membrane inhibitor of reactive lysis and the erythrocyte phenotypes of paroxysmal nocturnal hemoglobinuria. *Journal of Clinical Investigation* **84**:1387.

1025. Meri, S., P. Mattila, and R. Renkonen. 1993. Regulation of CD59 expression on the human endothelial cell line EA.hy 926. *European Journal of Immunology* **23**:2511.

1026. Holguin, M. H., C. B. Martin, J. H. Weis, and C. J. Parker. 1993. Enhanced expression of the complement regulatory protein, membrane inhibitor of reactive lysis (CD59), is regulated at the level of transcription. *Blood* **82**:968.

1027. Marchbank, K. J., C. W. van den Berg, and B. P. Morgan. 1997. Mechanisms of complement resistance induced by non-lethal complement attack and by growth arrest. *Immunology* **90**:647.

1028. Shibata, T., and T. Kohsaka. 1995. Effects of complement activation on the expression of CD59 by human mesangial cells. *Journal of Immunology* **155**:403.

1029. Bjorge, L., J. Hakulinen, T. Wahlstrom, R. Matre, and S. Meri. 1997.

Complement-regulatory proteins in ovarian malignancies. *International Journal of Cancer* **70**:14.

1030. Bjorge, L., C. A. Vedeler, E. Ulvestad, and R. Matre. 1994. Expression and function of CD59 on colonic adenocarcinoma cells. *European Journal of Immunology* **24**:1597.

1031. Niehans, G. A., D. L. Cherwitz, N. A. Staley, D. J. Knapp, and A. P. Dalmasso. 1996. Human carcinomas variably express the complement inhibitory proteins CD46 (membrane cofactor protein), CD55 (decay-accelerating factor), and CD59 (protectin). *American Journal of Pathology* **149**:129.

1032. Hatanaka, M., T. Seya, M. Matsumoto, T. Hara, M. Nonaka, N. Inoue, J. Takeda, and A. Shimizu. 1996. Mechanisms by which the surface expression of the glycosyl-phosphatidylinositol-anchored complement regulatory proteins decay-accelerating factor (CD55) and CD59 is lost in human leukaemia cell lines. *Biochemical Journal* **314**:969.

1033. Lehto, T., E. Honkanen, A. M. Teppo, and S. Meri. 1995. Urinary excretion of protectin (CD59), complement SC5b-9 and cytokines in membranous glomerulonephritis. *Kidney International* **47**:1403.

1034. Hakulinen, J., and S. Meri. 1995. Shedding and enrichment of the glycolipid-anchored complement lysis inhibitor protectin (CD59) into milk fat globules. *Immunology* **85**:495.

1035. Bjorge, L., T. S. Jensen, C. A. Vedeler, E. Ulvestad, E. K. Kristoffersen, and R. Matre. 1993. Soluble CD59 in pregnancy and infancy [letter]. *Immunology Letters* **36**:233.

1036. Davies, A., and P. J. Lachmann. 1993. Membrane defence against complement lysis: the structure and biological properties of CD59. *Immunologic Research* **12**:258.

1037. Rollins, S. A., and P. J. Sims. 1990. The complement-inhibitory activity of CD59 resides in its capacity to block incorporation of C9 into membrane C5b-9. *Journal of Immunology* **144**:3478.

1038. Ninomiya, H., and P. J. Sims. 1992. The human complement regulatory protein CD59 binds to the alpha-chain of C8 and to the 'b' domain of C9. *Journal of Biological Chemistry* **267**:13675.

1039. Sawada, R., K. Ohashi, H. Anaguchi, H. Okazaki, M. Hattori, N. Minato, and M. Naruto. 1990. Isolation and expression of the full-length cDNA encoding CD59 antigen of human lymphocytes. *DNA and Cell Biology* **9**:213.

1040. Sawada, R., K. Ohashi, K. Okano, M. Hattori, N. Minato, and M. Naruto. 1989. Complementary DNA sequence and deduced peptide sequence for CD59/MEM-43 antigen, the human homologue of murine lymphocyte antigen Ly-6C. *Nucleic Acids Research* **17**:6728.

1041. Sugita, Y., Y. Nakano, E. Oda, K. Noda, T. Tobe, N. H. Miura, and M. Tomita. 1993. Determination of carboxyl-terminal residue and disulfide bonds of MACIF (CD59), a glycosyl-phosphatidylinositol-anchored membrane protein. *Journal of Biochemistry* **114**:473.

1042. Shakin-Eshleman, S. H., S. L. Spitalnik, and L. Kasturi. 1996. The amino acid at the X position of an Asn-X-Ser sequon is an important determinant of N-linked core-glycosylation efficiency. *Journal of Biological Chemistry* **271**:6363.

1043. Kasturi, L., H. Chen, and S. H. Shakin-Eshleman. 1997. Regulation of N-linked core glycosylation: use of a site-directed mutagenesis approach to identify Asn-Xaa-Ser/Thr sequons that are poor oligosaccharide acceptors. *Biochemical Journal* **323**:415.

1044. Fleming, T. J., C. O'hUigin, and T. R. Malek. 1993. Characterization of two novel Ly-6 genes. Protein sequence and potential structural similarity to alpha-bungarotoxin and other neurotoxins. *Journal of Immunology* **150**:5379.

1045. Palfree, R. G. 1996. Ly-6-domain proteins – new insights and new members: a C-terminal Ly-6 domain in sperm acrosomal protein SP-10. *Tissue Antigens* **48**:71.

1046. Philbrick, W. M., R. G. Palfree, S. E. Maher, M. M. Bridgett, S. Sirlin, and A. L. Bothwell. 1990. The CD59 antigen is a structural homologue of murine Ly-6 antigens but lacks interferon inducibility. *European Journal of Immunology* **20**:87.

1047. Bickmore, W. A., D. Longbottom, K. Oghene, J. M. Fletcher, and V. van Heyningen. 1993. Colocalization of the human CD59 gene to 11p13 with the MIC11 cell surface antigen. *Genomics* **17**:129.

1048. Davies, A., A. B. Wilson, J. C. Bramley, C. Willers, V. Van Heyningen, W. A. Bickmore, and P. J. Lachmann. 1995. Identification of MIC 11 antigen as an epitope of the CD59 molecule. *Immunology* **85**:220.

1049. Tone, M., L. A. Walsh, and H. Waldmann. 1992. Gene structure of human CD59 and demonstration that discrete mRNAs are generated by alternative polyadenylation. *Journal of Molecular Biology* **227**:971.

1050. Petranka, J. G., D. E. Fleenor, K. Sykes, R. E. Kaufman, and W. F. Rosse. 1992. Structure of the CD59-encoding gene: further evidence of a relationship to murine lymphocyte antigen Ly-6 protein. *Proceedings of the National Academy of Sciences of the United States of America* **89**:7876.

1051. Holguin, M. H., C. B. Martin, T. Eggett, and C. J. Parker. 1996. Analysis of the gene that encodes the complement regulatory protein, membrane inhibitor of reactive lysis (CD59). Identification of an alternatively spliced exon and characterization of the transcriptional regulatory regions of the promoter. *Journal of Immunology* **157**:1659.

1052. Yamashina, M., E. Ueda, T. Kinoshita, T. Takami, A. Ojima, H. Ono, H. Tanaka, N. Kondo, T. Orii, N. Okada, *et al.* 1990. Inherited complete deficiency of 20-kilodalton homologous restriction factor (CD59) as a cause of paroxysmal nocturnal hemoglobinuria. *New England Journal of Medicine* **323**:1184.

1053. Motoyama, N., N. Okada, M. Yamashina, and H. Okada. 1992. Paroxysmal nocturnal hemoglobinuria due to hereditary nucleotide deletion in the HRF20 (CD59) gene. *European Journal of Immunology* **22**:2669.

1054. Ninomiya, H., B. H. Stewart, S. A. Rollins, J. Zhao, A. L. Bothwell, and P. J. Sims. 1992. Contribution of the N-linked carbohydrate of erythrocyte antigen CD59 to its complement-inhibitory activity. *Journal of Biological Chemistry* **267**:8404.

1055. Akami, T., K. Arakawa, M. Okamoto, K. Akioka, I. Fujiwara, I. Nakai, M. Mitsuo, R. Sawada, M. Naruto, and T. Oka. 1994. Enhancement of the complement regulatory function of CD59 by site-directed mutagenesis at the N-glycosylation site. *Transplantation Proceedings* **26:**1256.

1056. Rother, R. P., J. Zhao, Q. Zhou, and P. J. Sims. 1996. Elimination of potential sites of glycosylation fails to abrogate complement regulatory function of cell surface CD59. *Journal of Biological Chemistry* **271:**23842.

1057. Bodian, D. L., S. J. Davis, B. P. Morgan, and N. K. Rushmere. 1997. Mutational analysis of the active site and antibody epitopes of the complement-inhibitory glycoprotein, CD59. *Journal of Experimental Medicine* **185:**507.

1058. Rushmere, N. K., S. Tomlinson, and B. P. Morgan. 1997. Expression of rat CD59: functional analysis confirms lack of species selectivity and reveals that glycosylation is not required for function. *Immunology* **90:**640.

1059. Suzuki, H., N. Yamaji, A. Egashira, K. Yasunaga, Y. Sugita, and Y. Masuho. 1996. Effect of the sugar chain of soluble recombinant CD59 on complement inhibitory activity. *FEBS Letters* **399:**272.

1060. Rudd, P. M., B. P. Morgan, M. R. Wormald, D. J. Harvey, C. W. van den Berg, S. J. Davis, M. A. Ferguson, and R. A. Dwek. 1997. The glycosylation of the complement regulatory protein, human erythrocyte CD59. *Journal of Biological Chemistry* **272:**7229.

1061. Menu, E., B. C. Tsai, A. L. Bothwell, P. J. Sims, and B. E. Bierer. 1994. CD59 costimulation of T cell activation. CD58 dependence and requirement for glycosylation. *Journal of Immunology* **153:**2444.

1062. Nakano, Y., T. Tozaki, N. Kikuta, T. Tobe, E. Oda, N. H. Miura, T. Sakamoto, and M. Tomita. 1995. Determination of the active site of CD59 with synthetic peptides. *Molecular Immunology* **32:**241.

1063. Fletcher, C. M., R. A. Harrison, P. J. Lachmann, and D. Neuhaus. 1993. Sequence-specific 1H-NMR assignments and folding topology of human CD59. *Protein Science* **2:**2015.

1064. Fletcher, C. M., R. A. Harrison, P. J. Lachmann, and D. Neuhaus. 1994. Structure of a soluble, glycosylated form of the human complement regulatory protein CD59. *Structure* **2:**185.

1065. Kieffer, B., P. C. Driscoll, I. D. Campbell, A. C. Willis, P. A. van der Merwe, and S. J. Davis. 1994. Three-dimensional solution structure of the extracellular region of the complement regulatory protein CD59, a new cell-surface protein domain related to snake venom neurotoxins. *Biochemistry* **33:**4471.

1066. Petranka, J., J. Zhao, J. Norris, N. B. Tweedy, R. E. Ware, P. J. Sims, and W. F. Rosse. 1996. Structure–function relationships of the complement regulatory protein, CD59. *Blood Cells, Molecules and Diseases* **22:**281.

1067. Yu, J., S. Dong, N. K. Rushmere, B. P. Morgan, R. Abagyan, and S. Tomlinson. 1997. Mapping the regions of the complement inhibitor CD59 responsible for its species selective activity. *Biochemistry* **36:**9423.

1068. Yu, J., R. Abagyan, S. Dong, A. Gilbert, V. Nussenzweig, and S. Tomlinson. 1997. Mapping the active site of CD59. *Journal of Experimental Medicine* **185:**745.

1069. Zhao, X. J., J. Zhao, Q. S. Zhou, and P. J. Sims. 1998. Identity of the residues responsible for the species-restricted complement inhibitory function of human CD59. *Journal of Biological Chemistry* **273**:10665.

1070. Tomlinson, S., M. B. Whitlow, and V. Nussenzweig. 1994. A synthetic peptide from complement protein C9 binds to CD59 and enhances lysis of human erythrocytes by C5b-9. *Journal of Immunology* **152**:1927.

1071. Lockert, D. H., K. M. Kaufman, C. P. Chang, T. Husler, J. M. Sodetz, and P. J. Sims. 1995. Identity of the segment of human complement C8 recognized by complement regulatory protein CD59. *Journal of Biological Chemistry* **270**:19723.

1072. Husler, T., D. H. Lockert, and P. J. Sims. 1996. Role of a disulfide-bonded peptide loop within human complement C9 in the species-selectivity of complement inhibitor CD59. *Biochemistry* **35**:3263.

1073. Husler, T., D. H. Lockert, K. M. Kaufman, J. M. Sodetz, and P. J. Sims. 1995. Chimeras of human complement C9 reveal the site recognized by complement regulatory protein CD59. *Journal of Biological Chemistry* **270**:3483.

1074. Whitlow, M. B., K. Iida, I. Stefanova, A. Bernard, and V. Nussenzweig. 1990. H19, a surface membrane molecule involved in T-cell activation, inhibits channel formation by human complement. *Cellular Immunology* **126**:176.

1075. Hahn, W. C., E. Menu, A. L. Bothwell, P. J. Sims, and B. E. Bierer. 1992. Overlapping but nonidentical binding sites on CD2 for CD58 and a second ligand CD59. *Science* **256**:1805.

1076. Deckert, M., J. Kubar, D. Zoccola, G. Bernard-Pomier, P. Angelisova, V. Horejsi, and A. Bernard. 1992. CD59 molecule: a second ligand for CD2 in T cell adhesion. *European Journal of Immunology* **22**:2943.

1077. Deckert, M., J. Kubar, and A. Bernard. 1992. CD58 and CD59 molecules exhibit potentializing effects in T cell adhesion and activation. *Journal of Immunology* **148**:672.

1078. van der Merwe, P. A., A. N. Barclay, D. W. Mason, E. A. Davies, B. P. Morgan, M. Tone, A. K. Krishnam, C. Ianelli, and S. J. Davis. 1994. Human cell-adhesion molecule CD2 binds CD58 (LFA-3) with a very low affinity and an extremely fast dissociation rate but does not bind CD48 or CD59. *Biochemistry* **33**:10149.

1079. Korty, P. E., C. Brando, and E. M. Shevach. 1991. CD59 functions as a signal-transducing molecule for human T cell activation. *Journal of Immunology* **146**:4092.

1080. Stefanova, I., V. Horejsi, I. J. Ansotegui, W. Knapp, and H. Stockinger. 1991. GPI-anchored cell-surface molecules complexed to protein tyrosine kinases. *Science* **254**:1016.

1081. Hatanaka, M., T. Seya, S. Miyagawa, M. Matsumoto, T. Hara, K. Tanaka, and A. Shimizu. 1998. Cellular distribution of a GPI-anchored complement regulatory protein CD59: homodimerization on the surface of HeLa and CD59-transfected CHO cells. *Journal of Biochemistry* **123**:579.

1082. van den Berg, C. W., T. Cinek, M. B. Hallett, V. Horejsi, and B. P. Morgan. 1995. Exogenous glycosyl phosphatidylinositol-anchored CD59 associates with kinases in membrane clusters on U937 cells and becomes Ca(2+)-signaling competent. *Journal of Cell Biology* **131**:669.

1083. Meri, S., B. P. Morgan, M. Wing, J. Jones, A. Davies, E. Podack, and P. J. Lachmann. 1990. Human protectin (CD59), an 18-20-kD homologous complement restriction factor, does not restrict perforin-mediated lysis. *Journal of Experimental Medicine* **172:**367.

1084. Tomita, A., E. L. Radike, and C. J. Parker. 1993. Isolation of erythrocyte membrane inhibitor of reactive lysis type II. Identification as glycophorin A. *Journal of Immunology* **151:**3308.

1085. Tomita, A., and C. J. Parker. 1994. Aberrant regulation of complement by the erythrocytes of hereditary erythroblastic multinuclearity with a positive acidified serum lysis test (HEMPAS). *Blood* **83:**250.

1086. Brauch, H., D. Roelcke, and U. Rother. 1983. Glycophorin A inhibits lysis by the complement attack phase. *Immunobiology* **165:**115.

1087. Okada, N., T. Yasuda, T. Tsumita, H. Shinomiya, S. Utsumi, and H. Okada. 1982. Regulation of glycophorin of complement activation via the alternative pathway. *Biochemical and Biophysical Research Communications* **108:**770.

1088. Okada, H., and H. Tanaka. 1983. Species-specific inhibition by glycophorins of complement activation via the alternative pathway. *Molecular Immunology* **20:**1233.

1089. Marshall, P., A. Hasegawa, E. A. Davidson, V. Nussenzweig, and M. Whitlow. 1996. Interaction between complement proteins C5b-7 and erythrocyte membrane sialic acid. *Journal of Experimental Medicine* **184:**1225.

1090. Ohanian, S. H., S. I. Schlager, and T. Borsos. 1978. Molecular interactions of cells with antibody and complement: influence of metabolic and physical properties of the target on the outcome of humoral immune attack. *Contemporary Topics in Molecular Immunology* **7:**153.

1091. Ohanian, S. H., S. I. Schlager, and S. Saha. 1982. Effect of lipids, structural precursors of lipids and fatty acids on complement-mediated killing of antibody-sensitized nucleated cells. *Molecular Immunology* **19:**535.

1092. Schlager, S. I., and S. H. Ohanian. 1980. Tumor cell lipid composition and sensitivity to humoral immune killing. II. Influence of plasma membrane and intracellular lipid and fatty acid content. *Journal of Immunology* **125:**508.

1093. Schlager, S. I., and S. H. Ohanian. 1980. Plasma membrane and intracellular lipid synthesis in tumor cells rendered resistant to humoral immune killing after treatment with hormones. *Journal of the National Cancer Institute* **64:**943.

1094. Goldberg, V., and H. Green. 1958. The cytotoxic action of immune gamma globulin and complement on Ehrlich ascites tumor cells. I. Ultrastructural studies. *Journal of Experimental Medicine* **109:**504.

1095. Green, H., Barrow, P., and B. Goldberg. 1959. Effect of antibody and complement on permeability control in ascites tumor cells and erythrocytes. *Journal of Experimental Medicine* **100:**699.

1096. Kaliner, M., and K. F. Austen. 1974. Adenosine 3′5′-monophosphate: inhibition of complement-mediated cell lysis. *Science* **183:**659.

1097. Boyle, M. D. P., S. H. Ohanian, and T. Borsos. 1976. Lysis of tumor cells by antibody and complement. VII. Complement-dependent [86]Rb release – a nonlethal event? *Journal of Immunology* **117:**1346.

1098. Boyle, M. D., S. H. Ohanian, and T. Borsos. 1976. Studies on the terminal stages of antibody-complement-mediated killing of a tumor cell. II. Inhibition of transformation of T to dead cells by 3'5' cAMP. *Journal of Immunology* **116**:1276.

1099. Morgan, B. P., and A. K. Campbell. 1985. The recovery of human polymorphonuclear leucocytes from sublytic complement attack is mediated by changes in intracellular free calcium. *Biochemical Journal* **231**:205.

1100. Campbell, A. K., and B. P. Morgan. 1985. Monoclonal antibodies demonstrate protection of polymorphonuclear leukocytes against complement attack. *Nature* **317**:164.

1101. Shin, M. L., and D. F. Carney. 1988. Cytotoxic action and other metabolic consequences of terminal complement proteins. *Progress in Allergy* **40**:44.

1102. Carney, D. F., C. L. Koski, and M. L. Shin. 1985. Elimination of terminal complement intermediates from the plasma membrane of nucleated cells: the rate of disappearance differs for cells carrying C5b-7 or C5b-8 or a mixture of C5b-8 with a limited number of C5b-9. *Journal of Immunology* **134**:1804.

1103. Carney, D. F., C. H. Hammer, and M. L. Shin. 1986. Elimination of terminal complement complexes in the plasma membrane of nucleated cells: influence of extracellular Ca2+ and association with cellular Ca2+. *Journal of Immunology* **137**:263.

1104. Sims, P. J., and T. Wiedmer. 1986. Repolarization of the membrane potential of blood platelets after complement damage: evidence for a Ca++ -dependent exocytotic elimination of C5b-9 pores. *Blood* **68**:556.

1105. Sims, P. J., E. M. Faioni, T. Wiedmer, and S. J. Shattil. 1988. Complement proteins C5b-9 cause release of membrane vesicles from the platelet surface that are enriched in the membrane receptor for coagulation factor Va and express prothrombinase activity. *Journal of Biological Chemistry* **263**:18205.

1106. Morgan, B. P. 1992. Effects of the membrane attack complex of complement on nucleated cells. *Current Topics in Microbiology and Immunology* **178**:115.

1107. Stein, J. M., and J. P. Luzio. 1991. Ectocytosis caused by sublytic autologous complement attack on human neutrophils. The sorting of endogenous plasma-membrane proteins and lipids into shed vesicles. *Biochemical Journal* **274**:381.

1108. Scolding, N. J., B. P. Morgan, W. A. Houston, C. Linington, A. K. Campbell, and D. A. Compston. 1989. Vesicular removal by oligodendrocytes of membrane attack complexes formed by activated complement. *Nature* **339**:620.

1109. Wiedmer, T., C. T. Esmon, and P. J. Sims. 1986. On the mechanism by which complement proteins C5b-9 increase platelet prothrombinase activity. *Journal of Biological Chemistry* **261**:14587.

1110. Wiedmer, T., B. Ando, and P. J. Sims. 1987. Complement C5b-9-stimulated platelet secretion is associated with a Ca2+-initiated activation of cellular protein kinases. *Journal of Biological Chemistry* **262**:13674.

1111. Osler, W. 1888. Hereditary angioneurotic oedema. *American Journal of Medical Science* **95**:362.

1112. Donaldson, V. H. 1979. Hereditary angioneurotic edema. *Disease-A-Month* **26:**1.

1113. Frank, M. M., J. A. Gelfand, and J. P. Atkinson. 1976. Hereditary angioedema: the clinical syndrome and its management. *Annals of Internal Medicine* **84:**580.

1114. Brackertz, D., E. Isler, and F. Kueppers. 1975. Half-life of C1INH in hereditary angioneurotic oedema (HAE). *Clinical Allergy* **5:**89.

1115. Agostoni, A. 1989. Inherited C1 inhibitor deficiency. *Complement and Inflammation* **6:**112.

1116. Aulak, K. S., and R. S. Harrison. 1990. Rapid and sensitive techniques for identification and analysis of 'reactive-centre' mutants of C1-inhibitor proteins contained in type II hereditary angio-oedema plasmas. *Biochemical Journal* **271:**565.

1117. Donaldson, V. H., R. A. Harrison, F. S. Rosen, D. H. Bing, G. Kindness, J. Canar, C. J. Wagner, and S. Awad. 1985. Variability in purified dysfunctional C1(−)-inhibitor proteins from patients with hereditary angioneurotic edema. Functional and analytical gel studies. *Journal of Clinical Investigation* **75:**124.

1118. Siddique, Z., A. R. McPhaden, and K. Whaley. 1995. Characterization of nucleotide sequence variants and disease-specific mutations involving the 3′ end of the C1-inhibitor gene in hereditary angio-oedema. *Human Heredity* **45:**98.

1119. Siddique, Z., A. R. McPhaden, D. McCluskey, and K. Whaley. 1992. A single base deletion from the C1-inhibitor gene causes type I hereditary angio-oedema. *Human Heredity* **42:**231.

1120. Tosi, M. 1993. Molecular genetics of C1-inhibitor and hereditary angio-oedaema. In *Complement in Health and Disease.* K. Whaley, M. Loos, J.M. Weiler, eds. Kluwer, Lancaster, p. 245.

1121. Shoemaker, L. R., S. J. Schurman, V. H. Donaldson, and A. E. Davis, 3rd. 1994. Hereditary angioneurotic oedema: characterization of plasma kinin and vascular permeability-enhancing activities. *Clinical and Experimental Immunology* **95:**22.

1122. Bork, K., and G. Witzke. 1989. Long-term prophylaxis with C1-inhibitor (C1 INH) concentrate in patients with recurrent angioedema caused by hereditary and acquired C1-inhibitor deficiency. *Journal of Allergy and Clinical Immunology* **83:**677.

1123. Waytes, A. T., F. S. Rosen, and M. M. Frank. 1996. Treatment of hereditary angioedema with a vapor-heated C1 inhibitor concentrate [see comments]. *New England Journal of Medicine* **334:**1630.

1124. Agostoni, A., and M. Cicardi. 1992. Hereditary and acquired C1-inhibitor deficiency: biological and clinical characteriztics in 235 patients. *Medicine* **71:**206.

1125. Rosse, W. F., G. L. Logue, H. R. Silberman, and M. M. Frank. 1976. The effect of synthetic androgens in hereditary angioneurotic edema: alteration of C1 inhibitor and C4 levels. *Transactions of the Association of American Physicians* **89:**122.

1126. Sheffer, A. L., D. T. Fearon, and K. F. Austen. 1979. Clinical and biochemical effects of impeded androgen (oxymetholone) therapy of hereditary angioedema. *Journal of Allergy and Clinical Immunology* **64:**275.

1127. Frank, M. M., J. S. Sergent, M. A. Kane, and D. W. Alling. 1972. Epsilon aminocaproic acid therapy of hereditary angioneurotic edema. A double-blind study. *New England Journal of Medicine* **286:**808.

1128. Sheffer, A. L., K. F. Austen, and F. S. Rosen. 1972. Tranexamic acid therapy in hereditary angioneurotic edema. *New England Journal of Medicine* **287:**452.

1129. Van Dellen, R. G. 1996. Long-term treatment of C1 inhibitor deficiency with epsilon-aminocaproic acid in two patients. *Mayo Clinic Proceedings* **71:**1175.

1130. Bain, B. J., D. Catovsky, and P. W. Ewan. 1993. Acquired angioedema as the presenting feature of lymphoproliferative disorders of mature B-lymphocytes. *Cancer* **72:**3318.

1131. Mathur, R., P. J. Toghill, and I. D. Johnston. 1993. Acquired C1 inhibitor deficiency with lymphoma causing recurrent angioedema. *Postgraduate Medical Journal* **69:**646.

1132. Wasserfallen, J. B., P. Spaeth, L. Guillou, and A. R. Pecoud. 1995. Acquired deficiency in C1-inhibitor associated with signet ring cell gastric adenocarcinoma: a probable connection of antitumor-associated antibodies, hemolytic anemia, and complement turnover. *Journal of Allergy and Clinical Immunology* **95:**124.

1133. Cicardi, M., A. Beretta, M. Colombo, D. Gioffre, M. Cugno, and A. Agostoni. 1996. Relevance of lymphoproliferative disorders and of anti-C1 inhibitor autoantibodies in acquired angio-oedema. *Clinical and Experimental Immunology* **106:**475.

1134. Cicardi, M., G. Bisiani, M. Cugno, P. Spath, and A. Agostoni. 1993. Autoimmune C1 inhibitor deficiency: report of eight patients. *American Journal of Medicine* **95:**169.

1135. Whaley, K., R. B. Sim, and S. He. 1996. Autoimmune C1-inhibitor deficiency [editorial]. *Clinical and Experimental Immunology* **106:**423.

1136. Jackson, J., R. B. Sim, K. Whaley, and C. Feighery. 1989. Autoantibody facilitated cleavage of C1-inhibitor in autoimmune angioedema. *Journal of Clinical Investigation* **83:**698.

1137. Mandle, R., C. Baron, E. Roux, R. Sundel, J. Gelfand, K. Aulak, A. E. Davis, 3rd, F. S. Rosen, and D. H. Bing. 1994. Acquired C1 inhibitor deficiency as a result of an autoantibody to the reactive center region of C1 inhibitor. *Journal of Immunology* **152:**4680.

1138. Schur, P. H., H. Borel, E. W. Gelfand, C. A. Alper, and F. S. Rosen. 1970. Selective gamma-g globulin deficiencies in patients with recurrent pyogenic infections. *New England Journal of Medicine* **283:**631.

1139. Alper, C. A., T. Boenisch, and L. Watson. 1972. Genetic polymorphism in human glycine-rich beta-glycoprotein. *Journal of Experimental Medicine* **135:**68.

1140. Vyse, T. J., P. J. Spath, K. A. Davies, B. J. Morley, P. Philippe, P. Athanassiou, C. M. Giles, and M. J. Walport. 1994. Hereditary complement factor I deficiency. *Quarterly Journal of Medicine* **87:**385.

1141. Thompson, R. A., and P. J. Lachmann. 1977. A second case of human C3b inhibitor (KAF) deficiency. *Clinical and Experimental Immunology* **27**:23.

1142. Figueroa, J. E., and P. Densen. 1991. Infectious diseases associated with complement deficiencies. *Clinical Microbiology Reviews* **4**:359.

1143. Bonnin, A. J., H. J. Zeitz, and A. Gewurz. 1993. Complement factor I deficiency with recurrent aseptic meningitis. *Archives of Internal Medicine* **153**:1380.

1144. Solal-Celigny, P., M. Laviolette, J. Hebert, P. C. Atkins, M. Sirois, G. Brun, G. Lehner-Netsch, and J. M. Delage. 1982. C3b inactivator deficiency with immune complex manifestations. *Clinical and Experimental Immunology* **47**:197.

1145. Moller Rasmussen, J., B. Teisner, H. H. Jepsen, S. E. Svehag, F. Knudsen, H. Kirstein, and M. Buhl. 1988. Three cases of factor I deficiency: the effect of treatment with plasma. *Clinical and Experimental Immunology* **74**:131.

1146. Vyse, T. J., B. J. Morley, I. Bartok, E. L. Theodoridis, K. A. Davies, A. D. Webster, and M. J. Walport. 1996. The molecular basis of hereditary complement factor I deficiency. *Journal of Clinical Investigation* **97**:925.

1147. Thompson, R. A., and M. H. Winterborn. 1981. Hypocomplementaemia due to a genetic deficiency of beta 1H globulin. *Clinical and Experimental Immunology* **46**:110.

1148. Levy, M., L. Halbwachs-Mecarelli, M. C. Gubler, G. Kohout, A. Bensenouci, P. Niadet, G. Hauptmann, and P. Lesavre. 1986. H deficiency in two brothers with atypical dense intramembranous disease. *Kidney International* **30**:949.

1149. Nielsen, H. E., C. Koch, P. Magnussen, and I. Lind. 1989. Complement deficiencies in selected groups of patients with meningococcal disease. *Scandinavian Journal of Infectious Diseases* **21**:389.

1150. Vogt, B. A., R. J. Wyatt, B. A. Burke, S. C. Simonton, and C. E. Kashtan. 1995. Inherited factor H deficiency and collagen type III glomerulopathy. *Pediatric Nephrology* **9**:11.

1151. Lopez-Larrea, C., M. A. Dieguez, A. Enguix, O. Dominguez, B. Marin, and E. Gomez. 1987. A familial deficiency of complement factor H. *Biochemistry Society Transactions* **15**:648.

1152. Hogasen, K., J. H. Jansen, T. E. Mollnes, J. Hovdenes, and M. Harboe. 1995. Hereditary porcine membranoproliferative glomerulonephritis type II is caused by factor H deficiency. *Journal of Clinical Investigation* **95**:1054.

1153. Lehner, P. J., K. A. Davies, M. J. Walport, A. P. Cope, R. Wurzner, A. Orren, B. P. Morgan, and J. Cohen. 1992. Meningococcal septicaemia in a C6-deficient patient and effects of plasma transfusion on lipopolysaccharide release. *Lancet* **340**:1379.

1154. Lemesle, F. G., J. C. Gris, J. F. Schwed, and C. Arich. 1993. Rapid recovery of acquired purpura fulminans in a patient with familial C4bBP deficiency. *Intensive Care Medicine* **19**:115.

1155. Sjoholm, A. G., J. H. Braconier, and C. Soderstrom. 1982. Properdin deficiency in a family with fulminant meningococcal infections. *Clinical and Experimental Immunology* **50**:291.

1156. Goonewardena, P., A. G. Sjoholm, L. A. Nilsson, and U. Pettersson.

1988. Linkage analysis of the properdin deficiency gene: suggestion of a locus in the proximal part of the short arm of the X chromosome. *Genomics* **2:**115.

1157. Wadelius, C., M. Pigg, M. Sundvall, A. G. Sjoholm, P. Goonewardena, E. J. Kuijper, C. C. Tijssen, A. Jansz, P. J. Spath, U. B. Schaad, et al. 1992. Linkage analysis in properdin deficiency families: refined location in proximal Xp. *Clinical Genetics* **42:**8.

1158. Ash, S., C. Johnson, M. Shohat, T. Shohat, and M. Schlesinger. 1994. Further mapping of the properdin deficiency gene in a Tunisian Jewish family – for genetic homogeneity. *Israel Journal of Medical Sciences* **30:**626.

1159. Kolble, K., A. J. Cant, A. C. Fay, K. Whaley, M. Schlesinger, and K. B. Reid. 1993. Carrier detection in families with properdin deficiency by microsatellite haplotyping. *Journal of Clinical Investigation* **91:**99.

1160. Fijen, C. A., R. van den Bogaard, M. R. Daha, J. Dankert, M. Mannens, and E. J. Kuijper. 1996. Carrier detection by microsatellite haplotyping in 10 properdin type 1-deficient families. *European Journal of Clinical Investigation* **26:**902.

1161. Gelfand, E. W., C. P. Rao, J. O. Minta, T. Ham, D. B. Purkall, and S. Ruddy. 1987. Inherited deficiency of properdin and C2 in a patient with recurrent bacteremia. *American Journal of Medicine* **82:**671.

1162. Holme, E. R., J. Veitch, A. Johnston, G. Hauptmann, B. Uring-Lambert, M. Seywright, V. Docherty, W. N. Morley, and K. Whaley. 1989. Familial properdin deficiency associated with chronic discoid lupus erythematosus. *Clinical and Experimental Immunology* **76:**76.

1163. Soderstrom, C., A. G. Sjoholm, R. Svensson, and S. Ostenson. 1989. Another Swedish family with complete properdin deficiency: association with fulminant meningococcal disease in one male family member. *Scandinavian Journal of Infectious Diseases* **21:**259.

1164. Fredrikson, G. N., J. Westberg, E. J. Kuijper, C. C. Tijssen, A. G. Sjoholm, M. Uhlen, and L. Truedsson. 1996. Molecular characterization of properdin deficiency type III: dysfunction produced by a single point mutation in exon 9 of the structural gene causing a tyrosine to aspartic acid interchange. *Journal of Immunology* **157:**3666.

1165. Westberg, J., G. N. Fredrikson, L. Truedsson, A. G. Sjoholm, and M. Uhlen. 1995. Sequence-based analysis of properdin deficiency: identification of point mutations in two phenotypic forms of an X-linked immunodeficiency. *Genomics* **29:**1.

1166. Ohi, H., and T. Yasugi. 1994. Occurrence of C3 nephritic factor and C4 nephritic factor in membranoproliferative glomerulonephritis (MPGN). *Clinical and Experimental Immunology* **95:**316.

1167. Mathieson, P. W., and D. K. Peters. 1993. Deficiency and depletion of complement in the pathogenesis of nephritis and vasculitis. *Kidney International* Supplement 42:S13.

1168. Fremeaux-Bacchi, V., L. Weiss, C. Demouchy, A. May, S. Palomera, and M. D. Kazatchkine. 1994. Hypocomplementaemia of poststreptococcal acute glomerulonephritis is associated with C3 nephritic factor (C3NeF) IgG autoantibody activity. *Nephrology, Dialysis, Transplantation* **9:**1747.

1169. Carmichael, A. J., and J. R. Marsden. 1993. Urticarial vasculitis: a presentation of C3 nephritic factor [letter]. *British Journal of Dermatology* **128**:589.

1170. Wayte, J., G. Bird, and J. D. Wilkinson. 1996. The clinical significance of partial lipoatrophy and C3 hypocomplementaemia: a report of two cases. *Clinical and Experimental Dermatology* **21**:131.

1171. Mathieson, P. W., R. Wurzner, D. B. Oliveria, P. J. Lachmann, and D. K. Peters. 1993. Complement-mediated adipocyte lysis by nephritic factor sera. *Journal of Experimental Medicine* **177**:1827.

1172. Kawagoe, K., D. Kitamura, M. Okabe, I. Taniuchi, M. Ikawa, T. Watanabe, T. Kinoshita, and J. Takeda. 1996. Glycosylphosphatidylinositol-anchor-deficient mice: implications for clonal dominance of mutant cells in paroxysmal nocturnal hemoglobinuria. *Blood* **87**:3600.

1173. Bessler, M., P. J. Mason, P. Hillmen, and L. Luzzatto. 1994. Mutations in the PIG-A gene causing partial deficiency of GPI-linked surface proteins (PNH II) in patients with paroxysmal nocturnal haemoglobinuria. *British Journal of Haematology* **87**:863.

1174. Ham, T. H. 1937. Chronic haemolytic anaemia with paroxysmal nocturnal haemoglobinuria. A study of the mechanism of haemolysis in relation to acid-base equilibrium. *New England Journal of Medicine*. **217**:915.

1175. Ham, T. H., and J.H. Dingle. 1939. Studies on destruction of red blood cells. II. Chronic haemolytic anaemia with paroxysmal nocturnal haemoglobinuria: certain immunological aspects of the haemolytic mechanism with special reference to seum complement. *Journal of Clinical Investigation*. **18**:657.

1176. Rosse, W. F., and C. J. Parker. 1985. Paroxysmal nocturnal haemoglobinuria. *Clinics in Haematology* **14**:105.

1177. Hall, S. E., and W. F. Rosse. 1996. The use of monoclonal antibodies and flow cytometry in the diagnosis of paroxysmal nocturnal hemoglobinuria. *Blood* **87**:5332.

1178. Lin, R. C., J. Herman, L. Henry, and G. L. Daniels. 1988. A family showing inheritance of the Inab phenotype. *Transfusion* **28**:427.

1179. Telen, M. J., and A. M. Green. 1989. The Inab phenotype: characterization of the membrane protein and complement regulatory defect. *Blood* **74**:437.

1180. Reid, M. E., G. Mallinson, R. B. Sim, J. Poole, V. Pausch, A. H. Merry, Y. W. Liew, and M. J. Tanner. 1991. Biochemical studies on red blood cells from a patient with the Inab phenotype (decay-accelerating factor deficiency). *Blood* **78**:3291.

1181. Merry, A. H., V. I. Rawlinson, M. Uchikawa, M. R. Daha, and R. B. Sim. 1989. Studies on the sensitivity to complement-mediated lysis of erythrocytes (Inab phenotype) with a deficiency of DAF (decay accelerating factor). *British Journal of Haematology* **73**:248.

1182. Holguin, M. H., C. B. Martin, N. J. Bernshaw, and C. J. Parker. 1992. Analysis of the effects of activation of the alternative pathway of complement on erythrocytes with an isolated deficiency of decay accelerating factor. *Journal of Immunology* **148**:498.

1183. Lublin, D. M., G. Mallinson, J. Poole, M. E. Reid, E. S. Thompson, B. R. Ferdman, M. J. Telen, D. J. Anstee, and M. J. Tanner. 1994. Molecular basis of reduced or absent expression of decay-accelerating factor in Cromer blood group phenotypes. *Blood* **84:**1276.

1184. Lublin, D. M., E. S. Thompson, A. M. Green, C. Levene, and M. J. Telen. 1991. Dr(a-) polymorphism of decay accelerating factor. Biochemical, functional, and molecular characterization and production of allele-specific transfectants. *Journal of Clinical Investigation* **87:**1945.

1185. Fukuda, H., T. Seya, T. Hara, M. Matsumoto, T. Kinoshita, and T. Masaoka. 1991. Deficiency of complement decay-accelerating factor (DAF, CD55) in non-Hodgkin's lymphoma. *Immunology Letters* **29:**205.

1186. Walport, M., and P. Lachmann. 1984. C3 receptors, complement deficiency and SLE. *British Journal of Rheumatology* **23:**3.

1187. Schifferli, J. A., Y. C. Ng, J. P. Paccaud, and M. J. Walport. 1989. The role of hypocomplementaemia and low erythrocyte complement receptor type 1 numbers in determining abnormal immune complex clearance in humans. *Clinical and Experimental Immunology* **75:**329.

1188. Lachmann, P. J., and M. J. Walport. 1987. Deficiency of the effector mechanisms of the immune response and autoimmunity. *Ciba Foundation Symposium* **129:**149.

1189. Pascual, M., C. Danielsson, G. Steiger, and J. A. Schifferli. 1994. Proteolytic cleavage of CR1 on human erythrocytes in vivo: evidence for enhanced cleavage in AIDS. *European Journal of Immunology* **24:**702.

1190. Ohsaka, A., K. Saionji, N. Watanabe, H. Yokomichi, Y. Sugahara, R. Nagayama, and J. Igari. 1994. Complement receptor type 1 (CR1) deficiency on neutrophils in myelodysplastic syndrome. *British Journal of Haematology* **88:**409.

1191. Rao, N., D. J. Ferguson, S. F. Lee, and M. J. Telen. 1991. Identification of human erythrocyte blood group antigens on the C3b/C4b receptor. *Journal of Immunology* **146:**3502.

1192. Moulds, J. M., M. W. Nickells, J. J. Moulds, M. C. Brown, and J. P. Atkinson. 1991. The C3b/C4b receptor is recognized by the Knops, McCoy, Swain-Langley, and York blood group antisera. *Journal of Experimental Medicine* **173:**1159.

1193. Petty, A. C., C. A. Green, J. Poole, and G. L. Daniels. 1997. Analysis of Knops blood group antigens on CR1 (CD35) by the MAIEA test and by immunoblotting. *Transfusion Medicine* **7:**55.

1194. Moulds, J. M., J. J. Moulds, M. Brown, and J. P. Atkinson. 1992. Antiglobulin testing for CR1-related (Knops/McCoy/Swain-Langley/York) blood group antigens: negative and weak reactions are caused by variable expression of CR1. *Vox Sanguinis* **62:**230.

1195. Moulds, J. M., and K. E. Rowe. 1996. Neutralization of Knops system antibodies using soluble complement receptor 1. *Transfusion* **36:**517.

1196. Seya, T., H. Tejima, H. Fukuda, T. Hara, M. Matsumoto, M. Hatanaka, Y. Sugita, and T. Masaoka. 1993. Acute promyelocytic leukemia with CD59 deficiency. *Leukemia Research* **17:**895.

1197. Hatanaka, M., T. Seya, M. Matsumoto, T. Hara, M. Nonaka, N. Inoue, J. Takeda, and A. Shimizu. 1996. Mechanisms by which the surface expression of the glycosyl-phosphatidylinositol-anchored complement regulatory proteins decay-accelerating factor (CD55) and CD59 is lost in human leukaemia cell lines. *Biochemical Journal* **314:**969.

1198. Harris, C. L., and B. P. Morgan. 1995. Characterization of a glycosyl-phosphatidylinositol anchor-deficient subline of Raji cells. An analysis of the functional importance of complement inhibitors on the Raji cell line. *Immunology* **86:**311.

1199. Kelly, R. W. 1995. Immunosuppressive mechanisms in semen: implications for contraception. *Human Reproduction* **10:**1686.

1200. Rooney, I. A., T. J. Oglesby, and J. P. Atkinson. 1993. Complement in human reproduction: activation and control. *Immunologic Research* **12:**276.

1201. Rooney, I. A. 1994. Complement, complement inhibitors and fertilization. In *New aspects of complement structure and function.* A. Erdei, ed. R.G. Landes, Austin, p. 117.

1202. Price, R. J., and B. Boettcher. 1979. The presence of complement in human cervical mucus and its possible relevance to infertility in women with complement-dependent sperm-immobilizing antibodies. *Fertility and Sterility* **32:**61.

1203. Mathur, S., H. O. Williamson, F. C. Derrick, P. R. Madyastha, J. T. D. Melchers, G. L. Holtz, E. R. Baker, C. L. Smith, and H. H. Fudenberg. 1981. A new microassay for spermocytotoxic antibody: comparison with passive hemagglutination assay for antisperm antibodies in couples with unexplained fertility. *Journal of Immunology* **126:**905.

1204. Haas, G. G., Jr., R. Weiss-Wik, and D. P. Wolf. 1982. Identification of anti-sperm antibodies on sperm of infertile men. *Fertility and Sterility* **38:**54.

1205. Perricone, R., N. Pasetto, C. De Carolis, E. Vaquero, E. Piccione, L. Baschieri, and L. Fontana. 1992. Functionally active complement is present in human ovarian follicular fluid and can be activated by seminal plasma. *Clinical and Experimental Immunology* **89:**154.

1206. D' Cruz, O. J., G. G. Haas, Jr., and H. Lambert. 1990. Evaluation of anti-sperm complement-dependent immune mediators in human ovarian follicular fluid. *Journal of Immunology* **144:**3841.

1207. Bronson, R. A., G. W. Cooper, and D. L. Rosenfeld. 1982. Sperm-specific isoantibodies and autoantibodies inhibit the binding of human sperm to the human zona pellucida. *Fertility and Sterility* **38:**724.

1208. Bronson, R. A., G. W. Cooper, and D. L. Rosenfeld. 1983. Complement-mediated effects of sperm head-directed human antibodies on the ability of human spermatozoa to penetrate zona-free hamster eggs. *Fertility and Sterility* **40:**91.

1209. Aitken, R. J., J. M. Parslow, T. B. Hargreave, and W. F. Hendry. 1988. Influence of antisperm antibodies on human sperm function. *British Journal of Urology* **62:**367.

1210. D' Cruz, O. J., G. G. Haas, Jr., B. L. Wang, and L. E. DeBault. 1991. Activation of human complement by IgG antisperm antibody and the demonstration of C3 and C5b-9-mediated immune injury to human sperm. *Journal of Immunology* **146:**611.

1211. Petersen, B. H., C. J. Lammel, D. P. Stites, and G. F. Brooks. 1980. Human seminal plasma inhibition of complement. *Journal of Laboratory and Clinical Medicine* **96**:582.

1212. Brooks, G. F., C. J. Lammel, B. H. Petersen, and D. P. Stites. 1981. Human seminal plasma inhibition of antibody complement-mediated killing and opsonization of Neisseria gonorrhoeae and other gram-negative organisms. *Journal of Clinical Investigation* **67**:1523.

1213. Tarter, T. H., and N. J. Alexander. 1984. Complement-inhibiting activity of seminal plasma. *American Journal of Reproductive Immunology* **6**:28.

1214. O' Bryan, M. K., H. W. Baker, J. R. Saunders, L. Kirszbaum, I. D. Walker, P. Hudson, D. Y. Liu, M. D. Glew, A. J. d'Apice, and B. F. Murphy. 1990. Human seminal clusterin (SP-40,40). Isolation and characterization. *Journal of Clinical Investigation* **85**:1477.

1215. Bronson, R. A., and K. T. Preissner. 1997. Measurement of vitronectin content of human spermatozoa and vitronectin concentration within seminal fluid. *Fertility and Sterility* **68**:709.

1216. Griswold, M. D., K. Roberts, and P. Bishop. 1986. Purification and characterization of a sulfated glycoprotein secreted by Sertoli cells. *Biochemistry* **25**:7265.

1217. Ronquist, G., and I. Brody. 1985. The prostasome: its secretion and function in man. *Biochimica et Biophysica Acta* **822**:203.

1218. Rooney, I. A., J. E. Heuser, and J. P. Atkinson. 1996. GPI-anchored complement regulatory proteins in seminal plasma. An analysis of their physical condition and the mechanisms of their binding to exogenous cells. *Journal of Clinical Investigation* **97**:1675.

1219. Seya, T., T. Hara, M. Matsumoto, H. Kiyohara, I. Nakanishi, T. Kinouchi, M. Okabe, A. Shimizu, and H. Akedo. 1993. Membrane cofactor protein (MCP, CD46) in seminal plasma and on spermatozoa in normal and 'sterile' subjects. *European Journal of Immunology* **23**:1322.

1220. Bozas, S. E., L. Kirszbaum, R. L. Sparrow, and I. D. Walker. 1993. Several vascular complement inhibitors are present on human sperm. *Biology of Reproduction* **48**:503.

1221. Fenichel, P., F. Cervoni, P. Hofmann, M. Deckert, C. Emiliozzi, B. L. Hsi, and B. Rossi. 1994. Expression of the complement regulatory protein CD59 on human spermatozoa: characterization and role in gametic interaction. *Molecular Reproduction and Development* **38**:338.

1222. Taylor, C. T., and P. M. Johnson. 1996. Complement-binding proteins are strongly expressed by human preimplantation blastocysts and cumulus cells as well as gametes. *Molecular Human Reproduction* **2**:52.

1223. Schena, F. P., C. Manno, L. Selvaggi, G. Loverro, S. Bettocchi, and L. Bonomo. 1982. Behaviour of immune complexes and the complement system in normal pregnancy and pre-eclampsia. *Journal of Clinical and Laboratory Immunology* **7**:21.

1224. Sinha, D., M. Wells, and W. P. Faulk. 1984. Immunological studies of human placentae: complement components in pre-eclamptic chorionic villi. *Clinical and Experimental Immunology* **56**:175.

1225. Tedesco, F., O. Radillo, G. Candussi, A. Nazzaro, T. E. Mollnes, and D. Pecorari. 1990. Immunohistochemical detection of terminal complement complex and S protein in normal and pre-eclamptic placentae. *Clinical and Experimental Immunology* **80**:236.

1226. Fenichel, P., M. Donzeau, F. Cervoni, Y. Menezo, and B. L. Hsi. 1995. Expression of complement regulatory proteins on human eggs and pre-implantation embryos. *American Journal of Reproductive Immunology* **33**:155.

1227. Roberts, J. M., C. T. Taylor, G. C. Melling, C. R. Kingsland, and P. M. Johnson. 1992. Expression of the CD46 antigen, and absence of class I MHC antigen, on the human oocyte and preimplantation blastocyst. *Immunology* **75**:202.

1228. Oglesby, T. J., J. E. Longwith, and P. C. Huettner. 1996. Human complement regulator expression by the normal female reproductive tract. *Anatomical Record* **246**:78.

1229. Dohr, G. 1987. HLA and TLX antigen expression on the human oocyte, zona pellucida and granulosa cells. *Human Reproduction* **2**:657.

1230. Huffaker, J., S. S. Witkin, L. Cutler, M. L. Druzin, and W. J. Ledger. 1989. Total complement activity in maternal sera, amniotic fluids and cord sera in women with premature labor, premature rupture of membranes or chorioamnionitis. *Surgery, Gynecology and Obstetrics* **168**:397.

1231. Rooney, I. A., and B. P. Morgan. 1992. Characterization of the membrane attack complex inhibitory protein CD59 antigen on human amniotic cells and in amniotic fluid. *Immunology* **76**:541.

1232. Rooney, I. A., and B. P. Morgan. 1990. Protection of human amniotic epithelial cells (HAEC) from complement-mediated lysis: expression on the cells of three complement inhibitory membrane proteins. *Immunology* **71**:308.

1233. Tsujimura, A., K. Shida, M. Kitamura, M. Nomura, J. Takeda, H. Tanaka, M. Matsumoto, K. Matsumiya, A. Okuyama, Y. Nishimune, M. Okabe, and T. Seya. 1998. Molecular cloning of a murine homologue of membrane cofactor protein (CD46): Preferential expression in testicular germ cells. *Biochemical Journal* **330**:163.

1234. Hosokawa, M., M. Nonaka, N. Okada, M. Nonaka, and H. Okada. 1996. Molecular cloning of guinea pig membrane cofactor protein: preferential expression in testis. *Journal of Immunology* **157**:4946.

1235. Cooper, N. R. 1991. Complement evasion strategies of microorganisms. *Immunology Today* **12**:327.

1236. Smith, G. L. 1993. Vaccinia virus glycoproteins and immune evasion. The sixteenth Fleming lecture. *Journal of General Virology* **74**:1725.

1237. Fishelson, Z. 1994. Complement-related proteins in pathogenic organisms. *Springer Seminars in Immunopathology* **15**:345.

1238. Reeder, J. C., and G. V. Brown. 1996. Antigenic variation and immune evasion in Plasmodium falciparum malaria. *Immunology and Cell Biology* **74**:546.

1239. Davis-Poynter, N. J., and H. E. Farrell. 1996. Masters of deception: a review of herpesvirus immune evasion strategies. *Immunology and Cell Biology* **74**:513.

1240. Allen, J. E., and R. M. Maizels. 1996. Immunology of human helminth infection. *International Archives of Allergy and Immunology* **109**:3.

1241. Fishelson, Z. 1994. Complement-related proteins in pathogenic organisms. *Springer Seminars in Immunopathology* **15**:345.

1242. Jokiranta, T. S., L. Jokipii, and S. Meri. 1995. Complement resistance of parasites [editorial]. *Scandinavian Journal of Immunology* **42**:9.

1243. Lachmann, P. J., and A. Davies. 1997. Complement and immunity to viruses. *Immunological Reviews* **159**:69.

1244. Clarkson, N. A., R. Kaufman, D. M. Lublin, T. Ward, P. A. Pipkin, P. D. Minor, D. J. Evans, and J. W. Almond. 1995. Characterization of the echovirus 7 receptor: domains of CD55 critical for virus binding. *Journal of Virology* **69**:5497.

1245. Powell, R. M., T. Ward, D. J. Evans, and J. W. Almond. 1997. Interaction between echovirus 7 and its receptor, decay-accelerating factor (CD55): evidence for a secondary cellular factor in A-particle formation. *Journal of Virology* **71**:9306.

1246. Bergelson, J. M., J. A. Cunningham, G. Droguett, E. A. Kurt-Jones, A. Krithivas, J. S. Hong, M. S. Horwitz, R. L. Crowell, and R. W. Finberg. 1997. Isolation of a common receptor for coxsackie B viruses and adenoviruses 2 and 5. *Science* **275**:1320.

1247. Shafren, D. R., D. T. Williams, and R. D. Barry. 1997. A decay-accelerating factor-binding strain of coxsackievirus B3 requires the coxsackievirus-adenovirus receptor protein to mediate lytic infection of rhabdomyosarcoma cells. *Journal of Virology* **71**:9844.

1248. Karp, C. L., M. Wysocka, L. M. Wahl, J. M. Ahearn, P. J. Cuomo, B. Sherry, G. Trinchieri, and D. E. Griffin. 1996. Mechanism of suppression of cell-mediated immunity by measles virus [published erratum appears in *Science* 1997 Feb 21;275(5303):1053]. *Science* **273**:228.

1249. Schneider-Schaulies, J., L. M. Dunster, R. Schwartz-Albiez, G. Krohne, and V. ter Meulen. 1995. Physical association of moesin and CD46 as a receptor complex for measles virus. *Journal of Virology* **69**:2248.

1250. Devaux, P., and D. Gerlier. 1997. Antibody cross-reactivity with CD46 and lack of cell surface expression suggest that moesin might not mediate measles virus binding. *Journal of Virology* **71**:1679.

1251. Doi, Y., M. Kurita, M. Matsumoto, T. Kondo, T. Noda, S. Tsukita, S. Tsukita, and T. Seya. 1998. Moesin is not a receptor for measles virus entry into mouse embryonic stem cells. *Journal of Virology* **72**:1586.

1252. Hsu, E. C., F. Sarangi, C. Iorio, M. S. Sidhu, S. A. Udem, D. L. Dillehay, W. Xu, P. A. Rota, W. J. Bellini, and C. D. Richardson. 1998. A single amino acid change in the hemagglutinin protein of measles virus determines its ability to bind CD46 and reveals another receptor on marmoset B cells. *Journal of Virology* **72**:2905.

1253. Bartz, R., R. Firsching, B. Rima, V. Ter Meulen, and J. Schneider-Schaulies. 1998. Differential receptor usage by measles virus strains. *Journal of General Virology* **79**:1015.

1254. Tanaka, K., M. Xie, and Y. Yanagi. 1998. The hemagglutinin of recent measles virus isolates induces cell fusion in a marmoset cell line, but not in other CD46-positive human and monkey cell lines, when expressed together with the F protein. *Archives of Virology* **143**:213.

1255. Hsu, E. C., R. E. Dorig, F. Sarangi, A. Marcil, C. Iorio, and C. D. Richardson. 1997. Artificial mutations and natural variations in the CD46 molecules from human and monkey cells define regions important for measles virus binding. *Journal of Virology* **71**:6144.

1256. Manchester, M., M. K. Liszewski, J. P. Atkinson, and M. B. Oldstone. 1994. Multiple isoforms of CD46 (membrane cofactor protein) serve as receptors for measles virus. *Proceedings of the National Academy of Sciences of the United States of America* **91**:2161.

1257. Kurita, M., Y. Yanagi, T. Hara, S. Nagasawa, M. Matsumoto, and T. Seya. 1995. Human lymphocytes are more susceptible to measles virus than granulocytes, which is attributable to the phenotypic differences of their membrane cofactor protein (CD46). *Immunology Letters* **48**:91.

1258. Buchholz, C. J., U. Schneider, P. Devaux, D. Gerlier, and R. Cattaneo. 1996. Cell entry by measles virus: long hybrid receptors uncouple binding from membrane fusion. *Journal of Virology* **70**:3716.

1259. Seya, T., M. Kurita, K. Iwata, Y. Yanagi, K. Tanaka, K. Shida, M. Hatanaka, M. Matsumoto, S. Jun, A. Hirano, S. Ueda, and S. Nagasawa. 1997. The CD46 transmembrane domain is required for efficient formation of measles-virus-mediated syncytium. *Biochemical Journal* **322**:135.

1260. Hirano, A., S. Yant, K. Iwata, J. Korte-Sarfaty, T. Seya, S. Nagasawa, and T. C. Wong. 1996. Human cell receptor CD46 is down regulated through recognition of a membrane-proximal region of the cytoplasmic domain in persistent measles virus infection. *Journal of Virology* **70**:6929.

1261. Yant, S., A. Hirano, and T. C. Wong. 1997. Identification of a cytoplasmic Tyr-X-X-Leu motif essential for down regulation of the human cell receptor CD46 in persistent measles virus infection. *Journal of Virology* **71**:766.

1262. Nemerow, G. R., R. A. Houghten, M. D. Moore, and N. R. Cooper. 1989. Identification of an epitope in the major envelope protein of Epstein–Barr virus that mediates viral binding to the lymphocyte-B EBV receptor (CR-2). *Cell* **56**:369.

1263. Spear, G. T., B. L. Sullivan, A. L. Landay, and T. F. Lint. 1990. Neutralization of human immunodeficiency virus type 1 by complement occurs by viral lysis. *Journal of Virology* **64**:5869.

1264. Spear, G. T., D. M. Takefman, B. L. Sullivan, A. L. Landay, and S. Zolla-Pazner. 1993. Complement activation by human monoclonal antibodies to human immunodeficiency virus. *Journal of Virology* **67**:53.

1265. Sullivan, B. L., E. J. Knopoff, M. Saifuddin, D. M. Takefman, M. N. Saarloos, B. E. Sha, and G. T. Spear. 1996. Susceptibility of HIV-1 plasma virus to complement-mediated lysis. Evidence for a role in clearance of virus in vivo. *Journal of Immunology* **157**:1791.

1266. Dierich, M. P., C. F. Ebenbichler, P. Marschang, G. Fust, N. M. Thielens, and G. J. Arlaud. 1993. HIV and human complement: mechanisms of interaction and biological implication. *Immunology Today* **14**:435.

1267. Speth, C., L. Kacani, and M. P. Dierich. 1997. Complement receptors in HIV infection. *Immunological Reviews* **159**:49.

1268. Montefiori, D. C. 1997. Role of complement and Fc receptors in the pathogenesis of HIV-1 infection. *Springer Seminars in Immunopathology* **18**:371.

1269. Ebenbichler, C. F., N. M. Thielens, R. Vornhagen, P. Marschang, G. J. Arlaud, and M. P. Dierich. 1991. Human immunodeficiency virus type 1 activates the classical pathway of complement by direct C1 binding through specific sites in the transmembrane glycoprotein gp41. *Journal of Experimental Medicine* **174**:1417.

1270. Thielens, N. M., I. M. Bally, C. F. Ebenbichler, M. P. Dierich, and G. J. Arlaud. 1993. Interaction of C1 with HIV-1. *Behring Institute Mitteilungen* **93**:165.

1271. Thielens, N. M., I. M. Bally, C. F. Ebenbichler, M. P. Dierich, and G. J. Arlaud. 1993. Further characterization of the interaction between the C1q subcomponent of human C1 and the transmembrane envelope glycoprotein gp41 of HIV-1. *Journal of Immunology* **151**:6583.

1272. Marschang, P., L. Gurtler, M. Totsch, N. M. Thielens, G. J. Arlaud, A. Hittmair, H. Katinger, and M. P. Dierich. 1993. HIV-1 and HIV-2 isolates differ in their ability to activate the complement system on the surface of infected cells. *AIDS* **7**:903.

1273. Haurum, J. S., S. Thiel, I. M. Jones, P. B. Fischer, S. B. Laursen, and J. C. Jensenius. 1993. Complement activation upon binding of mannan-binding protein to HIV envelope glycoproteins. *AIDS* **7**:1307.

1274. Susal, C., M. Kirschfink, M. Kropelin, V. Daniel, and G. Opelz. 1994. Complement activation by recombinant HIV-1 glycoprotein gp120. *Journal of Immunology* **152**:6028.

1275. Tremblay, M., S. Meloche, R. P. Sekaly, and M. A. Wainberg. 1990. Complement receptor 2 mediates enhancement of human immunodeficiency virus 1 infection in Epstein–Barr virus-carrying B cells. *Journal of Experimental Medicine* **171**:1791.

1276. Boyer, V., C. Delibrias, N. Noraz, E. Fischer, M. D. Kazatchkine, and C. Desgranges. 1992. Complement receptor type 2 mediates infection of the human CD4-negative Raji B-cell line with opsonized HIV. *Scandinavian Journal of Immunology* **36**:879.

1277. Cheng-Mayer, C., J. T. Rutka, M. L. Rosenblum, T. McHugh, D. P. Stites, and J. A. Levy. 1987. Human immunodeficiency virus can productively infect cultured human glial cells. *Proceedings of the National Academy of Sciences of the United States of America* **84**:3526.

1278. Folks, T. M., S. W. Kessler, J. M. Orenstein, J. S. Justement, E. S. Jaffe, and A. S. Fauci. 1988. Infection and replication of HIV-1 in purified progenitor cells of normal human bone marrow. *Science* **242**:919.

1279. Reisinger, E. C., W. Vogetseder, D. Berzow, D. Kofler, G. Bitterlich, H. A. Lehr, H. Wachter, and M. P. Dierich. 1990. Complement-mediated enhancement of HIV-1 infection of the monoblastoid cell line U937. *AIDS* **4**:961.

1280. Boyer, V., C. Desgranges, M. A. Trabaud, E. Fischer, and M. D. Kazatchkine. 1991. Complement mediates human immunodeficiency virus type 1 infection of a human T cell line in a CD4- and antibody-independent fashion. *Journal of Experimental Medicine* **173**:1151.

1281. Gras, G. S., and D. Dormont. 1991. Antibody-dependent and antibody-independent complement-mediated enhancement of human immunodeficiency virus type 1 infection in a human, Epstein–Barr virus-transformed B-lymphocytic cell line. *Journal of Virology* **65**:541.

1282. June, R. A., S. Z. Schade, M. J. Bankowski, M. Kuhns, A. McNamara, T. F. Lint, A. L. Landay, and G. T. Spear. 1991. Complement and antibody mediate enhancement of HIV infection by increasing virus binding and provirus formation. *AIDS* **5**:269.

1283. Joling, P., L. J. Bakker, J. A. Van Strijp, T. Meerloo, L. de Graaf, M. E. Dekker, J. Goudsmit, J. Verhoef, and H. J. Schuurman. 1993. Binding of human immunodeficiency virus type-1 to follicular dendritic cells in vitro is complement dependent. *Journal of Immunology* **150**:1065.

1284. Schmitz, J., J. van Lunzen, K. Tenner-Racz, G. Grossschupff, P. Racz, H. Schmitz, M. Dietrich, and F. T. Hufert. 1994. Follicular dendritic cells retain HIV-1 particles on their plasma membrane, but are not productively infected in asymptomatic patients with follicular hyperplasia. *Journal of Immunology* **153**:1352.

1285. Pantaleo, G., C. Graziosi, L. Butini, P. A. Pizzo, S. M. Schnittman, D. P. Kotler, and A. S. Fauci. 1991. Lymphoid organs function as major reservoirs for human immunodeficiency virus. *Proceedings of the National Academy of Sciences of the United States of America* **88**:9838.

1286. Pantaleo, G., C. Graziosi, J. F. Demarest, L. Butini, M. Montroni, C. H. Fox, J. M. Orenstein, D. P. Kotler, and A. S. Fauci. 1993. HIV infection is active and progressive in lymphoid tissue during the clinically latent stage of disease. *Nature* **362**:355.

1287. Montefiori, D. C., R. J. Cornell, J. Y. Zhou, J. T. Zhou, V. M. Hirsch, and P. R. Johnson. 1994. Complement control proteins, CD46, CD55, and CD59, as common surface constituents of human and simian immunodeficiency viruses and possible targets for vaccine protection. *Virology* **205**:82.

1288. Saifuddin, M., C. J. Parker, M. E. Peeples, M. K. Gorny, S. Zolla-Pazner, M. Ghassemi, I. A. Rooney, J. P. Atkinson, and G. T. Spear. 1995. Role of virion-associated glycosylphosphatidylinositol-linked proteins CD55 and CD59 in complement resistance of cell line-derived and primary isolates of HIV-1. *Journal of Experimental Medicine* **182**:501.

1289. Spear, G. T., N. S. Lurain, C. J. Parker, M. Ghassemi, G. H. Payne, and M. Saifuddin. 1995. Host cell-derived complement control proteins CD55 and CD59 are incorporated into the virions of two unrelated enveloped viruses. Human T cell leukemia/lymphoma virus type I (HTLV-I) and human cytomegalovirus (HCMV). *Journal of Immunology* **155**:4376.

1290. Spiller, O. B., B. P. Morgan, F. Tufaro, and D. V. Devine. 1996. Altered expression of host-encoded complement regulators on human cytomegalovirus-infected cells. *European Journal of Immunology* **26**:1532.

1291. Spiller, O. B., S. M. Hanna, D. V. Devine, and F. Tufaro. 1997. Neutralization of cytomegalovirus virions: the role of complement. *Journal of Infectious Diseases* **176**:339.

1292. Edwards, M. S., D. L. Kasper, H. J. Jennings, C. J. Baker, and A. Nicholson-Weller. 1982. Capsular sialic-acid prevents activation of the alternative complement pathway by type-III, group-B streptococci. *Journal of Immunology* **128**:1278.

1293. Tomlinson, S., L. Pontes De Carvalho, F. Vandekerckhove, and V. Nussenzweig. 1992. Resialylation of sialidase-treated sheep and human erythrocytes by Trypanosoma cruzi trans-sialidase: restoration of complement resistance of desialylated sheep erythrocytes. *Glycobiology* **2**:549.

1294. Tomlinson, S., L. C. Pontes de Carvalho, F. Vandekerckhove, and V. Nussenzweig. 1994. Role of sialic acid in the resistance of Trypanosoma cruzi trypomastigotes to complement. *Journal of Immunology* **153**:3141.

1295. Ram, S., A. K. Sharma, S. D. Simpson, S. Gulati, D. P. McQuillen, M. K. Pangburn, and P. A. Rice. 1998. A novel sialic acid binding site on factor H mediates serum resistance of sialylated Neisseria gonorrhoeae. *Journal of Experimental Medicine* **187**:743.

1296. Shchelkunov, S. N., V. M. Blinov, A. V. Totmenin, S. S. Marennikova, A. A. Kolykhalov, I. V. Frolov, V. E. Chizhikov, V. V. Gytorov, P. V. Gashikov, E. F. Belanov, and *et al.* 1993. Nucleotide sequence analysis of variola virus HindIII M, L, I genome fragments. *Virus Research* **27**:25.

1297. Kotwal, G. J., and B. Moss. 1988. Vaccinia virus encodes a secretory polypeptide structurally related to complement control proteins. *Nature* **335**:176.

1298. Kotwal, G. J., S. N. Isaacs, R. McKenzie, M. M. Frank, and B. Moss. 1990. Inhibition of the complement cascade by the major secretory protein of vaccinia virus. *Science* **250**:827.

1299. McKenzie, R., G. J. Kotwal, B. Moss, C. H. Hammer, and M. M. Frank. 1992. Regulation of complement activity by vaccinia virus complement-control protein. *Journal of Infectious Diseases* **166**:1245.

1300. Sahu, A., S. N. Isaacs, A. M. Soulika, and J. D. Lambris. 1998. Interaction of vaccinia virus complement control protein with human complement proteins: factor I-mediated degradation of C3b to iC3b(1) inactivates the alternative complement pathway. *Journal of Immunology* **160**:5596.

1301. Isaacs, S. N., G. J. Kotwal, and B. Moss. 1992. Vaccinia virus complement-control protein prevents antibody-dependent complement-enhanced neutralization of infectivity and contributes to virulence. *Proceedings of the National Academy of Sciences of the United States of America* **89**:628.

1302. Engelstad, M., S. T. Howard, and G. L. Smith. 1992. A constitutively expressed vaccinia gene encodes a 42-kDa glycoprotein related to complement control factors that forms part of the extracellular virus envelope. *Virology* **188**:801.

1303. Isaacs, S. N., E. J. Wolffe, L. G. Payne, and B. Moss. 1992. Characterization of a vaccinia virus-encoded 42-kilodalton Class I membrane glycoprotein component of the extracellular virus envelope. *Journal of Virology* **66**:7217.

1304. Miller, C. G., S. N. Shchelkunov, and G. J. Kotwal. 1997. The cowpox virus-encoded homolog of the vaccinia virus complement control protein is an inflammation modulatory protein. *Virology* **229**:126.

1305. Massung, R. F., J. J. Esposito, L. I. Liu, J. Qi, T. R. Utterback, J. C. Knight, L. Aubin, T. E. Yuran, J. M. Parsons, V. N. Loparev, and et al. 1993. Potential virulence determinants in terminal regions of variola smallpox virus genome. *Nature* **366**:748.

1306. Albrecht, J. C., J. Nicholas, D. Biller, K. R. Cameron, B. Biesinger, C.

Newman, S. Wittmann, M. A. Craxton, H. Coleman, B. Fleckenstein, et al. 1992. Primary structure of the herpesvirus saimiri genome. *Journal of Virology* **66:**5047.

1307. Albrecht, J. C., J. Nicholas, K. R. Cameron, C. Newman, B. Fleckenstein, and R. W. Honess. 1992. Herpesvirus saimiri has a gene specifying a homologue of the cellular membrane glycoprotein CD59. *Virology* **190:**527.

1308. Rother, R. P., S. A. Rollins, W. L. Fodor, J. C. Albrecht, E. Setter, B. Fleckenstein, and S. P. Squinto. 1994. Inhibition of complement-mediated cytolysis by the terminal complement inhibitor of herpesvirus saimiri. *Journal of Virology* **68:**730.

1309. Albrecht, J. C., and B. Fleckenstein. 1992. New member of the multigene family of complement control proteins in herpesvirus saimiri. *Journal of Virology* **66:**3937.

1310. Randall, R. E., C. Newman, and R. W. Honess. 1984. Isolation and characterization of monoclonal antibodies to structural and nonstructural herpesvirus saimiri proteins. *Journal of Virology* **52:**872.

1311. Fodor, W. L., S. A. Rollins, S. Bianco-Caron, R. P. Rother, E. R. Guilmette, W. V. Burton, J. C. Albrecht, B. Fleckenstein, and S. P. Squinto. 1995. The complement control protein homolog of herpesvirus saimiri regulates serum complement by inhibiting C3 convertase activity. *Journal of Virology* **69:**3889.

1312. Cines, D. B., A. P. Lyss, M. Bina, R. Corkey, N. A. Kefalides, and H. M. Friedman. 1982. Fc and C3 receptors induced by herpes simplex virus on cultured human endothelial cells. *Journal of Clinical Investigation* **69:**123.

1313. Friedman, H. M., G. H. Cohen, R. J. Eisenberg, C. A. Seidel, and D. B. Cines. 1984. Glycoprotein C of herpes simplex virus 1 acts as a receptor for the C3b complement component on infected cells. *Nature* **309:**633.

1314. McNearney, T. A., C. Odell, V. M. Holers, P. G. Spear, and J. P. Atkinson. 1987. Herpes simplex virus glycoproteins gC-1 and gC-2 bind to the third component of complement and provide protection against complement-mediated neutralization of viral infectivity. *Journal of Experimental Medicine* **166:**1525.

1315. Fries, L. F., H. M. Friedman, G. H. Cohen, R. J. Eisenberg, C. H. Hammer, and M. M. Frank. 1986. Glycoprotein C of herpes simplex virus 1 is an inhibitor of the complement cascade. *Journal of Immunology* **137:**1636.

1316. Eisenberg, R. J., M. Ponce de Leon, H. M. Friedman, L. F. Fries, M. M. Frank, J. C. Hastings, and G. H. Cohen. 1987. Complement component C3b binds directly to purified glycoprotein C of herpes simplex virus types 1 and 2. *Microbial Pathogenesis* **3:**423.

1317. Harris, S. L., I. Frank, A. Yee, G. H. Cohen, R. J. Eisenberg, and H. M. Friedman. 1990. Glycoprotein C of herpes simplex virus type 1 prevents complement-mediated cell lysis and virus neutralization. *Journal of Infectious Diseases* **162:**331.

1318. Hidaka, Y., Y. Sakai, Y. Toh, and R. Mori. 1991. Glycoprotein C of herpes simplex virus type 1 is essential for the virus to evade antibody-independent complement-mediated virus inactivation and lysis of virus-infected cells. *Journal of General Virology* **72:**915.

1319. Gerber, S. I., B. J. Belval, and B. C. Herold. 1995. Differences in the role of glycoprotein C of HSV-1 and HSV-2 in viral binding may contribute to serotype differences in cell tropism. *Virology* **214:**29.

1320. Hung, S. L., C. Peng, I. Kostavasili, H. M. Friedman, J. D. Lambris, R. J. Eisenberg, and G. H. Cohen. 1994. The interaction of glycoprotein C of herpes simplex virus types 1 and 2 with the alternative complement pathway. *Virology* **203:**299.

1321. Hung, S. L., S. Srinivasan, H. M. Friedman, R. J. Eisenberg, and G. H. Cohen. 1992. Structural basis of C3b binding by glycoprotein C of herpes simplex virus. *Journal of Virology* **66:**4013.

1322. Kostavasil, I., A. Sahu, H. M. Friedman, R. J. Eisenberg, G. H. Cohen, and J. D. Lambris. 1997. Mechanism of complement inactivation by glycoprotein C of herpes simplex virus. *Journal of Immunology* **158:**1763.

1323. Seidel-Dugan, C., M. Ponce de Leon, H. M. Friedman, L. F. Fries, M. M. Frank, G. H. Cohen, and R. J. Eisenberg. 1988. C3b receptor activity on transfected cells expressing glycoprotein C of herpes simplex virus types 1 and 2. *Journal of Virology* **62:**4027.

1324. Tal-Singer, R., C. Seidel-Dugan, L. Fries, H. P. Huemer, R. J. Eisenberg, G. H. Cohen, and H. M. Friedman. 1991. Herpes simplex virus glycoprotein C is a receptor for complement component iC3b. *Journal of Infectious Diseases* **164:**750.

1325. Kubota, Y., T. A. Gaither, J. Cason, J. J. O' Shea, and T. J. Lawley. 1987. Characterization of the C3 receptor induced by herpes simplex virus type 1 infection of human epidermal, endothelial, and A431 cells. *Journal of Immunology* **138:**1137.

1326. Mold, C., B. M. Bradt, G. R. Nemerow, and N. R. Cooper. 1988. Epstein–Barr virus regulates activation and processing of the third component of complement. *Journal of Experimental Medicine* **168:**949.

1327. Rimoldi, M. T., A. Sher, S. Heiny, A. Lituchy, C. H. Hammer, and K. Joiner. 1988. Developmentally regulated expression by Trypanosoma cruzi of molecules that accelerate the decay of complement C3 convertases. *Proceedings of the National Academy of Sciences of the United States of America* **85:**193.

1328. Joiner, K. A., W. D. daSilva, M. T. Rimoldi, C. H. Hammer, A. Sher, and T. L. Kipnis. 1988. Biochemical characterization of a factor produced by trypomastigotes of Trypanosoma cruzi that accelerates the decay of complement C3 convertases. *Journal of Biological Chemistry* **263:**11327.

1329. Tambourgi, D. V., T. L. Kipnis, W. D. da Silva, K. A. Joiner, A. Sher, S. Heath, B. F. Hall, and G. B. Ogden. 1993. A partial cDNA clone of trypomastigote decay-accelerating factor (T-DAF), a developmentally regulated complement inhibitor of Trypanosoma cruzi, has genetic and functional similarities to the human complement inhibitor DAF. *Infection and Immunity* **61:**3656.

1330. Norris, K. A., B. Bradt, N. R. Cooper, and M. So. 1991. Characterization of a Trypanosoma cruzi C3 binding protein with functional and genetic similarities to the human complement regulatory protein, decay-accelerating factor. *Journal of Immunology* **147:**2240.

1331. Norris, K. A., and J. E. Schrimpf. 1994. Biochemical analysis of the membrane and soluble forms of the complement regulatory protein of Trypanosoma cruzi. *Infection and Immunity* **62:**236.

1332. Norris, K. A. 1998. Stable transfection of Trypanosoma cruzi epimastigotes with the trypomastigote-specific complement regulatory protein cDNA confers complement resistance. *Infection and Immunity* **66:**2460.

1333. Van Voorhis, W. C., L. Barrett, R. Koelling, and A. G. Farr. 1993. FL-160 proteins of Trypanosoma cruzi are expressed from a multigene family and contain two distinct epitopes that mimic nervous tissues. *Journal of Experimental Medicine* **178:**681.

1334. Norris, K. A., J. E. Schrimpf, and M. J. Szabo. 1997. Identification of the gene family encoding the 160-kilodalton Trypanosoma cruzi complement regulatory protein. *Infection and Immunity* **65:**349.

1335. Fischer, E., M. A. Ouaissi, P. Velge, J. Cornette, and M. D. Kazatchkine. 1988. gp 58/68, a parasite component that contributes to the escape of the trypomastigote form of T. cruzi from damage by the human alternative complement pathway. *Immunology* **65:**299.

1336. Iida, K., M. B. Whitlow, and V. Nussenzweig. 1989. Amastigotes of Trypanosoma cruzi escape destruction by the terminal complement components. *Journal of Experimental Medicine* **169:**881.

1337. Marikovsky, M., M. Parizade, R. Arnon, and Z. Fishelson. 1990. Complement regulation on the surface of cultured schistosomula and adult worms of Schistosoma mansoni. *European Journal of Immunology* **20:**221.

1338. Parizade, M., R. Arnon, P. J. Lachmann, and Z. Fishelson. 1994. Functional and antigenic similarities between a 94-kD protein of Schistosoma mansoni (SCIP-1) and human CD59. *Journal of Experimental Medicine* **179:**1625.

1339. Pearce, E. J., B. F. Hall, and A. Sher. 1990. Host-specific evasion of the alternative complement pathway by schistosomes correlates with the presence of a phospholipase C-sensitive surface molecule resembling human decay accelerating factor. *Journal of Immunology* **144:**2751.

1340. Horta, M. F., F. J. Ramalho-Pinto, and M. F. Horta. 1991. Role of human decay-accelerating factor in the evasion of Schistosoma mansoni from the complement-mediated killing in vitro. *Journal of Experimental Medicine* **174:**1399.

1341. Braga, L. L., H. Ninomiya, J. J. McCoy, S. Eacker, T. Wiedmer, C. Pham, S. Wood, P. Sims, and W. A. Petri. 1992. Inhibition of complement membrane attack complex by the galactose-specific adhesin of *Entamoeba histolytica. Journal of Clinical Investigation* **90:**1131.

1342. Wexler, D. E., R. D. Nelson, and P. P. Cleary. 1983. Human neutrophil chemotactic response to group A streptococci: bacteria-mediated interference with complement-derived chemotactic factors. *Infection and Immunity* **39:**239.

1343. Wexler, D. E., and P. P. Cleary. 1985. Purification and characteriztics of the streptococcal chemotactic factor inactivator. *Infection and Immunity* **50:**757.

1344. Wexler, D. E., D. E. Chenoweth, and P. P. Cleary. 1985. Mechanism of action of the group A streptococcal C5a inactivator. *Proceedings of the National Academy of Sciences of the United States of America* **82**:8144.

1345. O' Connor, S. P., and P. P. Cleary. 1986. Localization of the streptococcal C5a peptidase to the surface of group A streptococci. *Infection and Immunity* **53**:432.

1346. Chen, C. C., and P. P. Cleary. 1990. Complete nucleotide sequence of the streptococcal C5a peptidase gene of *Streptococcus pyogenes*. *Journal of Biological Chemistry* **265**:3161.

1347. Hill, H. R., J. F. Bohnsack, E. Z. Morris, N. H. Augustine, C. J. Parker, P. P. Cleary, and J. T. Wu. 1988. Group B streptococci inhibit the chemotactic activity of the fifth component of complement. *Journal of Immunology* **141**:3551.

1348. Bohnsack, J. F., K. W. Mollison, A. M. Buko, J. C. Ashworth, and H. R. Hill. 1991. Group B streptococci inactivate complement component C5a by enzymic cleavage at the C-terminus. *Biochemical Journal* **273**:635.

1349. Bohnsack, J. F., X. N. Zhou, P. A. Williams, P. P. Cleary, C. J. Parker, and H. R. Hill. 1991. Purification of the proteinase from group B streptococci that inactivates human C5a. *Biochimica et Biophysica Acta* **1079**:222.

1350. Bohnsack, J. F., X. N. Zhou, J. N. Gustin, C. E. Rubens, C. J. Parker, and H. R. Hill. 1992. Bacterial evasion of the antibody response: human IgG antibodies neutralize soluble but not bacteria-associated group B streptococcal C5a-ase. *Journal of Infectious Diseases* **165**:315.

1351. Cleary, P. P., J. Handley, A. N. Suvorov, A. Podbielski, and P. Ferrieri. 1992. Similarity between the group B and A streptococcal C5a peptidase genes. *Infection and Immunity* **60**:4239.

1352. Chmouryguina, I., A. Suvorov, P. Ferrieri, and P. P. Cleary. 1996. Conservation of the C5a peptidase genes in group A and B streptococci. *Infection and Immunity* **64**:2387.

1353. Oda, T., Y. Kojima, T. Akaike, S. Ijiri, A. Molla, and H. Maeda. 1990. Inactivation of chemotactic activity of C5a by the serratial 56-kilodalton protease. *Infection and Immunity* **58**:1269.

1354. Okada, N., H. Tanaka, H. Takizawa, and H. Okada. 1995. A monoclonal antibody that blocks the complement regulatory activity of guinea pig erythrocytes and characterization of the antigen involved as guinea pig decay-accelerating factor. *Journal of Immunology* **154**:6103.

1355. Sugita, Y., M. Uzawa, and M. Tomita. 1987. Isolation of decay-accelerating factor (DAF) from rabbit erythrocyte membranes. *Journal of Immunological Methods* **104**:123.

1356. Kameyoshi, Y., M. Matsushita, and H. Okada. 1989. Murine membrane inhibitor of complement which accelerates decay of human C3 convertase. *Immunology* **68**:439.

1357. Nickells, M. W., and J. P. Atkinson. 1990. Characterization of CR1- and membrane cofactor protein-like proteins of two primates. *Journal of Immunology* **144**:4262.

1358. Nonaka, M., T. Miwa, N. Okada, M. Nonaka, and H. Okada. 1995. Multiple isoforms of guinea pig decay-accelerating factor (DAF) generated by alternative splicing. *Journal of Immunology* **155**:3037.

1359. Wang, G., M. Nonaka, C. He, N. Okada, I. Nakashima, and H. Okada. 1998. Functional differences among multiple isoforms of guinea pig decay-accelerating factor. *Journal of Immunology* **160**:3014.

1360. Spicer, A. P., M. F. Seldin, and S. J. Gendler. 1995. Molecular cloning and chromosomal localization of the mouse decay-accelerating factor genes. Duplicated genes encode glycosylphosphatidylinositol-anchored and transmembrane forms. *Journal of Immunology* **155**:3079.

1361. Fukuoka, Y., A. Yasui, N. Okada, and H. Okada. 1996. Molecular cloning of murine decay accelerating factor by immunoscreening. *International Immunology* **8**:379.

1362. Song, W. C., C. Deng, K. Raszmann, R. Moore, R. Newbold, J. A. McLachlan, and M. Negishi. 1996. Mouse decay-accelerating factor: selective and tissue-specific induction by estrogen of the gene encoding the glycosylphosphatidylinositol-anchored form. *Journal of Immunology* **157**:4166.

1363. Manthei, U., M. W. Nickells, S. H. Barnes, L. L. Ballard, W. Y. Cui, and J. P. Atkinson. 1988. Identification of a C3b/iC3 binding protein of rabbit platelets and leukocytes. A CR1-like candidate for the immune adherence receptor. *Journal of Immunology* **140**:1228.

1364. Seya, T., M. Okada, K. Hazeki, and S. Nagasawa. 1990. Regulatory system of guinea-pig complement C3b: two factor I-cofactor proteins on guinea-pig peritoneal granulocytes. *Biochemical and Biophysical Research Communications* **170**:504.

1365. van den Berg, C. W., J. M. Perez de la Lastra, D. Llanes, and B. P. Morgan. 1997. Purification and characterization of the pig analogue of human membrane cofactor protein (CD46/MCP). *Journal of Immunology* **158**:1703.

1366. Toyomura, K., T. Fujimura, H. Murakami, T. Natsume, T. Shigehisa, N. Inoue, J. Takeda, and T. Kinoshita. 1997. Molecular cloning of a pig homologue of membrane cofactor protein (CD46). *International Immunology* **9**:869.

1367. Nickells, M. W., V. B. Subramanian, L. Clemenza, and J. P. Atkinson. 1995. Identification of complement receptor type 1-related proteins on primate erythrocytes. *Journal of Immunology* **154**:2829.

1368. Birmingham, D. J., X. P. Shen, D. Hourcade, M. W. Nickells, and J. P. Atkinson. 1994. Primary sequence of an alternatively spliced form of CR1. Candidate for the 75,000 M(r) complement receptor expressed on chimpanzee erythrocytes. *Journal of Immunology* **153**:691.

1369. Birmingham, D. J., C. M. Logar, X. P. Shen, and W. Chen. 1996. The baboon erythrocyte complement receptor is a glycophosphatidylinositol-linked protein encoded by a homologue of the human CR1-like genetic element. *Journal of Immunology* **157**:2586.

1370. Arnaiz-Villena, A., and P. Sheldon. 1975. Autoimmune New Zealand mouse. Alterations in C3 receptors, complement-induced clumps and follicular adherence of immune complexes. *Immunology* **29**:1103.

1371. Rabellino, E. M., G. D. Ross, H. T. Trang, N. Williams, and D. Metcalf. 1978. Membrane receptors of mouse leukocytes. II. Sequential expression of membrane receptors and phagocytic capacity during leukocyte differentiation. *Journal of Experimental Medicine* **147**:434.

1372. Edberg, J. C., L. Tosic, and R. P. Taylor. 1989. Immune adherence and the processing of soluble complement-fixing antibody/DNA immune complexes in mice. *Clinical Immunology and Immunopathology* **51**:118.

1373. Kinoshita, T., S. Lavoie, and V. Nussenzweig. 1985. Regulatory proteins for the activated third and fourth components of complement (C3b and C4b) in mice. II. Identification and properties of complement receptor type 1 (CR1). *Journal of Immunology* **134**:2564.

1374. Holers, V. M., T. Kinoshita, and H. Molina. 1992. The evolution of mouse and human complement C3-binding proteins: divergence of form but conservation of function. *Immunology Today* **13**:231.

1375. Molina, H., T. Kinoshita, K. Inoue, J. C. Carel, and V. M. Holers. 1990. A molecular and immunochemical characterization of mouse CR2. Evidence for a single gene model of mouse complement receptors 1 and 2. *Journal of Immunology* **145**:2974.

1376. Molina, H., W. Wong, T. Kinoshita, C. Brenner, S. Foley, and V. M. Holers. 1992. Distinct receptor and regulatory properties of recombinant mouse complement receptor 1 (CR1) and Crry, the two genetic homologues of human CR1. *Journal of Experimental Medicine* **175**:121.

1377. Kalli, K. R., and D. T. Fearon. 1994. Binding of C3b and C4b by the CR1-like site in murine CR1. *Journal of Immunology* **152**:2899.

1378. Molina, H., T. Kinoshita, C. B. Webster, and V. M. Holers. 1994. Analysis of C3b/C3d binding sites and factor I cofactor regions within mouse complement receptors 1 and 2. *Journal of Immunology* **153**:789.

1379. Kozono, Y., R. C. Duke, M. S. Schleicher, and V. M. Holers. 1995. Co-ligation of mouse complement receptors 1 and 2 with surface IgM rescues splenic B cells and WEHI-231 cells from anti-surface IgM-induced apoptosis. *European Journal of Immunology* **25**:1013.

1380. Molina, H., V. M. Holers, B. Li, Y. Fung, S. Mariathasan, J. Goellner, J. Strauss-Schoenberger, R. W. Karr, and D. D. Chaplin. 1996. Markedly impaired humoral immune response in mice deficient in complement receptors 1 and 2. *Proceedings of the National Academy of Sciences of the United States of America* **93**:3357.

1381. Takahashi, K., Y. Kozono, T. J. Waldschmidt, D. Berthiaume, R. J. Quigg, A. Baron, and V. M. Holers. 1997. Mouse complement receptors type 1 (CR1;CD35) and type 2 (CR2;CD21): expression on normal B cell subpopulations and decreased levels during the development of autoimmunity in MRL/lpr mice. *Journal of Immunology* **159**:1557.

1382. Quigg, R. J., and V. M. Holers. 1995. Characterization of rat complement receptors and regulatory proteins. CR2 and Crry are conserved, and the C3b receptor of neutrophils and platelets is distinct from CR1. *Journal of Immunology* **155**:1481.

1383. Quigg, R. J., M. L. Galishoff, A. E. d. Sneed, and D. Kim. 1993. Isolation and characterization of complement receptor type 1 from rat glomerular epithelial cells. *Kidney International* **43**:730.

1384. Quigg, R. J., J. J. Alexander, C. F. Lo, A. Lim, C. He, and V. M. Holers. 1997. Characterization of C3-binding proteins on mouse neutrophils and platelets. *Journal of Immunology* **159**:2438.

1385. Wong, W. W., and D. T. Fearon. 1985. p65: a C3b-binding protein on murine cells that shares antigenic determinants with the human C3b receptor (Cr1) and is distinct from murine C3b receptor. *Journal of Immunology* **134**:4048.

1386. Aegerter-Shaw, M., J. L. Cole, L. B. Klickstein, W. W. Wong, D. T. Fearon, P. A. Lalley, and J. H. Weis. 1987. Expansion of the complement receptor gene family. Identification in the mouse of two new genes related to the CR1 and CR2 gene family. *Journal of Immunology* **138**:3488.

1387. Li, B., C. Sallee, M. Dehoff, S. Foley, H. Molina, and V. M. Holers. 1993. Mouse Crry/p65. Characterization of monoclonal antibodies and the tissue distribution of a functional homologue of human MCP and DAF. *Journal of Immunology* **151**:4295.

1388. Foley, S., B. Li, M. Dehoff, H. Molina, and V. M. Holers. 1993. Mouse Crry/p65 is a regulator of the alternative pathway of complement activation. *European Journal of Immunology* **23**:1381.

1389. Kim, Y. U., T. Kinoshita, H. Molina, D. Hourcade, T. Seya, L. M. Wagner, and V. M. Holers. 1995. Mouse complement regulatory protein Crry/p65 uses the specific mechanisms of both human decay-accelerating factor and membrane cofactor protein. *Journal of Experimental Medicine* **181**:151.

1390. Kurtz, C. B., M. S. Paul, M. Aegerter, J. J. Weis, and J. H. Weis. 1989. Murine complement receptor gene family. II. Identification and characterization of the murine homolog (Cr2) to human CR2 and its molecular linkage to Crry. *Journal of Immunology* **143**:2058.

1391. Paul, M. S., M. Aegerter, K. Cepek, M. D. Miller, and J. H. Weis. 1990. The murine complement receptor gene family. III. The genomic and transcriptional complexity of the Crry and Crry-ps genes. *Journal of Immunology* **144**:1988.

1392. Takizawa, H., N. Okada, and H. Okada. 1994. Complement inhibitor of rat cell membrane resembling mouse Crry/p65. *Journal of Immunology* **152**:3032.

1393. Sakurada, C., H. Seno, N. Dohi, H. Takizawa, M. Nonaka, N. Okada, and H. Okada. 1994. Molecular cloning of the rat complement regulatory protein, 512 antigen. *Biochemical and Biophysical Research Communications* **198**:819.

1394. Quigg, R. J., C. F. Lo, J. J. Alexander, A. E. Sneed, 3rd, and G. Moxley. 1995. Molecular characterization of rat Crry: widespread distribution of two alternative forms of Crry mRNA. *Immunogenetics* **42**:362.

1395. Funabashi, K., N. Okada, S. Matsuo, T. Yamamoto, B. P. Morgan, and H. Okada. 1994. Tissue distribution of complement regulatory membrane proteins in rats. *Immunology* **81**:444.

1396. Lasser, E. C., J. H. Lang, S. G. Lyon, A. E. Hamblin, and M. Howard. 1981. Glucocorticoid-induced elevations of C1-esterase inhibitor: a mechanism for protection against lethal dose range contrast challenge in rabbits. *Investigative Radiology* **16**:20.

1397. Al-Abdullah, I. H., R. B. Sim, J. Sheil, and J. F. Greally. 1984. The effect of danazol on the production of C1 inhibitor in the guinea pig. *Complement* **1**:27.

1398. Schwogler, S., M. Odenthal, T. Knittel, K. H. Meyer zum Buschenfelde, and G. Ramadori. 1992. Fat-storing cells of the rat liver synthesize and secrete C1-esterase inhibitor; modulation by cytokines. *Hepatology* **16**:794.

1399. van den Berg, C. W., P. C. Aerts, and H. van Dijk. 1992. C1-inhibitor prevents PEG fractionation-induced, EDTA-resistant activation of mouse complement. *Molecular Immunology* **29**:363.

1400. Van Nostrand, W. E., and D. D. Cunningham. 1987. Purification of a proteinase inhibitor from bovine serum with C1-inhibitor activity. *Biochimica et Biophysica Acta* **923**:167.

1401. Russell, J. A., K. Whaley, and S. Heaphy. 1997. The sequence of a cDNA encoding functional murine C1-inhibitor protein. *Biochimica et Biophysica Acta* **1352**:156.

1402. Fischer, M. B., A. P. Prodeus, A. Nicholson-Weller, M. Ma, J. Murrow, R. R. Reid, H. B. Warren, A. L. Lage, F. D. Moore, Jr., F. S. Rosen, and M. C. Carroll. 1997. Increased susceptibility to endotoxin shock in complement C3- and C4-deficient mice is corrected by C1 inhibitor replacement. *Journal of Immunology* **159**:976.

1403. Kaidoh, T., and I. Gigli. 1987. Phylogeny of C4b-C3b cleaving activity: similar fragmentation patterns of human C4b and C3b produced by lower animals. *Journal of Immunology* **139**:194.

1404. Kaidoh, T., and I. Gigli. 1989. Phylogeny of the plasma regulatory proteins of the complement system. *Progress in Clinical and Biological Research* **297**:199.

1405. Grossberger, D., A. Marcuz, L. Du Pasquier, and J. D. Lambris. 1989. Conservation of structural and functional domains in complement component C3 of Xenopus and mammals. *Proceedings of the National Academy of Sciences of the United States of America* **86**:1323.

1406. Kunnath-Muglia, L. M., G. H. Chang, R. B. Sim, A. J. Day, and R. A. Ezekowitz. 1993. Characterization of Xenopus laevis complement factor I structure – conservation of modular structure except for an unusual insert not present in human factor I. *Molecular Immunology* **30**:1249.

1407. Minta, J. O., M. J. Wong, C. A. Kozak, L. M. Kunnath-Muglia, and G. Goldberger. 1996. cDNA cloning, sequencing and chromosomal assignment of the gene for mouse complement factor I (C3b/C4b inactivator): identification of a species specific divergent segment in factor I. *Molecular Immunology* **33**:101.

1408. Zipfel, P. F., C. Kemper, A. Dahmen, and I. Gigli. 1996. Cloning and recombinant expression of a barred sand bass (Paralabrax nebulifer) cDNA. The encoded protein displays structural homology and immunological crossreactivity to human complement/cofactor related plasma proteins. *Developmental and Comparative Immunology* **20**:407.

1409. Soames, C. J., A. J. Day, and R. B. Sim. 1996. Prediction from sequence comparisons of residues of factor H involved in the interaction with complement component C3b. *Biochemical Journal* **315**:523.

1410. D'Cruz, O. J., and N. K. Day. 1985. Structural and functional similarities between the major hemolymph protein of fall armyworm and cat C4 binding protein of the complement system. *Developmental and Comparative Immunology* **9**:541.

1411. Barnum, S. R., T. Kristensen, D. D. Chaplin, M. F. Seldin, and B. F. Tack. 1989. Molecular analysis of the murine C4b-binding protein gene. Chromosome assignment and partial gene organization. *Biochemistry* **28**:8312.

1412. Hillarp, A., H. Wiklund, A. Thern, and B. Dahlback. 1997. Molecular cloning of rat C4b binding protein alpha- and beta-chains: structural and functional relationships among human, bovine, rabbit, mouse, and rat proteins. *Journal of Immunology* **158**:1315.

1413. Hughes, T. R., S. J. Piddlesden, J. D. Williams, R. A. Harrison, and B. P. Morgan. 1992. Isolation and characterization of a membrane protein from rat erythrocytes which inhibits lysis by the membrane attack complex of rat complement. *Biochemical Journal* **284**:169.

1414. Hughes, T. R., S. Meri, M. Davies, J. D. Williams, and B. P. Morgan. 1993. Immunolocalization and characterization of the rat analogue of human CD59 in kidney and glomerular cells. *Immunology* **80**:439.

1415. Rushmere, N. K., R. A. Harrison, C. W. van den Berg, and B. P. Morgan. 1994. Molecular cloning of the rat analogue of human CD59: structural comparison with human CD59 and identification of a putative active site. *Biochemical Journal* **304**:595.

1416. van den Berg, C. W., and B. P. Morgan. 1994. Complement-inhibiting activities of human CD59 and analogues from rat, sheep, and pig are not homologously restricted. *Journal of Immunology* **152**:4095.

1417. van den Berg, C. W., R. A. Harrison, and B. P. Morgan. 1993. The sheep analogue of human CD59: purification and characterization of its complement inhibitory activity. *Immunology* **78**:349.

1418. van den Berg, C. W., R. A. Harrison, and B. P. Morgan. 1995. A rapid method for the isolation of analogues of human CD59 by preparative SDS-PAGE: application to pig CD59. *Journal of Immunological Methods* **179**:223.

1419. Fodor, W. L., S. A. Rollins, S. Bianco-Caron, W. V. Burton, E. R. Guilmette, R. P. Rother, G. B. Zavoico, and S. P. Squinto. 1995. Primate terminal complement inhibitor homologues of human CD59. *Immunogenetics* **41**:51.

1420. Powell, M. B., K. J. Marchbank, N. K. Rushmere, C. W. van den Berg, and B. P. Morgan. 1997. Molecular cloning, chromosomal localization, expression, and functional characterization of the mouse analogue of human CD59. *Journal of Immunology* **158**:1692.

1421. Hinchliffe, S. J., N. K. Rushmere, S. M. Hanna, and B. P. Morgan. 1998. Molecular cloning and functional characterization of the pig analogue of CD59: relevance to xenotransplantation. *Journal of Immunology* **160**:3924.

1422. Fritz, I. B., K. Burdzy, B. Setchell, and O. Blaschuk. 1983. Ram rete testis fluid contains a protein (clusterin) which influences cell–cell interactions in vitro. *Biology of Reproduction* **28**:1173.

1423. Matsuo, S., H. Nishikage, F. Yoshida, A. Nomura, S. J. Piddlesden, and B. P. Morgan. 1994. Role of CD59 in experimental glomerulonephritis in rats. *Kidney International* **46**:191.

1424. Dickneite, G. 1993. Influence of C1-inhibitor on inflammation, edema and shock. *Behring Institute Mitteilungen* **93**:299.

1425. Vesentini, S., L. Benetti, C. Bassi, A. Bonora, A. Campedelli, G. Zamboni, P. Castelli, and P. Pederzoli. 1993. Effects of choline-esterase inhibitor in experimental acute pancreatitis in rats. Preliminary results. *International Journal of Pancreatology* **13**:217.

1426. Niederau, C., R. Brinsa, M. Niederau, R. Luthen, G. Strohmeyer, and L. D. Ferrell. 1995. Effects of C1-esterase inhibitor in three models of acute pancreatitis. *International Journal of Pancreatology* **17**:189.

1427. Buerke, M., T. Murohara, and A. M. Lefer. 1995. Cardioprotective effects of a C1 esterase inhibitor in myocardial ischemia and reperfusion. *Circulation* **91**:393.

1428. Horstick, G., A. Heimann, O. Götze, G. Hafner, O. Berg, P. Boehmer, P. Becker, H. Darius, H. J. Rupprecht, M. Loos, S. Bhakdi, J. Meyer, and O. Kempski. 1997. Intracoronary application of C1 esterase inhibitor improves cardiac function and reduces myocardial necrosis in an experimental model of ischemia and reperfusion. *Circulation* **95**:701.

1429. Hack, C. E., A. C. Ogilvie, B. Eisele, A. J. Eerenberg, J. Wagstaff, and L. G. Thijs. 1993. C1-inhibitor substitution therapy in septic shock and in the vascular leak syndrome induced by high doses of interleukin-2. *Intensive Care Medicine* **19**:S19.

1430. Ogilvie, A. C., J. W. Baars, A. J. Eerenberg, C. E. Hack, H. M. Pinedo, L. G. Thijs, and J. Wagstaff. 1994. A pilot study to evaluate the effects of C1 esterase inhibitor on the toxicity of high-dose interleukin 2. *British Journal of Cancer* **69**:596.

1431. Atkinson, J. P. 1996. Impact of the discovery of the membrane regulators of complement. *Research in Immunology* **147**:95.

1432. Weisman, H. F., T. Bartow, M. K. Leppo, H. C. Marsh, Jr., G. R. Carson, M. F. Concino, M. P. Boyle, K. H. Roux, M. L. Weisfeldt, and D. T. Fearon. 1990. Soluble human complement receptor type 1: in vivo inhibitor of complement suppressing post-ischemic myocardial inflammation and necrosis. *Science* **249**:146.

1433. Weisman, H. F., T. Bartow, M. K. Leppo, M. P. Boyle, H. C. Marsh, Jr., G. R. Carson, K. H. Roux, M. L. Weisfeldt, and D. T. Fearon. 1990. Recombinant soluble CR1 suppressed complement activation, inflammation, and necrosis associated with reperfusion of ischemic myocardium. *Transactions of the Association of American Physicians* **103**:64.

1434. Moore, F. D., Jr. 1994. Therapeutic regulation of the complement system in acute injury states. *Advances in Immunology* **56**:267.

1435. Kalli, K. R., P. Hsu, and D. T. Fearon. 1994. Therapeutic uses of recombinant complement protein inhibitors. *Springer Seminars in Immunopathology* **15**:417.

1436. Piddlesden, S. J., M. K. Storch, M. Hibbs, A. M. Freeman, H. Lassmann, and B. P. Morgan. 1994. Soluble recombinant complement receptor 1 inhibits inflammation and demyelination in antibody-mediated demyelinating experimental allergic encephalomyelitis. *Journal of Immunology* **152**:5477.

1437. Goodfellow, R. M., A. S. Williams, J. L. Levin, B. D. Williams, and B. P. Morgan. 1997. Local therapy with soluble complement receptor 1 (sCR1) suppresses inflammation in rat monoarticular arthritis. *Clinical and Experimental Immunology* **110**:45.

1438. Moran, P., H. Beasley, A. Gorrell, E. Martin, P. Gribling, H. Fuchs, N. Gillett, L. E. Burton, and I. W. Caras. 1992. Human recombinant soluble decay accelerating factor inhibits complement activation in vitro and in vivo. *Journal of Immunology* **149**:1736.

1439. Christiansen, D., J. Milland, B. R. Thorley, I. F. McKenzie, P. L. Mottram, L. J. Purcell, and B. E. Loveland. 1996. Engineering of recombinant soluble CD46: an inhibitor of complement activation. *Immunology* **87**:348.

1440. Christiansen, D., J. Milland, B. R. Thorley, I. F. McKenzie, and B. E. Loveland. 1996. A functional analysis of recombinant soluble CD46 in vivo and a comparison with recombinant soluble forms of CD55 and CD35 in vitro. *European Journal of Immunology* **26**:578.

1441. Iwata, K., T. Seya, H. Ariga, and S. Nagasawa. 1994. Expression of a hybrid complement regulatory protein, membrane cofactor protein decay accelerating factor on Chinese hamster ovary. Comparison of its regulatory effect with those of decay accelerating factor and membrane cofactor protein. *Journal of Immunology* **152**:3436.

1442. Higgins, P. J., J. L. Ko, R. Lobell, C. Sardonini, M. K. Alessi, and C. G. Yeh. 1997. A soluble chimeric complement inhibitory protein that possesses both decay-accelerating and factor I cofactor activities. *Journal of Immunology* **158**:2872.

1443. Davies, A. 1996. Policing the membrane: cell surface proteins which regulate complement. *Research in Immunology* **147**:82.

1444. Sugita, Y., K. Ito, K. Shiozuka, H. Suzuki, H. Gushima, M. Tomita, and Y. Masuho. 1994. Recombinant soluble CD59 inhibits reactive haemolysis with complement. *Immunology* **82**:34.

1445. Meri, S., T. Lehto, C. W. Sutton, J. Tyynela, and M. Baumann. 1996. Structural composition and functional characterization of soluble CD59: heterogeneity of the oligosaccharide and glycophosphoinositol (GPI) anchor revealed by laser-desorption mass spectrometric analysis. *Biochemical Journal* **316**:923.

1446. Fodor, W. L., S. A. Rollins, E. R. Guilmette, E. Setter, and S. P. Squinto. 1995. A novel bifunctional chimeric complement inhibitor that regulates C3 convertase and formation of the membrane attack complex. *Journal of Immunology* **155**:4135.

1447. Scesney, S. M., S. C. Makrides, M. L. Gosselin, P. J. Ford, B. M. Andrews, E. G. Hayman, and H. C. Marsh, Jr. 1996. A soluble deletion mutant of the human complement receptor type 1, which lacks the C4b binding site, is a selective inhibitor of the alternative complement pathway. *European Journal of Immunology* **26**:1729.

1448. Gralinski, M. R., B. C. Wiater, A. N. Assenmacher, and B. R. Lucchesi. 1996. Selective inhibition of the alternative complement pathway by sCR1[desLHR-A] protects the rabbit isolated heart from human complement-mediated damage. *Immunopharmacology* **34**:79.

1449. Akahori, T., Y. Yuzawa, K. Nishikawa, T. Tamatani, R. Kannagi, M. Miyasaka, H. Okada, N. Hotta, and S. Matsuo. 1997. Role of a sialyl Lewis(x)-like epitope selectively expressed on vascular endothelial cells in local skin inflammation of the rat. *Journal of Immunology* **158**:5384.

1450. Lowe, J. B., and P. A. Ward. 1997. Therapeutic inhibition of carbohydrate-protein interactions in vivo. *Journal of Clinical Investigation* **99**:822.

1451. Makrides, S. C., P. A. Nygren, B. Andrews, P. J. Ford, K. S. Evans, E. G. Hayman, H. Adari, M. Uhlen, and C. A. Toth. 1996. Extended in vivo half-life of human soluble complement receptor type 1 fused to a serum albumin-binding receptor. *Journal of Pharmacology and Experimental Therapeutics* **277**:534.

1452. Kalli, K. R., P. Hsu, T. J. Bartow, J. M. Ahearn, A. K. Matsumoto, L. B. Klickstein, and D. T. Fearon. 1991. Mapping of the C3b-binding site of CR1 and construction of a (CR1)2- F(ab')2 chimeric complement inhibitor. *Journal of Experimental Medicine* **174**:1451.

1453. Kalli, K. R., and D. T. Fearon. 1994. Binding of C3b and C4b by the CR1-like site in murine CR1. *Journal of Immunology* **152**:2899.

1454. Kalli, K. R., P. Hsu, and D. T. Fearon. 1994. Therapeutic uses of recombinant complement protein inhibitors. *Springer Seminars in Immunopathology* **15**:417.

1455. Wang, Y., Q. Hu, J. A. Madri, S. A. Rollins, A. Chodera, and L. A. Matis. 1996. Amelioration of lupus-like autoimmune disease in NZB/WF1 mice after treatment with a blocking monoclonal antibody specific for complement component C5. *Proceedings of the National Academy of Sciences of the United States of America* **93**:8563.

1456. Kroshus, T. J., S. A. Rollins, A. P. Dalmasso, E. A. Elliott, L. A. Matis, S. P. Squinto, and R. M. Bolman, 3rd. 1995. Complement inhibition with an anti-C5 monoclonal antibody prevents acute cardiac tissue injury in an ex vivo model of pig-to-human xenotransplantation. *Transplantation* **60**:1194.

1457. Rollins, S. A., L. A. Matis, J. P. Springhorn, E. Setter, and D. W. Wolff. 1995. Monoclonal antibodies directed against human C5 and C8 block complement-mediated damage of xenogeneic cells and organs. *Transplantation* **60**:1284.

1458. Wang, Y., S. A. Rollins, J. A. Madri, and L. A. Matis. 1995. Anti-C5 monoclonal antibody therapy prevents collagen-induced arthritis and ameliorates established disease. *Proceedings of the National Academy of Sciences of the United States of America* **92**:8955.

1459. Thomas, T. C., S. A. Rollins, R. P. Rother, M. A. Giannoni, S. L. Hartman, E. A. Elliott, S. H. Nye, L. A. Matis, S. P. Squinto, and M. J. Evans. 1996. Inhibition of complement activity by humanized anti-C5 antibody and single-chain Fv. *Molecular Immunology* **33**:1389.

1460. Würzner, R., M. Schulze, L. Happe, A. Franzke, F. A. Bieber, M. Oppermann, and O. Götze. 1991. Inhibition of terminal complement complex formation and cell lysis by monoclonal antibodies. *Complement and Inflammation* **8**:328.

1461. Evans, M. J., S. A. Rollins, D. W. Wolff, R. P. Rother, A. J. Norin, D. M. Therrien, G. A. Grijalva, J. P. Mueller, S. H. Nye, S. P. Squinto, and J. A.

Wilkins. 1995. In vitro and in vivo inhibition of complement activity by a single-chain Fv fragment recognizing human C5. *Molecular Immunology* **32**:1183.

1462. Thomas, T. C., S. A. Rollins, R. P. Rother, M. A. Giannoni, S. L. Hartman, E. A. Elliott, S. H. Nye, L. A. Matis, S. P. Squinto, and M. J. Evans. 1996. Inhibition of complement activity by humanized anti-C5 antibody and single-chain Fv. *Molecular Immunology* **33**:1389.

1463. Amsterdam, E. A., G. L. Stahl, H. L. Pan, S. V. Rendig, M. P. Fletcher, and J. C. Longhurst. 1995. Limitation of reperfusion injury by a monoclonal antibody to C5a during myocardial infarction in pigs. *American Journal of Physiology* **268**:H448.

1464. Morgan, E. L., J. A. Ember, S. D. Sanderson, W. Scholz, R. Buchner, R. D. Ye, and T. E. Hugli. 1993. Anti-C5a receptor antibodies. Characterization of neutralizing antibodies specific for a peptide, C5aR-(9-29), derived from the predicted amino-terminal sequence of the human C5a receptor. *Journal of Immunology* **151**:377.

1465. Mulligan, M. S., E. Schmid, B. Beck-Schimmer, G. O. Till, H. P. Friedl, R. B. Brauer, T. E. Hugli, M. Miyasaka, R. L. Warner, K. J. Johnson, and P. A. Ward. 1996. Requirement and role of C5a in acute lung inflammatory injury in rats. *Journal of Clinical Investigation* **98**:503.

1466. Reemstma, K., B. H. McCracken, J. U. Schlegel, and M. Pearl. 1964. Heterotransplantation of the kidney: two clinical experiences. *Science* **143**:700.

1467. Starzl, T. E., T. L. Marchioro, G. Peters, C. H., Kirkpatrick, W. E. C. Wilson, and K. A. Porter. 1964. Renal heterotransplantation from baboon to man: experience with 6 cases. *Transplantation* **2**:752.

1468. Bailey, L. L., S. L. Nehlsen-Cannarella, W. Concepcion, and W. B. Jolley. 1985. Baboon-to-human cardiac xenotransplantation in a neonate. *Journal of the American Medical Association* **254**:3321.

1469. Auchincloss, H. J. 1988. Xenogenic transplantation: a review. *Transplantation* **46**:1.

1470. Pierson, R. N. 3rd, D. J. White, and J. Wallwork. 1993. Ethical considerations in clinical cardiac xenografting [letter; comment]. *Journal of Heart and Lung Transplantation* **12**:876.

1471. Niekrasz, M., Y. Ye, and L.L. Rolf. 1992. The pig as organ donor for man. *Transplantation Proceedings* **24**:625.

1472. Calne, R. Y. 1970. Organ transplantation between widely disparate species. *Transplantation Proceedings* **2**:550.

1473. Stevens, R. B., and J. L. Platt. 1992. The pathogenesis of hyperacute xenograft rejection. *American Journal of Kidney Diseases* **20**:414.

1474. Hammer, C., M. Suckfull, and D. Saumweber. 1992. Evolutionary and immunological aspects of xenotransplantation. *Transplantation Proceedings* **24**:2397.

1475. Hammer, C. 1994. Fundamental problems of xenotransplantation. *Pathologie Biologie* **42**:203.

1476. Schilling, A., W. Land, E. Pratschke, K. Pielsticker, and W. Brendel. 1976. Dominant role of complement in the hyperacute xenograft rejection reaction. *Surgery, Gynecology and Obstetrics* **142**:29.

1477. Oglesby, T. J., D. White, I. Tedja, K. Liszewski, L. Wright, J. Van den Bogarde, and J. P. Atkinson. 1991. Protection of mammalian cells from complement-mediated lysis by transfection of human membrane cofactor protein and decay-accelerating factor. *Transactions of the Association of American Physicians* **104**:164.

1478. Fodor, W. L., B. L. Williams, L. A. Matis, J. A. Madri, S. A. Rollins, J. W. Knight, W. Velander, and S. P. Squinto. 1994. Expression of a functional human complement inhibitor in a transgenic pig as a model for the prevention of xenogeneic hyperacute organ rejection. *Proceedings of the National Academy of Sciences of the United States of America* **91**:11153.

1479. Kroshus, T. J., R. M. Bolman, III, A. P. Dalmasso, S. A. Rollins, E. R. Guilmette, B. L. Williams, S. P. Squinto, and W. L. Fodor. 1996. Expression of human CD59 in transgenic pig organs enhances organ survival in an ex vivo xenogeneic perfusion model. *Transplantation* **61**:1513.

1480. Diamond, L. E., K. R. McCurry, E. R. Oldham, M. Tone, H. Waldmann, J. L. Platt, and J. S. Logan. 1995. Human CD59 expressed in transgenic mouse hearts inhibits the activation of complement. *Transplant Immunology* **3**:305.

1481. Byrne, G. W., K. R. McCurry, D. Kagan, C. Quinn, M. J. Martin, J. L. Platt, and J. S. Logan. 1995. Protection of xenogeneic cardiac endothelium from human complement by expression of CD59 or DAF in transgenic mice. *Transplantation* **60**:1149.

1482. Kagan, D., J. Platt, J. Logan, and G. W. Byrne. 1994. Expression of complement regulatory factors using heterologous promoters in transgenic mice. *Transplantation Proceedings* **26**:1242.

1483. Cozzi, E., G. A. Langford, A. Richards, K. Elsome, R. Lancaster, P. Chen, N. Yannoutsos, and D. J. White. 1994. Expression of human decay accelerating factor in transgenic pigs. *Transplantation Proceedings* **26**:1402.

1484. Byrne, G., K. McCurry, M. Martin, J. Platt, and J. Logan. 1996. Development and analysis of transgenic pigs expressing the human complement regulatory protein CD59 and DAF. *Transplantation Proceedings* **28**:759.

1485. Carrington, C. A., A. C. Richards, E. Cozzi, G. Langford, N. Yannoutsos, and D. J. White. 1995. Expression of human DAF and MCP on pig endothelial cells protects from human complement. *Transplantation Proceedings* **27**:321.

1486. Dunning, J., P. C. Braidley, J. Wallwork, and D. J. White. 1994. Analysis of hyperacute rejection of pig hearts by human blood using an ex vivo perfusion model. *Transplantation Proceedings* **26**:1016.

1487. Pohlein, C., A. Pascher, M. Storck, V. K. Young, W. Konig, D. Abendroth, M. Wick, J. Thiery, D. J. White, and C. Hammer. 1996. Transgenic human DAF-expressing porcine livers: their function during hemoperfusion with human blood. *Transplantation Proceedings* **28**:770.

1488. McCurry, K. R., L. E. Diamond, D. L. Kooyman, G. W. Byrne, M. J. Martin, J. S. Logan, and J. L. Platt. 1996. Human complement regulatory proteins expressed in transgenic swine protect swine xenografts from humoral injury. *Transplantation Proceedings* **28**:758.

1489. Diamond, L. E., K. R. McCurry, M. J. Martin, S. B. McClellan, E. R.

Oldham, J. L. Platt, and J. S. Logan. 1996. Characterization of transgenic pigs expressing functionally active human CD59 on cardiac endothelium. *Transplantation* **61**:1241.

1490. McCurry, K. R., D. L. Kooyman, C. G. Alvarado, A. H. Cotterell, M. J. Martin, J. S. Logan, and J. L. Platt. 1995. Human complement regulatory proteins protect swine-to-primate cardiac xenografts from humoral injury. *Nature Medicine* **1**:423.

1491. Vaughan, H. A., B. E. Loveland, and M. S. Sandrin. 1994. Gal alpha(1,3)Gal is the major xenoepitope expressed on pig endothelial cells recognized by naturally occurring cytotoxic human antibodies. *Transplantation* **58**:879.

1492. Vaughan, H. A., I. F. McKenzie, and M. S. Sandrin. 1995. Biochemical studies of pig xenoantigens detected by naturally occurring human antibodies and the galactose alpha(1-3)galactose reactive lectin. *Transplantation* **59**:102.

1493. Sandrin, M. S., and I. F. McKenzie. 1994. Gal alpha (1,3)Gal, the major xenoantigen(s) recognised in pigs by human natural antibodies. *Immunological Reviews* **141**:169.

1494. Sandrin, M. S., W. L. Fodor, E. Mouhtouris, N. Osman, S. Cohney, S. A. Rollins, E. R. Guilmette, E. Setter, S. P. Squinto, and I. F. McKenzie. 1995. Enzymatic remodelling of the carbohydrate surface of a xenogenic cell substantially reduces human antibody binding and complement-mediated cytolysis. *Nature Medicine* **1**:1261.

1495. Rajasinghe, H. A., V. M. Reddy, W. W. Hancock, M. H. Sayegh, and F. L. Hanley. 1996. Key role of the alternate complement pathway in hyperacute rejection of rat hearts transplanted into fetal sheep. *Transplantation* **62**:407.

1496. Platt, J. L. 1996. The immunological barriers to xenotransplantation. *Critical Reviews in Immunology* **16**:331.

1497. White, D. 1996. Alteration of complement activity: a strategy for xenotransplantation. *Trends in Biotechnology* **14**:3.

1498. Pruitt, S. K., W. M. Baldwin, 3rd, H. C. Marsh, Jr., S. S. Lin, C. G. Yeh, and R. R. Bollinger. 1991. The effect of soluble complement receptor type 1 on hyperacute xenograft rejection. *Transplantation* **52**:868.

1499. Pruitt, S. K., A. D. Kirk, R. R. Bollinger, H. C. Marsh, Jr., B. H. Collins, J. L. Levin, J. R. Mault, J. S. Heinle, S. Ibrahim, A. R. Rudolph, et al. 1994. The effect of soluble complement receptor type 1 on hyperacute rejection of porcine xenografts. *Transplantation* **57**:363.

1500. Ryan, U. S. 1995. Complement inhibitory therapeutics and xenotransplantation. *Nature Medicine* **1**:967.

1501. Candinas, D., B. A. Lesnikoski, S. C. Robson, S. M. Scesney, I. Otsu, T. Myiatake, H. C. Marsh, Jr., U. S. Ryan, W. W. Hancock, and F. H. Bach. 1996. Soluble complement receptor type 1 and cobra venom factor in discordant xenotransplantation. *Transplantation Proceedings* **28**:581.

1502. Hayashi, S., M. Ito, M. Yasutomi, Y. Namii, I. Yokoyama, K. Uchida, and H. Takagi. 1995. Evidence that donor pretreatment with FK506 has a synergistic effect on graft prolongation in hamster-to-rat heart xenotransplantation. *Journal of Heart and Lung Transplantation* **14**:579.

1503. Tanaka, M., N. Murase, Q. Ye, W. Miyazaki, M. Nomoto, H. Miyazawa, R. Manez, Y. Toyama, A. J. Demetris, S. Todo, and T. E. Starzl. 1996. Effect of anticomplement agent K76 COOH on hamster-to-rat and guinea pig-to-rat heart xenotransplantation. *Transplantation* **62**:681.

1504. Magee, J. C., B. H. Collins, R. C. Harland, B. J. Lindman, R. R. Bollinger, M. M. Frank, and J. L. Platt. 1995. Immunoglobulin prevents complement-mediated hyperacute rejection in swine-to-primate xenotransplantation. *Journal of Clinical Investigation* **96**:2404.

1505. Platt, J. L. 1994. A perspective on xenograft rejection and accommodation. *Immunological Reviews* **141**:127.

1506. Dorling, A., C. Stocker, T. Tsao, D. O. Haskard, and R. I. Lechler. 1996. In vitro accommodation of immortalized porcine endothelial cells: resistance to complement mediated lysis and down-regulation of VCAM expression induced by low concentrations of polyclonal human IgG antipig antibodies. *Transplantation* **62**:1127.

1507. Yamakawa, M., K. Yamada, T. Tsuge, H. Ohrui, T. Ogata, M. Dobashi, and Y. Imai. 1994. Protection of thyroid cancer cells by complement-regulatory factors. *Cancer* **73**:2808.

1508. Hakulinen, J., and S. Meri. 1994. Expression and function of the complement membrane attack complex inhibitor protectin (CD59) on human breast cancer cells. *Laboratory Investigation* **71**:820.

1509. Varsano, S., I. Frolkis, L. Rashkovsky, D. Ophir, and Z. Fishelson. 1996. Protection of human nasal respiratory epithelium from complement-mediated lysis by cell-membrane regulators of complement activation. *American Journal of Respiratory Cell and Molecular Biology* **15**:731.

1510. Harris, C. L., K. S. Kan, G. T. Stevenson, and B. P. Morgan. 1997. Tumour cell killing using chemically engineered antibody constructs specific for tumour cells and the complement inhibitor CD59. *Clinical and Experimental Immunology* **107**:364.

1511. Stoiber, H., C. Ebenbichler, R. Schneider, J. Janatova, and M. P. Dierich. 1995. Interaction of several complement proteins with gp120 and gp41, the two envelope glycoproteins of HIV-1. *AIDS* **9**:19.

1512. Dierich, M. P., H. Stoiber, and A. Clivio. 1996. A 'complement-ary' AIDS vaccine. *Nature Medicine* **2**:153.

1513. Xiao, F., M. J. Eppihimer, B. H. Willis, and D. L. Carden. 1997. Complement-mediated lung injury and neutrophil retention after intestinal ischemia-reperfusion. *Journal of Applied Physiology* **82**:1459.

1514. Hill, J., T. F. Lindsay, F. Ortiz, C. G. Yeh, H. B. Hechtman, and F. D. Moore, Jr. 1992. Soluble complement receptor type 1 ameliorates the local and remote organ injury after intestinal ischemia-reperfusion in the rat. *Journal of Immunology* **149**:1723.

1515. Pemberton, M., G. Anderson, V. Vetvicka, D. E. Justus, and G. D. Ross. 1993. Microvascular effects of complement blockade with soluble recombinant CR1 on ischemia/reperfusion injury of skeletal muscle. *Journal of Immunology* **150**:5104.

1516. Chavez-Cartaya, R. E., G. P. DeSola, L. Wright, N. V. Jamieson, and D. J. White. 1995. Regulation of the complement cascade by soluble complement receptor type 1. Protective effect in experimental liver ischemia and reperfusion. *Transplantation* **59**:1047.

1517. Mulligan, M. S., C. G. Yeh, A. R. Rudolph, and P. A. Ward. 1992. Protective effects of soluble CR1 in complement- and neutrophil-mediated tissue injury. *Journal of Immunology* **148**:1479.

1518. Pratt, J. R., M. J. Hibbs, A. J. Laver, R. A. Smith, and S. H. Sacks. 1996. Effects of complement inhibition with soluble complement receptor-1 on vascular injury and inflammation during renal allograft rejection in the rat. *American Journal of Pathology* **149**:2055.

1519. Kaczorowski, S. L., J. K. Schiding, C. A. Toth, and P. M. Kochanek. 1995. Effect of soluble complement receptor-1 on neutrophil accumulation after traumatic brain injury in rats. *Journal of Cerebral Blood Flow and Metabolism* **15**:860.

1520. Couser, W. G., R. J. Johnson, B. A. Young, C. G. Yeh, C. A. Toth, and A. R. Rudolph. 1995. The effects of soluble recombinant complement receptor 1 on complement-mediated experimental glomerulonephritis. *Journal of the American Society of Nephrology* **5**:1888.

1521. Cheung, A. K., C. J. Parker, and M. Hohnholt. 1994. Soluble complement receptor type 1 inhibits complement activation induced by hemodialysis membranes in vitro. *Kidney International* **46**:1680.

1522. Himmelfarb, J., E. McMonagle, D. Holbrook, and C. Toth. 1995. Soluble complement receptor 1 inhibits both complement and granulocyte activation during ex vivo hemodialysis. *Journal of Laboratory and Clinical Medicine* **126**:392.

1523. Nishizawa, H., H. Yamada, H. Miyazaki, M. Ohara, K. Kaneko, T. Yamakawa, J. Wiener-Kronish, and I. Kudoh. 1996. Soluble complement receptor type 1 inhibited the systemic organ injury caused by acid instillation into a lung. *Anesthesiology* **85**:1120.

1524. Piddlesden, S. J., S. Jiang, J. L. Levin, A. Vincent, and B. P. Morgan. 1996. Soluble complement receptor 1 (sCR1) protects against experimental autoimmune myasthenia gravis. *Journal of Neuroimmunology* **71**:173.

1525. Jung, S., K. V. Toyka, and H. P. Hartung. 1995. Soluble complement receptor type 1 inhibits experimental autoimmune neuritis in Lewis rats. *Neuroscience Letters* **200**:167.

1526. Block, V. T., M. R. Daha, O. Tijsma, C. L. Harris, B. P. Morgan, G. J. Fleuren, and A. Gorter. 1998. A bispecific monoclonal antibody directed against both the membrane-bound complement regulator CD55 and the renal tumor-associated antigen G250 enhances C3 deposition and tumor cell lysis by complement. *Journal of Immunology* **160**:3437.

1527. Warwicker, P., T. H. J. Goodship, R. L. Donne, Y. Pirson, A. Nicholls, R. M. Ward, P. Turnpenny, and J. A. Goodship. 1998. Genetic studies into inherited and sporadic hemolytic uremic syndrome. *Kidney International* **53**:836–844.

# INDEX